# Advances in Mathematics Education

More information about this series at www.springer.com/series/8392

Birgit Pepin • Bettina Roesken-Winter
Editors

# From beliefs to dynamic affect systems in mathematics education

## Exploring a mosaic of relationships and interactions

Springer

*Editors*
Birgit Pepin
Sør-Trøndelag University College
Trondheim, Norway

Bettina Roesken-Winter
Professional School of Education
Department of Mathematics
Humboldt-Universität zu Berlin
Berlin, Germany

ISSN 1869-4918 ISSN 1869-4926 (electronic)
ISBN 978-3-319-06807-7 ISBN 978-3-319-06808-4 (eBook)
DOI 10.1007/978-3-319-06808-4
Springer Cham Heidelberg New York Dordrecht London

Library of Congress Control Number: 2014953525



Printed on acid-free paper

Springer is part of Springer Science+Business Media (www.springer.com)

# Foreword

## From *Hidden Dimensions* to *Dynamic Systems* in Affect Research

**Abstract** In this preface, I turn briefly to the genesis of an influential forerunner to the present volume on affect and mathematics education, namely, *Beliefs: A Hidden Variable in Mathematics Education?*, published over a decade ago. Some brief snapshots of the contents of the earlier manuscripts are presented and areas then identified in need of further, or more nuanced, exploration are highlighted. Foregrounding these issues serves as an expedient link between the present volume and the earlier work.

### *Introduction*

The *Mathematisches Forschungsinstitut Oberwolfach* (The Mathematical Research Institute of Oberwolfach) is idyllically situated in Germany's Black Forest. Over the years, the Institute has hosted many week-long workshops in which specialist mathematics topics are explored by experts in the field. "The Institute brings people together for a short but intense period, providing them with ideal conditions under which to pursue research activities which will *influence and stimulate the future development of the field*" (Mathematisches Forschungsinstitut Oberwolfach n.d., emphasis added).

The week of November 21 to November 27 in 1999 marked a particularly important time in the mathematics education research calendar. That week, uniquely, the issues for debate and exploration at the Institute were not confined to the field of mathematics per se. Instead the focus was firmly on a topic of great relevance to the field of mathematics education: *Mathematical beliefs and their impact on teaching and learning of mathematics.*

Consideration of issues raised during the rich and diverse flow of presentations and discussions extended well beyond the life of the workshop. The tradition of many of the previous, and indeed subsequent, workshops held at Oberwolfach was maintained. So, seemingly inevitably, follow up activities were pursued, culminating in the preparation of a volume designed to add to existing knowledge and literature in the field explored so intensively during the time spent at the Institute.

The gestation period between conception and birth of the book about the influence and impact of beliefs and aspects of the teaching and learning of mathematics was substantial. Core ideas initially presented and discussed in Oberwolfach were added to. Finally, in 2002, the manuscript *Beliefs: A Hidden Variable in Mathematics Education?* (Leder et al. 2002) was ready for publication.

## *The Foundations: Inspection*

What, more than a decade ago, caught the attention of researchers and practitioners concerned with beliefs and their impact on the teaching and learning of mathematics? The "then" state of the art, as captured in Leder et al. (2002), was covered in three distinct but overlapping sections: *Beliefs: Conceptualization and measurement; Teachers' beliefs;* and *Students' beliefs.* Inevitably, as in the current volume, affective components beyond beliefs also attracted much attention.

### Beliefs: Conceptualization and Measurement

"The contributions in the first section," Leder et al. (2002, p. 4) clarified,

> are particularly concerned with examining what is meant by mathematical beliefs and how they differ from other, related concepts. The authors draw extensively on existing literature, highlight consensus and confusion in the ways various terms have been used in earlier work, and indicate directions for further research without – collectively – offering a unified view on the main theoretical concepts explored: belief, conception, and knowledge. The overall thrust is on the diversity of different starting points which typically correspond to different emphases.

Reports of empirical research studies were largely absent in this section. Capturing the essence of beliefs was a common theme – a task made more difficult because of the multiple and often loose usage of the term "belief". This practice, it was frequently argued, should be challenged. That those working in the field are unlikely to embrace a single definition of the term belief was also generally agreed. So, it was asked, both explicitly and implicitly, should the field settle for another option: to aim for a commonly accepted set of definitions of beliefs, depending on the setting, the context, and the audience being addressed? How realistic, and achievable, is a different, related, and persistently elusive goal: to use more refined and more comprehensive instruments to tap beliefs, use more careful terminology to describe and characterize beliefs, and distinguish beliefs from other terms often used virtually interchangeably?

The magnitude of this goal was sketched evocatively by one of the reviewers of *Beliefs: A Hidden Variable in Mathematics Education?*

"The first stumbling block," wrote Mason (2004, p. 347) "is to work out what beliefs actually are, and where they fit into an entire alphabet of associated interlinked terms." These he listed as:

> A is for attitudes, affect, aptitude, and aims; B is for beliefs; C is for constructs, conceptions, and concerns; D is for demeanor and dispositions; E is for emotions, empathies, and expectations; F is for feelings; G is for goals and gatherings; H is for habits and habitus; I is for intentions, interests, and intuitions; J is for justifications and judgements; K is for knowing; L is for leanings; M is for meaning-to; N is for norms; O is for orientations and objectives; P is for propensities, perspectives, and predispositions; Q is for quirks and quiddity; R is for recognitions and resonances; S is for sympathies and sensations; T is for tendencies and truths; U is for understandings and undertakings; V is for values and views; W is for wishes, warrants, words, and weltanschauung; X is for xenophilia (perhaps); Y is for yearnings and yens; and Z is for zeitgeist and zeal. (Mason 2004, p. 347)

Whether written partly in jest or not, Mason's creative beliefs-alphabet captured an important challenge faced by researchers and followers of the field: the need to combat the loose and colloquial usage of the term belief and to adopt more constrained terminology.

What other directions were nominated in the book's introductory chapters for further or more nuanced explorations? Beliefs are not observable but are typically inferred from observations, from responses to interview probes, or answers to questionnaires. How can we be sure that the inferences we draw are accurate – a question that is posed frequently in the book and in the wider research community? Might other measures, physiological or neurological and/or laboratory based, provide new insights? How easy is it to make individuals aware of their beliefs about themselves and about mathematics, and how might these beliefs be changed if that is required for optimum mathematics learning?

As noted by D. McLeod and S. McLeod (2002), the chapters in this section "are not the last word on attempts to define beliefs.... Writers in the future will shape the research using terminology that meets their needs" (p. 119). That many have already done so is evident from the contributions in this and the other two parts of Leder et al. (2002). The extent to which later researchers have also adopted this pragmatic approach can be judged from inspection of the present, new volume, on research on beliefs and affect more broadly.

## Teachers' Beliefs

The synthesis of the second section, *Teachers' beliefs*, read in part:

> All the authors see a cyclical relationship between changing beliefs and changing practices; wherever one starts they affect each other. ... My reading of these chapters suggests that

> Chapman,[1] Llinares, and Philippou and Christou come down on the side of changing beliefs leading to change in practices, Lloyd and Hart on the side of changing practices leading to changes in beliefs, and the review chapter by Wilson and Cooney remains with the dialectic.... The chapters provide rich evidence of changing practices in teaching mathematics as seen through a lens of discourse of beliefs as mental objects that are both cognitive and affective. They are constituted in teachers' prior experiences and they need to become the subject of reflection and analysis. A whole range of activities can bring about change towards reform and that change will come about as beliefs change... The picture is rightly a complex one. (Lerman 2002, p. 235)

Exploring teachers' beliefs and their development are important topics in themselves. How these beliefs affect their instructional strategies and in class behaviors were dominant themes addressed, in different ways, by the contributors in this section. Typically, however, the proposed implications and applications for constructive practice have been tentative and nebulous. To quote Wilson and Cooney (2002, p. 145): "understanding context and developing alternative explanations for phenomena require researchers to dig deeply. ... The human condition is always beset with a strange mixture of rationality and irrationality that defy sharp lines of demarcation." That interviews and observations provide insights into the ways in which teachers behave and make sense of their world is clearly acknowledged. But, and under what circumstances, is the information thus gained sufficient to predict future actions? Much in this complex field remains untilled. Welcome additional insights are provided in the present volume.

Experimental data, generally based on work with small samples, were reported in several chapters in this section. While instructive, such findings are normally of questionable generalizability and unlikely to generate new theories that stretch the boundaries of the field. What other avenues could be fruitful?

An alternate, or perhaps more appropriately thought of as a complementary approach to the examination of changing classroom practices and the role of teachers' beliefs in this process, was advocated by Lerman (2002). Instead of regarding beliefs and belief systems as psychological constructs, as, he argued, is done by the authors of the chapters in this section, and indeed often in research about beliefs and belief systems more widely, why not look "at the issue of changing practices in mathematics teaching and learning (by) applying a *sociological* gaze rather than a psychological one" (p. 235, emphasis in the original). In parenthesis, it is worth noting that most of the authors of the work contained in the next section in the book "clearly acknowledge the tremendous influence socio-cultural context has on the formation of beliefs" (Lester 2002, p. 353). Readers of the current volume can themselves determine whether, and if so how, Lerman's challenge is taken up in the current publication.

---

[1] Details of the work of the authors mentioned in this quotation can be found in *Beliefs: A hidden variable in mathematics education?* They are not detailed in the reference list.

### Students' Beliefs

Research on students' beliefs is relatively new. "It is only during the past 15–20 years that beliefs research has come to be viewed as an essential ingredient in mathematics education research programs" according to Lester (2002, p. 346). Core issues investigated by the contributors to the final section of Leder et al. (2002) were summarized by Lester as follows:

- Do students' beliefs influence their interest in and motivation to learn mathematics?
- Why don't students worry about whether their solutions to word problems make sense?
- Do students' beliefs influence their ability to connect real-world and school mathematics?
- How do classrooms influence the development of beliefs?
- How do students' intuitive beliefs about mathematical operations affect their thinking processes?

Once again a multiplicity of methods and settings were adopted to explore students' beliefs about mathematics and the learning of mathematics. When included, experimental data were generally again confined to small samples. Inevitably, there were shortcomings in the various studies reported: the scope of instruments used to gather students' beliefs was necessarily constrained, and the need to infer beliefs from students' answers or actions remained a hurdle.

At the time, and also with hindsight, the findings reported – singly and collectively – were, and are, more appropriately considered as markers of progress in the field than as end points of research. Lester's (2002) concluding comments were indeed apt.

> I am skeptical (he wrote) about the credibility and reliability of the data presented by these five reports because I doubt that they accurately indicate what the students really believe.... I do not think most students really think much about what they believe about mathematics and as a result are not very aware of their beliefs. So, although I think these researchers are leading the way in efforts to develop good methodological tools for studying students' beliefs, a considerable amount of work remains to be done. (p. 353)

## *Building on the Foundations*

Some years ago, Freud (1940/1986, p. 286) wrote: "the concept of the unconscious has long been knocking at the gates of psychology and asking to be let in. Philosophy and literature have often toyed with it, but science could find no use for it." This contention is now heavily challenged, and indeed refuted, by the active and sustained research agenda on the interaction between affect and the teaching and learning of mathematics, spawned in the later part of the twentieth century and continuing unabated in the twenty-first century. Despite this activity, there continues to be room for painstakingly crafted, theoretically driven explorations and new innovative techniques. The contents of the current volume reflect, in varying degrees, continuing efforts to extend the boundaries of the field.

Monash University
Melbourne, Australia

Gilah C. Leder

## References

Freud, S. (1986). *Some elementary lessons in psycho-analysis. The standard edition of the complete psychological works of Sigmund Freund* (Vol. XXIII). London: Hogarth Press (Original work published in 1940).

Leder, G. C., Pehkonen, E., & Törner, G. (Eds.). (2002). *Beliefs: A hidden variable in mathematics education?* Dordrecht: Kluwer Academic Publishers.

Lerman, S. (2002). Situating research on mathematics teachers' beliefs on change. In G. C. Leder, E. Pehkonen, & G. Törner (Eds.), *Beliefs: A hidden variable in mathematics education?* (pp. 233–243). Dordrecht: Kluwer Academic Publishers.

Lester, F. K. Jr. (2002). Implications of research on students' beliefs for classroom practice. In G. C. Leder, E. Pehkonen, & G. Törner (Eds.), *Beliefs: A hidden variable in mathematics education?* (pp. 345–353). Dordrecht: Kluwer Academic Publishers.

McLeod, D. B., & Mcleod, S. H. (2002). Synthesis – Beliefs and mathematics education: Implications for learning, teaching, and research. In G. C. Leder, E. Pehkonen, & G. Törner (Eds.), *Beliefs: A hidden variable in mathematics education?* (pp. 115–123). Dordrecht: Kluwer Academic Publishers.

Mason, J. (2004). Are beliefs believable? *Mathematical Thinking and Learning, 6*(3), 343–351.

Mathematisches Forschungsinstitut Oberwolfach. (n.d.). Retrieved from http://www.mfo.de/about-the-institute

Wilson, M., & Cooney, T. J. (2002). Mathematics teacher change and development. The role of beliefs. In G. C. Leder, E. Pehkonen, & G. Törner (Eds.), *Beliefs: A hidden variable in mathematics education?* (pp. 127–147). Dordrecht: Kluwer Academic Publishers.

# Contents

# Introduction

The title of the book *From Beliefs to Dynamic Affect Systems in Mathematics Education: Exploring a Mosaic of Relationships and Interactions* was purposefully chosen. First, it has been our intention to bridge the gap between 'then and now': from the so-called 'beliefs literature' to recent developments in the field of affect research. Gilah Leder does this in her Foreword, where she delineates the advances in the field starting from her co-edited seminal book (Leder et al. 2002). Second, with the use of the term 'dynamic affect systems' we link to the view that some of the notions and categories in affect research can be regarded as 'systems' (see Schlögelmann 2003; for an overview of research into affect, see Evans 2000); in particular, beliefs are frequently referred to as belief systems (e.g. Aguirre and Speer 2000). Moreover, these systems are 'dynamic' as can be seen from, for instance, Goldin (2001):

1. Emotions (rapidly changing states of feeling, mild to very intense, that are usually local or embedded in a context);
2. Attitudes (moderately stable predispositions toward ways of feeling in classes of situations, involving a balance of affect and cognition);
3. Beliefs (internal representations to which the believer attributes truth, validity, or applicability, usually stable and highly cognitive, may be highly structured);
4. Values, ethics, and morals (deeply held preferences), sometimes characterized as 'personal truth', stable, highly affective as well as cognitive, may also be highly structured). (p. 3)

What does that mean? A short excursion into systems thinking (and 'complex systems') may provide some parallels to dynamic affect systems. According to the literature (e.g. Capra 1997), a 'system' is composed of interrelated parts or components (structures) that cooperate in processes (behaviour, actions). Systems thinking is based on the belief that the component parts of a system can best be understood in the context of relationships with each other and with other systems rather than in isolation – in other words, on the belief that 'the whole is greater than the sum of its parts'. It is argued that the only way to fully understand why a problem or element occurs and persists is to understand the parts in relation to the whole (this is in contrast to Descartian thinking which is reductionist) and hence proposes to view systems in a holistic manner. This means that systems thinking concerns an understanding of a system by examining the linkages, interactions and relationships between the elements that compose the entirety of the system. Linking this to 'affect systems', we can see why there have been so many debates over the last few decades about what constitutes 'emotions', 'attitudes', 'beliefs' and 'values', and there are no clear definitions available to date. Could it be that emotions, attitudes, beliefs and values each constitute a system (e.g. in an individual, or in a collective/group) and that these systems are indeed inter-related, or 'nested' within any one person/group, albeit nurtured by the context?

Moreover, systems thinking attempts to illustrate how small catalytic events that are separated by distance and time can be the cause of significant changes in particular systems: 'complex adaptive systems' is the literature name for what we called 'dynamic systems', referring to their dynamic nature. These are diverse and made up of multiple interconnected elements, and dynamic in that they have the capacity to change and learn from experience. Examples of dynamic systems may be any human collective activity, such as problem solving, in a cultural and social system, such as a mathematics classroom in a school. An example of dynamic affect systems may be the affect system of a group of pupils or teachers working on an unusual mathematical problem (e.g. Stylianides and Stylianides 2011; Pepin 2012) as a group or individually. The interesting features of such dynamic systems are the following:

- Due to the strong coupling between components in such systems, a failure in one or more components can lead to cascading failures, which may have catastrophic consequences on the functioning of the system: for an individual pupil, this may mean the development of a negative attitude due to a breakdown of understanding, because one component was not understood, and hence the affective side (of 'failure' in one part) may bring the whole affect (and learning) system to a halt.
- The 'history' of such dynamic systems may be important: because these systems are dynamic, they may change over time (e.g. different emotional experiences with different mathematical topic areas), and prior states may have an influence on present states.
- Dynamic systems may be nested (as also indicated earlier): the components of a dynamic system may themselves be dynamic systems. For example, a classroom (as a dynamic affect system) is made up of pupils, who each have their dynamic affect system/s.
- Dynamic affect systems may show emergent phenomena: that is to say that while the results may be sufficiently determined by the activity of the systems' basic constituents, they may have properties that can only be studied at a higher 'emergent' level. For example, a group of pupils may show a very different dynamic affect system, as a group, than the individual pupils would show on their own.
- A small perturbation/intervention of the dynamic affect system may cause a large effect (e.g. the butterfly effect), a proportional effect, or even no effect at all. This is due to the non-linearity of dynamic systems.

- Affect relationships contain feedback loops: both negative (damping) and positive (amplifying) feedback are always found in dynamic systems. The effects of a person's behaviour, for example, are fed back to the person in such a way that the person herself/himself may change.

Throughout the book, the authors have provided results of their affect studies which may be explainable using the 'dynamic affect system lens', which, further investigated, discussed and tested, may serve as a theoretical framework for analysis. For example, Hannula has developed a sophisticated theoretical frame, and it can be argued that there are clear links to systems thinking. Moreover, several of these characteristics of dynamic affect systems can be evidenced in selected chapters, and we now turn to the sections and chapters to provide an overview of the book.

The book provides a developing perspective on the issue of affect in the mathematics classroom in terms of three dimensions: (1) theoretical lenses in affect research; (2) relevance in the field – affective systems of individuals and collectives; and (3) methodological issues in affect research.

The first section is dedicated to present theoretical frames – we call them 'theoretical lenses' – to study the mosaic of relationships and interactions in the field of affect. Accordingly, the contribution by Skott revisits belief research through 'system affect lenses' whilst criticizing the dominant conceptualization of individual functioning in beliefs studies – favouring a participatory approach taken from social practice theory. In the same way, Radford advocates a cultural conception of emotions and their role in thinking in general and mathematical thinking in particular. The chapter by Di Martino and Zan traces the 'story' of the construct of attitude and provides a theoretical discussion on crucial aspects that help in understanding the mosaic of relationships and interactions within the affect field. Moreover, the theory section also explores the networking of theories to gain a deeper understanding of specific affect constructs as, for instance, described by Rolka and Rösken-Winter, and contributes to re-conceptualizing constructs such as teacher efficacy with respect to its multidimensionality, as suggested by Philippou and Pantziara.

In the second section, empirical research studies enrich the theory by providing relevant findings in terms of developing deeper understandings of individuals' and collectives' affective systems in mathematics education. Here, pupil and teacher beliefs and affect systems have been examined more closely. Eichler and Erens, for instance, explore the belief systems of secondary mathematics teachers as part of their mathematics-related affect. With regard to problem solving, the two contributions by Depaepe, De Corte and Verschaffel and by Gómez-Chacón capture the complex interaction of teacher and student beliefs. Depaepe, De Corte and Verschaffel review theoretical analyses and empirical studies that have focused on major aspects of teachers' instructional practices that affect students' non-realistic approaches to and beliefs about word problem solving. Gómez-Chacón studies the interaction between affect and cognition in problem solving while using technology in teaching mathematics. On a larger scale, three chapters discuss different relationships between affective variables. Ding, Pepin and Jones study the attitudes towards mathematics of Shanghai students in lower secondary schools and pay attention to their emotional disposition and perceived competence in mathematics. The chapter by Blömeke and Kaiser explores the effects of motivation on the belief systems of future mathematics teachers, while Forgasz, Leder, Mittelberg, Tan and Murimo present findings from recent research studies that point out how gender issues associated with a range of affective variables affect learning outcomes.

The third section is concerned with the methodological tools used, and needed, in affect research. How can the different methodological designs contribute data which help us to develop better understandings of teachers' and pupils' affect systems for teaching and learning mathematics and in which ways are knowledge and affect related? The first chapter of this section, by Chen and Leung, clarifies two major methodological issues in studying teachers' beliefs: analysing data and drawing conclusions on teachers' beliefs. The contribution by Bofah and Hannula discusses how an instrument developed in a specific context can be applied to a different culture, and how cross-validation can be established. Further, Andrà addresses methodological issues on a meta-level, networking and comparing different methodologies used in empirical research in the field of affect, and eventually bringing them together in an overarching conceptual framework. Finally, two contributions consider affective variables in connection with different facets of knowledge. Tsamir, Tirosh, Levenson, Tabach and Barkai investigate pre-school teachers' knowledge and self-efficacy, perceived to be required for teaching geometry, and highlight the need to design instruments which are sensitive to revealing aspects of self-efficacy in relation to different mathematical objects. Kuntze and Dreher dedicate their chapter to an investigation of the role of awareness of affective aspects and how it relates to teachers' Pedagogical Content Knowledge (PCK).

Furthermore, each section has been commented upon by an internationally renowned expert in the respective field. Hence, the commentary papers by Clarke, Hannula and Ruthven highlight the value of the individual chapters and the connections between them. In search of an organizational referent for the discussion on the theoretical lenses section, Clarke suggests the triadic conjunction of virtues: faith, hope and charity. To capture the relevance of affective systems and social factors in affect research, in the second section, Hannula introduces a meta-theoretical framework for research on mathematics-related affect. Reacting to contributions in the third section on methodological issues in affect research, Ruthven highlights the shared approach of using Likert-item questionnaires and how different reliable and validated measures can be combined.

In the closing chapter, Schoenfeld pursues two goals: first, he identifies how theory and methods may be productive in generating rigorous explanations of people's beliefs (or affect) and how these may influence their in-the-moment decision making; second, he addresses questions on what it takes to have a positive impact on people's beliefs.

Sør-Trøndelag University College
Trondheim, Norway
Birgit Pepin

Humboldt-Universität zu Berlin
Berlin, Germany
Bettina Roesken-Winter

## References

Aguirre, J., & Speer, N. M. (2000). Examining the relationship between beliefs and goals in teacher practice. *Journal of Mathematical Behaviour, 18*(3), 327–356.

Capra, F. (1997). *The web of Life*. London: HarperCollins.

Evans, J. (2000). *Adults' mathematical thinking and emotions*. London: Routledge Falmer.

Goldin, G. A. (2001). *Affect, meta-affect, and mathematical beliefs structures*. Pre-print, Rutgers University.

Leder, G. C., Pehkonen, E., & Törner, G. (2002). *Beliefs: A hidden variable in mathematics education?* Dordrecht: Kluwer.

Pepin, B. (2012). Affective systems of Norwegian mathematics students/teachers concerning 'unusual' problem solving. In *Proceedings of ICME 12*, Seoul, South Korea.

Schloegelmann, W. (2003). Can neuro-science help us to better understand affective reactions in mathematics learning? In *Proceedings of CERME 3*, Bellaria, Italy.

Stylianides, G., & Stylianides, A. (2011). An intervention of students' problem solving beliefs. In *Proceedings of CERME 7*, Rzeszów, Poland.

# Part I
# Theoretical Lenses in Affect Research

The first part of the book focuses on theoretical frames and lenses in affect research in mathematics education. Previously, no other book has brought together the different and intricate lenses used in this field of research: in this part the most recent and innovative are provided which build on seminal work in the field. Particular views require particular approaches, whether seen from the collective point of view, or from the teacher's or the pupil's viewpoint.

# Towards a Participatory Approach to 'Beliefs' in Mathematics Education

**Jeppe Skott**

**Abstract** Over the last three decades research in beliefs, and affect more generally, has developed into a significant field of study. It attempts to make sense of teachers' and students' understandings of mathematics, of its teaching and learning, and of themselves as doers, teachers, and learners of mathematics and of how these understandings relate to classroom practice. Studies of these issues have been published widely and in the most prestigious journals and book series. However, belief research is still confronted with significant conceptual and methodological problems. I suggest that this is at least in part due to the dominant conceptualization of individual functioning in belief research, one that is based on acquisitionism with its emphasis on human action as an enactment of previously reified mental entities. In the present chapter I build on social practice theory and symbolic interactionism to rephrase key issues of belief research, especially that of the relationship between beliefs and practice, in more participatory terms. The suggestion is to shift the focus from beliefs to the pre-reified processes that are said to give rise to them. This leads to more dynamic understandings of learning and lives in mathematics classrooms and serves to overcome some of the conceptual and methodological problems of the field.

**Keywords** Belief research • Mathematics teachers • Dynamic views of beliefs • Acquisition • Patterns of Participation (PoP)

Over the last three decades large numbers of studies have investigated the character of students' and teachers' beliefs about mathematics, about mathematics as taught in school, about the teaching and learning of the mathematics, and about themselves as learners, teachers, and doers of mathematics. Some studies focus on the development or relative stability of beliefs, for instance as they relate to the education of prospective or practising teachers, while others address the question of a possible correlation between students' and teachers' beliefs and the teaching-learning

J. Skott (✉)
Department of Mathematics Education, The Linnaeus University, Växjö, Sweden

Department of Education, Aarhus University, Copenhagen, Denmark
e-mail: jeppe.skott@edu.au.dk



B. Pepin, B. Roesken-Winter (eds.), *From beliefs to dynamic affect systems in mathematics education*, Advances in Mathematics Education,
DOI 10.1007/978-3-319-06808-4_1

practices that unfold in mathematics classrooms. As it relates to the students, the field is in the latter case concerned with how beliefs structure the students' approach to mathematics, sometimes in more domain-specific fields such as statistics or proof, and even "determine the way they engage in mathematical learning and problem solving" (De Corte et al. 2002, p. 298). In relation to teachers, the research interest in beliefs is part of an effort to supplement the focus on their knowledge of the contents and of the related educational issues with a more meta-cognitive and affective perspective. It is often assumed that the structuring effect of their beliefs on behaviour is as significant as suggested by de Corte et al. for the students (Schoenfeld 1992). The premise of the field of beliefs, then, was – and to some extent still is – that beliefs, understood as relatively stable, reified mental constructs, significantly influence students' and teachers' behaviour, also if they run counter to curricular intentions developed for instance in the research community. The promise of the field was – and still is – to solve, or at least alleviate the 'problems of implementation', i.e. the lack of congruity between such intentions and instructional practice, by changing the beliefs of prospective and practising teachers.

This research effort has contributed with more profound understandings of what Goldin (2002) calls affective/cognitive configurations and of the role they play for students and teachers engaged in classroom interaction. However, belief research is not an unproblematic endeavour. The key concept of the field, the one of beliefs, is ill-defined, and the methodological problem of getting access these elusive constructs is unresolved. Further, it has turned out to be difficult to fulfil the promise of making significant contributions to the current reform agenda, in spite of comprehensive development and research efforts to do so. There are two sides to this, as it is no easy task to facilitate belief change, and even when beliefs attributed to the teacher on the basis of questionnaires and interviews are in line with reform intentions, classroom practices do not always comply. The latter of these problems is the background to the development of interpretations of the beliefs-practice quandary that are less causal and more dynamic than the ones that dominate traditional belief research (Op't Eynde et al. 2006; Schoenfeld 2011a, b). They suggest that the impact of mathematics related beliefs may be moulded by other mental constructs the role and significance of which are modified by contextual constraints. These interpretations, then, emphasise the dynamic relationships among mental reifications in the form of beliefs and knowledge, and between such reifications and classroom processes.

In line with more traditional approaches to beliefs most of these dynamic interpretations rely on acquisitionist, especially constructivist, interpretations of human functioning. Students and teachers are expected to come to hold or possess reified mental entities, beliefs, through processes of assimilation and accommodation as they engage in social interaction. Subsequently they are to enact these reifications, though possibly in modified form due to contextual constraints. Acquisitionism, however, has been challenged in recent years, for instance by the more participatory approach adopted in most studies of identity (Hodgen and Askew 2007; Horn et al. 2008; Ma and Singer-Gabella 2011). In line with this, I suggest interpreting students' and teachers' affectively laden action and meaning-making as shifting modes of participation in different social practices, rather than as contextually constrained release of mental reifications. The argument is that the general lack of confirmation of the congruity thesis in belief research, i.e. the thesis of close correspondence between beliefs and practice, may be addressed not by suggesting that the role and significance of beliefs is contextually constrained, but by shifting the emphasis from mental reifications to the social processes on which they are assumed to be based. The use of 'beliefs' (with inverted commas) in the title of this chapter is to indicate that I address affective issues normally dealt with in the field of beliefs, but that I suggest minimizing the emphasis on mental reifications and conceptualising affective issues in more processual terms.

To make my argument I focus primarily, but not exclusively, on teachers' beliefs. I begin by discussing the concept of beliefs and argue that there seems to be some agreement about a core of the concept in spite of the lack of an agreed-upon definition. Next, I outline and categorise some of the dynamic approaches to beliefs and link them to aspects of this core (section "Dynamic views of teachers' beliefs"). Following from that, I discuss the acquisitionist underpinnings that orient the larger part of the beliefs literature, including at least some the more dynamic approaches, and argue that the concept of beliefs is used about objectifications, i.e. about reified mental entities assumed to have explanatory power for practice (section "Belief research and acquisitionism – or why believe in beliefs?"). I build on Sfard (2008) to outline some of the drawbacks of such an approach and argue that an over-reliance on objectifications is somewhat ironic in relation to beliefs. This is the backdrop for the suggestion to adopt a more participatory stance in the form of a conceptual framework in the making that I call *Patterns of Participation* (sections "PoP – towards a participatory account" and "Using PoP for empirical purposes"). My colleagues and I have argued elsewhere that PoP addresses some of the conceptual and methodological problems of belief research (Palmér 2013; Skott 2013; Skott et al. 2011); in the present context I highlight how it differs from dynamic approaches developed within the field.

## The Concept and Expected Functions of Beliefs

One of the challenges of belief research is that the key concept of the field, the one of beliefs, is not easily defined. Some scholars, both in mathematics education and beyond, engage in lengthy discussions of the concept, while others define it only implicitly and in use. The latter approach may be based on one or more rationales. It may be implied that there is a core or an essence to the notion of beliefs that is generally accepted, even though it is difficult to phrase a definition that captures all aspects of the concept and delineates its borders sufficiently clearly vis-à-vis related ones. It may also be based on the recognition that empirical work on a concept that is initially ill-defined may invite dialogue, which in turn serves to specify the concept in question in greater detail. And it may simply acknowledge that explicit definitions do not carry unequivocal meanings and may be interpreted in a multitude of ways, even if agreement is reached. Each of these rationales suggests that a further search for an agreed-upon definition is not worth the effort, at least at present, but that continued research in the field is, even though we may not be able to specify the contents of the concept of beliefs as clearly as we would like.

I have argued elsewhere that whether beliefs are defined explicitly or not, there seem to be four key aspects to the concept (Skott 2014). First, beliefs are used about mental constructs that are subjectively true for the person in question. This implies that beliefs are characterised by individual conviction, but also that the individual holding them may accept alternatives as reasonable and justifiable. Second, there is an element of affect to beliefs. Beliefs, then, are value-laden and characterised by a certain degree of commitment. Third, beliefs are considered relatively stable. The individual is expected to carry his or her beliefs in and out of different settings without changing them significantly, and belief change is expected to occur only as a result of substantial, new personal experiences. Fourth, and as argued above, beliefs are expected to significantly influence individuals' perceptions and interpretations of experiential encounters as well as their contributions to the practices in which they engage. In fact, it is unlikely that research on teachers' beliefs would have attracted more than minimal attention, if they were not believed (!) to impact practice.

To sum up, the notion of beliefs is used in the literature about mental reifications that are acquired on the basis of comprehensive, previous social experiences and that are characterised by considerable degrees of conviction, commitment, stability, and impact. The core of the beliefs concept may, then, be defined as subjectively true, value-laden mental constructs that are the relatively stable results of substantial prior experiences and that have significant impact on practice. I do not mean to imply that this definition is helpful when describing an everyday use of the notion (e.g. *I believe it is going to rain tomorrow*). However, the four key aspects may be said to constitute the core of the concept as understood in mainstream belief research. In fact, I use the phrase of *mainstream belief research* to designate approaches that focus on beliefs about mathematics and its teaching and learning and explicitly or implicitly endorse all four.

The last of the four characteristics of the core of beliefs suggests that once established these reified mental constructs serve two functions. The flow downwards in Fig. 1 emphasises the reification process itself and the subsequent, (semi-)causal relationship between beliefs on the one hand and instructional behaviour and classroom practice on the other. However, beliefs also serve a function for the reverse movement, i.e. for guiding perception and interpretation and turning immediate social encounters into more coherent life experiences. In this sense, they are an assimilatory filter that shapes the flow backwards in Fig. 1.

In the literature, however, the thesis that there is congruity between beliefs and behaviour has been challenged as much as confirmed (Fives and Buehl 2012). This obviously calls for an explanation in view of the premise of the field. One response is to capitalise on the conceptual and methodological problems of belief research and argue that the methods used in the field do not provide access to what people really believe or at least not to beliefs that matter for the situation at hand.

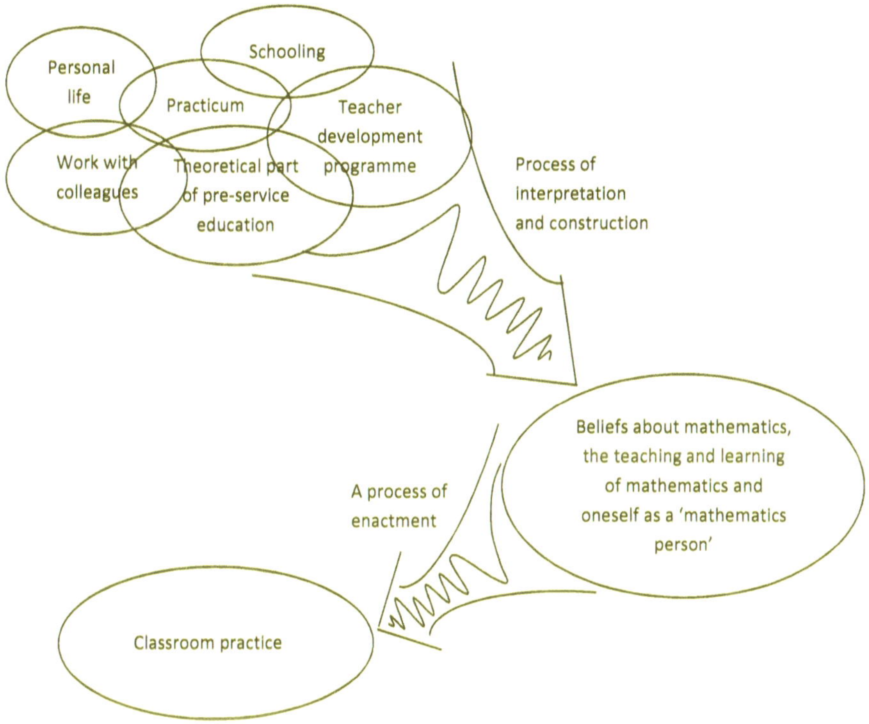


**Fig. 1** Beliefs in mainstream belief research

Consequently it is suggested that other types of analyses are needed (Speer 2008), or that beliefs are held in "clusters" (Green 1971) or "bundles" (Aguirre and Speer 2000) that have different relations to instruction and are structured so that the beliefs one does get access to for instance in interviews are different from the ones that manifest themselves in the classroom. These explanations are fully compatible with the claim that beliefs explain behaviour. Another set of responses, the ones discussed in the present context, modify this claim and suggest that 'context', in one or other interpretation of the term, may be a constraint on the opportunities for 'belief enactment', and that a more dynamic approach is needed to understanding the functioning of the individual in that 'context'.

In relation to the students, such dynamics is apparent for instance in the work of Malmivuori (2006). She discusses *self-systems*, i.e. relatively stable mental structures encompassing knowledge of mathematics, beliefs about the subject and about self in mathematics, affective schemata, and habitual behavioural patterns in mathematical situations. In Malmivuori's analysis, self-systems are "the basis for the functioning of students' [...] metacognitive, cognitive, and affective capacity used in mathematical thinking", but their role is conditioned by situation-specific factors (p. 151). Also working with students, Goldin et al. (2011) introduce *engagement structures* that have beliefs and values as one of ten inherent components or strands, and which they describe as "*behavioural/affective/social constellation[s]* situated in the person" (p. 548; emphasis in original). Examples of engagement structures include "Get the work done", i.e. completing assignments by following instructions; "Look how smart I am", i.e. impressing others with one's mathematical performance; and "Stay out of trouble", i.e. avoiding interactions that may cause conflict or distress. It is a main aim for Goldin et al. to describe the "*particulars* of *how* beliefs, values, emotional feeling, and social situations interact in a structured way to influence in-the-moment engagement with mathematics" (p. 552; emphasis in original). Engagement structures are embedded in people, but activated in particular situations and as such descriptive of a person's state. Beliefs, in contrast, are taken as traits that in the particular situation may motivate involvement in certain engagement structures, but inhibit the activation of others.

There is dual dynamic involved in the view of beliefs in these studies. First, they both acknowledge that immediate social interaction and the related contingencies play a role for the extent to which beliefs inform students' participation in the classroom. This dynamic relates to the person-context interface. Second, the studies consider an internal dynamic in the form of shifting relationships between the elements of self-systems (Malmivuori 2006) or engagement structures (Goldin et al. 2011). While these studies modify the assumption of a direct causality between beliefs and behaviour, they still expect relatively high levels of stability and impact of students' and teachers' mathematics related beliefs, and in this sense they are in line with mainstream belief research. Similar conceptualisations may be found in studies of teachers' beliefs.

## Dynamic Views of Teachers' Beliefs

One may differentiate at least four possible contextual and dynamic categories of perspectives on the role of teachers' beliefs for practice, which I label *enactment*, *activation*, *situatedness*, and *emergence*. They relate differently to the last two of the four characteristics of the core of the beliefs concept, the ones of relative stability and of expected impact. I build on different studies to elaborate on the distinctions between the categories. However, my intention is not primarily to 'locate' these studies in particular categories, but to use them as starting points for specifying the character of the categories themselves.

One dynamic interpretation, labelled *enactment*, is compatible with mainstream belief research in the sense defined above. Schoenfeld's recent work may be taken as a starting point for a description of this category (Schoenfeld 2011a, b). He subsumes beliefs under a broader concept of 'orientations' and takes what he calls resources (most notably knowledge) and goals into account.

Schoenfeld emphasises the role of planned behaviour in instruction, as "the vast majority of a teacher's actions in the classroom are shaped by the teacher's agenda", for instance as it materialises in the lesson plan (Schoenfeld 2011a, p. 9). Further he suggests that teachers base their behaviour on combinations of their goals, resources, and orientations both when teaching goes according to plan and when planned action is disrupted by contingencies, such as an unexpected response or suggestion from a student. There is in Schoenfeld's interpretation a dynamic relationship between the orientations, resources, and goals brought to the situation by the teacher and the goals that are pursued at the instant. Schoenfeld, then, assigns a significant role to orientations, most notably beliefs, as the individual teacher's choice of action among a range of possible options "depends on that teacher's orientation […] and what resources the teacher can bring to bear in support of the option he or she has chosen" (Schoenfeld 2011a, p. 13).

The dynamics between beliefs and behaviour suggested by Schoenfeld resemble the ones described by Malmivuori (2006) and Goldin et al. (2011). He acknowledges the significance of social interaction for the degree and character of belief enactment, and points to an internal dynamic that links emerging goals to shifting relationships between the goals, orientations, and resources brought to the classroom (Schoenfeld 2011a, b). This is all in line with the constructivist tenet that social interaction is a significant source of perturbations to what is conceived as the relatively autonomous functioning of the individual. In this interpretation, Schoenfeld assigns considerable stability and impact to teachers' beliefs. The position may be described as one of contextually constrained enactment of mathematics related beliefs.

In the cases Schoenfeld works with, the orientations that are enacted are closely related to the contents. This may be because of his emphasis on planned behaviour and because the examples he presents are from high school or college education and/or conducted by very experienced mathematics educators. A second dynamic perspective, the one of belief *activation*, is developed by others, who work with teachers with weaker backgrounds in mathematics and/or mathematics education. They have found that content-related beliefs may lose part of their significance or may be transformed in the educational process. Sztajn (2003), for instance, compares two elementary school teachers, Teresa and Julie. They both consider themselves in line with current reform initiatives, although their interpretations of the reform differ, and they are both convinced that their instructional approaches comply with these recommendations. However, these approaches are very different, and in Sztajn's interpretation the differences are not adequately accounted for by differences in the teachers' beliefs about the reform. Instead, she suggests that the teachers base instructional decisions on broader aspects of their students' lives than those related to their mathematical learning, and in particular it seems significant that Teresa and Julie teach children from very different socioeconomic backgrounds. Teresa teaches in a relatively poor neighbourhood, and in Sztajn's interpretation she emphasises rules and drill, as she seeks to "transform lower socioeconomic students into good citizens" (p. 69). In contrast, Julie works in a well-to-do area and teaches "higher-order thinking through educationally rich projects" in order to make schooling a good experience for the students (p. 69). Sztajn's point, then, is that beliefs beyond mathematics, especially a broad view of students' needs, play prominently in instruction, as teachers make ideological decisions "about what within the reform rhetoric fits particular children" (p. 70).

Also, my own previous work has questioned the extent to which teachers base instructional decisions on their beliefs about mathematics (Skott 2001, 2009). In one study, the teacher, Christopher, presented *school mathematics images* (SMIs) in interviews and questionnaires that were deemed highly compatible with aspects of the reform discourse (Skott 2001). However, in specific classroom episodes Christopher reacted to different groups of students in ways that appeared to be in mutual conflict with one another, and often also in conflict with his SMIs. Rather than interpreting these conflicts as expressions of teacher inconsistency, I saw them as cases in which Christopher's reformist intentions were to different degrees dominated by other concerns such as supporting the students' self-confidence and ensuring their position in the classroom community. In another study, a novice teacher, Larry, presents beliefs that are highly inspired by the reform, but he gets his first teaching position at a very conservative private school (Skott 2009). The tension between Larry's intention of supporting the students' own investigations and the school's emphasis on testing their command over standard procedures highlights the need for a contextualised views of beliefs that relate Larry's contributions to classroom interaction to three actual and virtual communities of practice, the ones of immediate classroom interaction, of the teachers at his school, and of his college education.

There is nothing in Schoenfeld's model of the moment-to-moment decision-making in teaching that is in principle at odds with the emphasis on broader educational issues in the studies by Sztajn and Skott. Both studies share Schoenfeld's view of teaching as goal-oriented, and the teachers may be seen as just bringing a broader set of orientations to the classroom, a set that encompasses educational issues beyond mathematics.

However, there is a difference in degree, if not in kind, between the types of dynamics involved in Schoenfeld's model (Schoenfeld 2011a) and my own previous work (Skott 2001, 2009). For Schoenfeld teaching is primarily a matter of planned enactment of orientations and resources, even though he does allow for contingencies. There is a stronger emphasis on the emergence of goals in the locally social in my own work. The two approaches share the view of beliefs as relatively stable constructs; the difference concerns the expected impact of those related to mathematics. Schoenfeld's model suggests that pre-existing, mathematics related beliefs are highly influential as teaching is basically a matter of enacting them (with due consideration of contingencies); in comparison I (in the studies above) suggest a stronger contextual dynamic that leaves it as a more open question if, what, and how beliefs are activated in classroom interaction.

Another interpretation, the *situated* perspective, suggests that mainstream belief research should not primarily be questioned on the expectation of impact of mathematics related beliefs, but on the one of their contextual stability. Beliefs, the argument goes, are situated or distributed, and there is little reason to expect that beliefs espoused in questionnaires and research interviews resemble the ones that are observed in classroom interaction. Hoyles (1992), for instance, suggests that once the situated character of beliefs is recognised, it becomes "self-evident that any individual can hold multiple (even contradictory) beliefs, and 'mismatch', 'transfer', and 'inconsistency' are irrelevant considerations" (p. 40). Looking back on her own previous study of a female teacher, Ms. X, who teaches a group of high ability girls, Hoyles raises the following questions:

> How far was Ms X's mathematical perspective constructed by the 'high ability' of the group? How far was her emphasis on effort related to her sex and the sex of the students? Was her particular blend of exposition/interaction partly a function of the age and specialism of the students? I would of course now answer these questions in the affirmative! (p. 40)

Somewhat in line with Hoyles, Lerman (2001) argues that although there is "a family resemblance between concepts, beliefs, and actions in one context and those in another, they are qualitatively different by virtue of those contexts" (p. 36). The beliefs observed in the classroom, then, are not unrelated to the ones that may dominate the interview situation, "but the classroom is its own setting" (p. 36).

The notion of situatedness clearly questions the assumption of belief stability across contexts, but not necessarily the one of belief impact. Indeed, if the situatedness of teachers' beliefs is used to explain why observed classroom practices differ when the same teacher works with different groups of students and why classroom practices differ from beliefs as inferred from research interviews or questionnaires, it is still implied that beliefs are an explanatory principle for practice. The differences are accounted for, not by suggesting that beliefs, still understood as reified prior experiences, do not matter for practice, but that the experiences gained in different settings are sufficiently different for the beliefs, to differ as well (Skott, 2014).

A fourth dynamic and contextual perspective on belief-practice relationships, the one of *emergence*, also involves a view of beliefs as in some sense situated. In this interpretation, however, beliefs are reflexively related to the classroom processes that evolve at the instant, and consequently less reified. This is the case for instance in work of Cobb and Yackel (1996), whose emphasis is on students rather than on teachers. Their framework includes the well-known concept of socio-mathematical norms, i.e. "the normative aspects of whole-class discussions that are specific to students' mathematical activity" (p. 178). Socio-mathematical norms are seen as collective counterparts to "mathematical beliefs and values". At a more general level of analysis, "classroom social norms" correspond to "beliefs about own role, others' role, and the general nature of mathematical activity in school", while at a more specific level "classroom mathematical practices" are seen as social correlates of individuals' "mathematical conceptions and activity".

Cobb and Yackel are explicit that they developed the framework for the purpose of accounting for and supporting the students' mathematical development, not to analyse the norms in their own right. The same is the case in Cobb's and his colleagues' subsequent use of the framework in development activities concerned with elementary students' learning of measurement (Stephan et al. 2003) and lower secondary students' work on data handling (Cobb et al. 2001, 2003). However, in order to make such accounts, presumably including accounts of students' beliefs, they found it necessary to include a social perspective and conduct analyses of classroom social norms, of socio-mathematical norms, and of classroom mathematical practices.

Two elements of the argument made by Cobb and his colleagues are particularly relevant in the present context. First, they argue that the relationship between beliefs and norms is reflexive rather than causal. This means for instance that “neither the social norms nor individual students’ belief are given primacy over the other [...] social norms and beliefs are seen to be reflexively related such that neither exists independently of the other” (Cobb and Yackel 1996, p. 178). Supposedly, the same holds also for teacher’s beliefs. Second, all norms are seen as established jointly by the students and the teacher. Classroom practices, then, are not interpreted as the teacher’s practices, but as the result of continuous renegotiation among all participants in the classroom community, in spite of the special role of the teacher in it.

As indicated above, Cobb and his colleagues were not primarily interested in teachers’ beliefs. However, the reflexivity between the social and the individual and the view of the classroom as a jointly emerging reality suggest a fourth possible category of perspectives on teachers’ beliefs, the one of emergence, according to which they are neither necessarily stable nor determiners of the practices that evolve in the classroom.

The point in this section is that it makes sense to discuss the dynamic character of beliefs in relation to the two dimensions of stability and impact. To make the point, I have used particular studies to elaborate on the meaning of each of the four cells in Table 1. I should reiterate, however, that this does not mean that these studies are necessarily ‘located’ in those cells. For instance, both Lerman and Hoyles argue for a ‘low-stability’ view of beliefs, but neither of them is explicit that situated beliefs are ‘high-impact’. However, I have used their studies to suggest what may characterise a ‘low-stability, high-impact’ approach to belief research. The studies mentioned in the cells in Table 1, then, are meant as reminders of the studies used to elaborate on the meaning of the particular cells, not necessarily as studies deemed exemplary for the cell in question.

The four cells in Table 1 all represent relatively dynamic interpretations of the belief-practice quandary. Possibly with the exception of the emergent perspective, however, they all interpret beliefs as individual reifications that significantly influence practice, although the beliefs in question are not necessarily related to mathematics (*activation*) and not necessarily stable across contexts (*situatedness*). This indicates that the acquisitionist underpinnings of mainstream belief research continue to orient the field, also when more dynamic interpretations are developed.

**Table 1** Dynamic perspectives on teachers’ beliefs

| | | Impact of teachers’ mathematics related beliefs on classroom practice | |
|---|---|---|---|
| | | High | Possibly low |
| Stability of teachers’ mathematics related beliefs across contexts | High | Enactment (Schoenfeld 2011a, b) | Activation (Sztajn 2003; Skott 2001) |
| | Possibly low | Situatedness (Hoyles 1992; Lerman 2001) | Emergence (Cobb and Yackel 1996) |

## Belief Research and Acquisitionism – Or Why Believe in Beliefs?

The notions of reification and objectification have been discussed by Sfard as part of her challenge to acquisition as a metaphor for learning and knowing (Sfard 2007, 2008). In her terminology, objectification is a two-stage process that transforms human engagement in discursive practices into apparently self-sustained, mental entities. The first stage is a reification, in which "sentences about processes and actions [are replaced by] propositions about states and objects" (Sfard 2008, p. 44). The second stage is an alienation in which reified objects get a life of their own, independently of the processes that initially gave rise to them.

Sfard's examples of objectification include the notions of number and of thinking. Number, she says, is a reification of a counting procedure (1, 2, 3, …) that is transformed into an adjective (three apples) and then into a noun-like entity that may be operated on (3 + 5 =). Finally, number becomes a mind independent, alienated object that has its own characteristics independently of any mental activity (whether 3517211 is a prime is independent of whether anybody cares to find out). As far as the other example, thinking, is concerned, Sfard argues that a discourse dominated by the acquisition metaphor makes us think "of knowledge as a kind of material, of human mind as a container, and of the learner as becoming an owner of the material stored in the container." (p. 49). The acts of knowing or coming to know in a particular situation are decontextualized and the content of knowing is considered an entity with a strong element of permanence. As a final example, one may use the concept of objectification on itself. Describing the stages of objectification, Sfard objectifies the process, and uses the term of objectification to point to an independent entity as well as to the process.

The core of the beliefs concept (cf. section "The concept and expected functions of beliefs") implies that beliefs are generally regarded as objectifications. Abelson (1986), for instance, suggests that beliefs resemble possessions that are acquired, kept, valued, and sometimes lost, and although people do not buy or sell beliefs, they often accept that beliefs come at a cost. Referring to Abelson's earlier work (Abelson 1979), Nespor (1987) develops a conceptualisation of beliefs on the basis of a qualitative study of teachers of different subjects. He suggests that beliefs are characterised among others by what he calls an "existential presumption". This is the tendency to phrase beliefs in terms of reifications that point to the existence or non-existence of the involved entities. The two mathematics teachers in Nespor's study, for example, explained students' (lack of) learning with reference to their "ability", "maturity", and "laziness". In Nespor's interpretation these terms not only describe the students' participation in the classroom, but function as "labels for entities thought to be embodied by the students" (p. 318). Nespor's point is that the teachers have acquired and now possess reified mental constructs that allow them to bring order to the poorly structured problems of practice by interpreting and explaining student action and subsequently to define their own tasks in the classroom.

Nespor's examples suggest that there are advantages and disadvantages to objectification. Similarly Sfard (2008) says that it increases the effectiveness of communication and constitutes the basis for accumulation of experience. However, objectified entities are the result of an ontological collapse, as the discursive construction of the object is disregarded, and the object itself is mistakenly conceived as belonging to a mind-independent, perceptually accessible reality. Consequently, objectified entities carry connotations of permanence and repetitiveness that may be unfounded, and they invite interpretations of the future in the image of the past. Sfard mentions mathematical inability and giftedness as examples to make her point.

In belief research, teachers' beliefs are viewed as the result of a two-stage objectification process on the part of the teachers themselves. However, the beliefs attributed to, or symbolically imposed upon, teachers are a result of a similar, but second-order objectification process on the part of an observer. The researcher working with beliefs attributes sets of reified priorities, beliefs, to teachers and uses them to explain instructional decisions. For example, Nespor's argument that the teachers in his study impose personality traits (ability; maturity; laziness) on the students to account for their behaviour may be turned on the field of beliefs itself to the extent that researchers in the field impose trait-like beliefs on the teachers (traditionalist; reform oriented; inconsistent) so as to explain *their* classroom behaviour.

The drawbacks of objectification that Sfard points to, then, are apparent in the dominant use of beliefs. The very idea that beliefs are a priori expected to influence classroom practice is a paradigm case of how the experiential base of an objectified entity is disregarded and of how the reifications themselves are expected to mould future activities. One may object that this is no more and no less than yet another example that people understand and act in and towards the world in ways that reflect the meaning they attribute to that world. However, mainstream belief research is based on the premise that pre-existing, de-contextualized, and temporally stable beliefs about mathematics and its teaching and learning are the main, if not the sole determiners of such meaning. The more participatory approach outlined below allows for the possibility that such meaning-making is more or at least differently dynamic than usually assumed in belief research, also in approaches that may be 'located' in cells [11], [12] and [21] in Table 1.

## PoP – Towards a Participatory Account

Recently, attempts have been made to challenge acquisitionism and develop or use more participatory accounts of human functioning. Sfard's work, referred to previously, is an ambitious attempt to develop such an account; Barwell (2013) draws on discursive psychology to make more locally social and dynamic analyses of what is normally discussed in terms of knowledge; and Wagner and Herbel-Eisenmann (2009) build on positioning theory to do so in the field of identity. In the field of affect few such attempts have been made, but Evans et al. (2006) and Horn (2007) are notable exceptions. In what follows I recapitulate my own attempt to build on social practice theory and symbolic interactionism to reconceptualise what is normally phrased in terms of beliefs, and develop a conceptual framework, called Patterns of Participation (PoP), that as far as possible avoids relying on objectifications when analysing classroom practices. In a more positive wording, PoP may be described as an approach to classroom interaction that views individuals' contributions in processual and participatory terms and interprets them as meaningful re-engagement in other past and present practices in view of the ones that unfold at the instant.

I suggested previously that Schoenfeld's model of teaching implies that belief enactment may be modified by two distinct types of dynamics, one that concerns the person-context interface and another – depending on the first – that refers to shifting relationships among different orientations, goals and resources, i.e. among different reifications. Similarly, there are dual dynamics involved also in 'belief activation', i.e. when belief selection is based primarily on immediate social interaction, and in what I described as the situated perspective (cf. Table 1). Although the three perspectives differ in their views of the character and functioning of the dynamic relationships, they all locate these in the bottom half of Fig. 1, i.e. between beliefs and classroom practice or among the beliefs themselves, possibly supplemented with other reifications in the form of knowledge and goals.

Reducing the emphasis on objectifications, PoP assigns greater significance to the relationship between the experiences at the top of Fig. 1 and classroom practice without relying on beliefs as an intermediary reification. However, merely transforming Fig. 1 by turning the top arrow clockwise and erasing beliefs would indicate an immediate, causal connection between prior experiences and classroom practice. This loses the potential of the more interactive interpretations of classroom processes that have been developed recently, including the ones outlined in section "Dynamic views of teachers' beliefs", and it may even be read as disregard for the attempts to understand individual meaning-making that have always fuelled belief research. The intention of PoP is exactly to focus on such meaning-making, but in contrast to belief research to suggest that it is to a lesser extent based on reifications (beliefs) than on dynamic re-engagement in the practices that in belief research are assumed to be the basis for them.

There is also a dual dynamic involved in PoP interpretations of classroom interaction, although it is somewhat different from the ones described previously. First, and somewhat in line with Cobb and Yackel (1996), classroom practices are viewed as social phenomena, not as an outcome of any individual's actions. In PoP they are seen as constituted in a process during which each individual continuously makes symbolic interpretations of others' actions as well as of others' (possible) reactions to one's own behaviour. This is inspired by symbolic interactionism, especially of its view of the self as consisting of two phases, an *I* and a *me* (Blumer 1969; Mead 1934). The *I* acts, but in the process the individual becomes an object to him- or herself, i.e. becomes the *me*. In the action, then, the individual takes the attitude of individual or generalised others and adjusts his or her actions accordingly. This is significant not least in relation to affective issues (Shott 1979). Second, behaviour is not seen as a release of reified mental entities, whether in the form of beliefs,

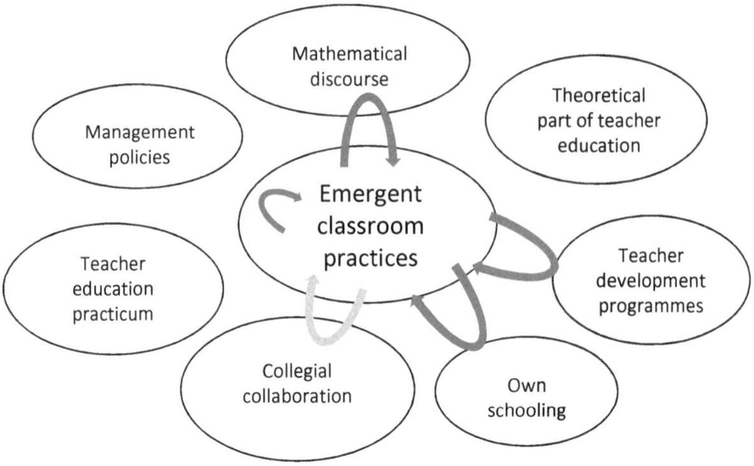


**Fig. 2** Drawing on other practices and figured worlds in classroom interaction

knowledge, or any combination of the two, but as an outcome of the teacher reengaging in a range of other social practices stemming for instance from team or department meetings, theoretical discussions in teacher education or development programmes, experiences from their own schooling, and many more.

Consider for example a teacher working with a group of students, who are trying to substantiate a mathematical conjecture, but finds it difficult to do so. The teacher's contributions to the interaction may change, if she, while engaging in a mathematical discourse in order to assist the students, also orients herself towards the reform, possibly as propagated in a recent teacher development programme; positions herself within a team of teachers, whose cooperation focuses on the well-being of individual students rather than on their subject matter learning; and manifests her own professional authority, as her mathematical competence was recently questioned in the class. In the interaction, then, i.e. as classroom practices emerge, the teacher draws upon and renegotiates the meaning of prior social practices (Fig. 2). This exemplifies that in PoP we look at classroom processes in an attempt to link the teacher's contributions to the interaction to other significant practices or to what Holland and her colleagues call figured worlds (Holland et al. 1998), i.e. collective as-if worlds in which "particular characters and actors are recognised, significance is assigned to certain acts, and particular outcomes are valued over others" (p. 52). The imaginary example above indicates that beyond mathematics such practices and figured worlds may include – among many more – a reform agenda, a team of cooperating teachers, and even schooling in a general sense. Schooling, for instance, qualifies as a significant figured world in relation to the challenge to the teacher's professional authority. Students and teachers are recognised as such due to their different positions in the interaction; significance is assigned to acts of teaching that position the teacher as knowledgeable and able to support students in solving the tasks at hand; and the valued outcomes include identifiable and recognisable shifts in understanding or proficiency on the part of the students. In the example the degree to which this sense of 'schooling' plays a role for the teacher may inform how she seeks to contribute to the students' reasoning about the task. I refer to two recent studies to indicate how PoP may be used for empirical purposes.

## Using PoP for Empirical Purposes

I indicated previously that belief research is faced with considerable methodological problems, as there is no easy access to the mental reifications assumed to reside entirely in people's minds. Consequently, a combination of for instance observations, interviews, and surveys is often used, based on the expectation that between them they shed sufficient light on what teachers and students 'really believe'. As I have argued elsewhere, however, methodical triangulation is itself problematic, as it takes contextual and temporal stability of beliefs for granted (Skott 2014).

Although teachers' participation in discourses and non-verbal practices is more readily observable than their beliefs, PoP-research is clearly confronted with its own methodological challenges (Skott 2013). As Fig. 2 indicates, the task is to interpret classroom action and meaning-making as they relate to the teachers' participation in other past and present practices that are not all equally accessible and the character and significance of which cannot be specified beforehand. To meet these challenges we use an approach inspired by recent developments of grounded theory (GT) in combination with methods that are often associated with educational ethnography (Charmaz 2006; Charmaz and Mitchell 2001). Initially data are generated from video recordings of considerable amounts of classroom teaching and from relatively open qualitative interviews, sometimes using stimulated recall. These methods are combined with informal observations of staffroom communication. The data are continuously coded, compared, and theorised in line with GT guidelines, but without the naturalistic connotations often associated with them. The analyses of the data lead to suggestions for practices and figured worlds beyond the classroom that are significant for the teacher's contributions to the ones that unfold within it. Subsequently new data are generated on these other practices and compared to those stemming from the classroom. In the case of Anna, discussed briefly below, it became apparent from the initial interviews that she relates closely to the functioning of "her team", i.e. the group of four teachers, who teach all subjects in grade 7 at her school. Consequently, I observed team meetings and conducted a group interview with the other team members, so as to get a sense of how the team negotiates the task of taking on the full responsibility for the year-group; of how Anna contributes to that negotiation; and of how the negotiation relates to Anna's shifting tales of herself as a professional at Northgate, including how she positions herself in the classroom.

It is apparent, then, that PoP makes use combinations of methods that are somewhat similar to the ones used in belief research, but does so for other reasons. In research on beliefs, multiple methods are used in attempts to gain access to the same mental constructs, a teacher's or student's beliefs. In PoP the intention is in some sense the opposite, i.e. as far as possible to get access to *different* practices and figured worlds. An open, qualitative interview may, for instance, shed light on aspects of how a teacher engages discursively with mathematics education, including if and how she relates to elements of the current reform; a set of classroom observations may suggest how she copes with the multiple challenges involved in classroom interaction in a particular context. Although her discursive engagement with the reform may matter for her contributions to classroom practice, the reform discourse and classroom interaction are viewed as decidedly different practices or figured worlds. The methodological decisions in PoP, then, are concerned with how best to develop an understanding of the range of different practices and figured worlds that are currently significant for the teacher in question, and of how they may inform one another, in particular how they relate to her contributions to classroom interaction. It follows that interviews, let alone surveys, are of limited value in PoP as the sole sources of data generation, as they provide little access to the range of practices beyond teachers' rhetorical commitment to specific discourses such as the reform.

To exemplify the methods as well as some results, I refer to the study of Anna, mentioned above (Skott 2013). Anna is a young, novice teacher, who teaches mathematics at Northgate Primary and Lower Secondary School in Denmark. She is selected for the study because of her mathematical and professional self-confidence and commitment at the time of her graduation. She explicitly considers herself a *mathematics* teacher, not just someone who teaches mathematics, and the initial interviews and observations suggest that one aspect of her dedication is linked to the current reform agenda, not least to investigations and student communication, which, she says, was promoted by her teacher education programme. The initial data also indicate that there are two other, but partly overlapping, practices and figured worlds beyond *mathematics* and *the reform* that are less immediately related to the contents of instruction, but that play prominently for Anna's tales of herself as young professional and for how she positions herself in the classroom. One of these is *teaming*. The initial interviews with Anna as well as the subsequent data from observations and interviews with the team suggest that the team does not plan instruction or teach together, at least not in the first 2 years after Anna's graduation. The team focuses on less content specific aspects, such as the social functioning of the classes and individual students' social and personal problems. This relates closely to the fourth figured world discerned from the initial data, one concerned with Anna's attempts to build trusting relationships with the students. The valued outcome of this world of *relationing* positions Anna as what she half-jokingly describes as being "somewhere between a mother and a friend" for the students. Between them the data generated with the different methods invite interpretations of the meanings these four practices and figured worlds have for Anna in different situations, including classroom contexts.

In the above analysis, then, the most prominent practices and figured worlds for Anna's meaning-making and instructional decisions in her classroom are *mathematics*, *the reform*, *teaming*, and *relationing*. However, the character and relative significance of each of them change as the interactions unfold. It is noticeable, for instance, that the aspect of the reform concerned with students' mathematical communication sometimes loses the qualifier of mathematical and becomes merely a matter of verbal exchanges. This may happen when Anna takes the attitude of what she at the instant considers vulnerable students, as the emphasis on communication in *the reform* is submerged by or embedded in *relationing* in her attempts to avoid jeopardising the students' self-confidence or her own relationship with them.

My colleagues and I also worked with another novice teacher, Susanne, who teaches mathematics in grade 5 at a school called Southern Heights (Skott et al. 2011). The students at the school have mixed social backgrounds, and as Susanne points out, a significant number of them come to school every morning "without breakfast and without a kiss and a hug and without all the other things that the rest of us consider matters of course" (p. 41).

Susanne draws heavily on what she describes as traditional teaching: "teaching-from-the-board and exercises" (p. 38). However, in a PoP interpretation, elements of other practices and figured worlds are renegotiated and inserted in isolated instances and on the fringes of the dominant instructional approaches. This is so for instance with the reform discourse on student understanding, i.e. "that doctrine that they need to understand and not just follow the rules" (p. 40). This discourse plays small but significantly different roles, when she introduces procedures for the students to copy and routinize, and when students, in spite of Susanne's emphasis on standard procedures, come up with unexpected suggestions for how to solve the tasks. Also, Susanne reinterprets what she considers successful initiatives on the part of the school to support children with social or personal problems. One such initiative is a special needs department for students with learning problems in particular subjects; another is the 'observation class', 'the obs', where students are sent, if they are unruly, but which Susanne, at least when she talks about the students in general, also sees as an opportunity for them to calm down, if they are under pressure. Susanne makes frequent use of both the special needs department and 'the obs'. Further, she seems inspired by these organisational measures at Southern Heights also in relation to other students in her class and asks groups of students to work on their own elsewhere or to work independently in the classroom, apparently in an attempt to create a sufficiently homogeneous group of students to work with herself. We suggest, however, that in the process this partial imitation of the school's segregation policies shifts its meaning for Susanne from taking care of students with problems to handling problematic students.

The practices and figured worlds that are significant differ in the two cases above and so do the ways in which their mutual relationships develop as classroom practices unfold. In spite of the differences, however, they both lend themselves to PoP interpretations.

## Concluding Remarks

Irrespective of the problems of defining and accessing mathematics related beliefs, empirical findings in belief research more than suggest that the expectation of a causal relationship between such reified constructs and behaviour needs to be modified. The section "Dynamic views of teachers' beliefs" outlines four such modifications that differ in their interpretations of the contextual and temporal stability of beliefs and of what and if beliefs are influential, even if stable. These responses, then, relate differently to the last two of the core aspects of beliefs as outlined in section "The concept and expected functions of beliefs", the ones of stability and impact.

In spite of the differences among these more dynamic interpretations, at least three of them conceptualise beliefs as objectifications. Section "Belief research and acquisitionism – or why believe in beliefs?" discusses some of the problems with this, including the somewhat ironic observation that the use of objectifications in the beliefs literature is characterised by a similar existential presumption to the one Nespor assigns to the beliefs of the teachers in his study. It is implicit in Nespor's discussion that teachers' beliefs that their students are able, mature or lazy do not qualify as reasonable explanations for the students' actions. One may wonder why belief research attributes similar explanatory power to the trait-like beliefs attributed for instance to teachers.

This is the backdrop to the presentation of PoP, a framework that suggests a negative answer to the question of whether we need to rely on objectified beliefs, when attempting to understand what roles the teacher plays in the practices that emerge in the classroom. This, however, needs an elaboration.

As Sfard points out, there are certainly advantages to the use of reifications, and we tend to make and use them in order to make sense of and function in complex situations. However, there are at least two reasons why belief research seems to rely too heavily on beliefs as reifications. First, researchers attribute beliefs to teachers and students, and in this sense the beliefs described in the literature are second-order reifications. More often than not the research participants do not explicate these reifications themselves. In spite of that they are expected to make sense of the world by using them. Second, even when research participants do describe their relationship to mathematics and its teaching and learning in reified terms, it is an empirical question what role the reifications play, if any. In the interviews with Anna and Susanne, they are both close to using reifications to describe themselves as teachers of mathematics, Susanne explicitly calling herself a traditionalist, and Anna emphasising that she is a *mathematics* teacher, who prioritises certain aspects of the reform. Doing so, they engage in a discursive construction of themselves as professionals, which may resemble or inform their contributions to unfolding classrooms events. Anna, for instance, time and again requires the students to find their own solution strategies and discuss their methods and results with one another. However, Anna's reengagement in the reform discourse in the classroom is often transformed as she appears to position herself in her team or among her colleagues in general or by her attempts to be "somewhere between a mother and a friend" for the students. As she makes sense of and contributes to emerging classroom practices, then, Anna takes the attitude of different individual and generalised others (e.g. students), including some that are not physically present (e.g. other team members) and others that are only established discursively (e.g. *the reform*).

PoP implies taking the dynamic perspective beyond an approach that looks at shifting relationships between pre-established reifications. Instead, it suggests attempting to understand how teachers draw on and renegotiate their participation in a range of other past and present practices and figured worlds as they engage in classroom interaction. This provides a differently dynamic perspective on what in beliefs terminology may be phrased as the beliefs-practice quandary.

## References


Abelson, R. P. (1979). Differences between belief and knowledge systems. *Cognitive Science, 3*(4), 355–366.

Abelson, R. P. (1986). Beliefs are like possessions. *Journal for the Theory of Social Behaviour, 16*(3), 223–250.

Aguirre, J., & Speer, N. M. (2000). Examining the relationship between beliefs and goals in teacher practice. *Journal of Mathematical Behavior, 18*(3), 327–356.

Barwell, R. (2013). Discursive psychology as an alternative perspective on mathematics teacher knowledge. *ZDM, 45*(5), 595–606.

Blumer, H. (1969). *Symbolic interactionism. Perspective and method.* Berkeley: University of Los Angeles Press.

Charmaz, K. (2006). *Constructing grounded theory. A practical guide through qualitative analysis.* London: Sage.

Charmaz, K., & Mitchell, R. G. (2001). Grounded theory in ethnography. In P. Atkinson, A. Coffey, S. Delamont, J. Lofland, & L. Lofland (Eds.), *Handbook of ethnography* (pp. 160–174). London: Sage.

Cobb, P., & Yackel, E. (1996). Constructivist, emergent, and sociocultural perspectives in the context of developmental research. *Educational Psychologist, 31*(3/4), 175–190.

Cobb, P., Stephan, M., McClain, K., & Gravemeijer, K. (2001). Participating in classroom mathematical practices. *The Journal of the Learning Sciences, 10*(1/2), 113–163.

Cobb, P., McClain, K., & Gravemeijer, K. (2003). Learning about statistical covariation. *Cognition and Instruction, 21*(1), 1–78.

De Corte, E., Op't Eynde, P., & Verschaffel, L. (2002). "Knowing what to believe": The relevance of students' mathematical beliefs for mathematics education. In B. K. Hofer & P. R. Pintrich (Eds.), *Personal epistemology. The psychology of beliefs about knowledge and knowing* (pp. 297–320). New York: Routledge.

Evans, J., Morgan, C., & Tsatsaroni, A. (2006). Discursive positioning and emotion in school mathematics practices. *Educational Studies in Mathematics, 63*(2), 209–226.

Fives, H., & Buehl, M. M. (2012). Spring cleaning for the messy construct of teachers' beliefs: What are they? Which have been examined? What can they tell us? In K. R. Harris, S. Graham, & T. Urdan (Eds.), *APA educational psychology handbook* (Individual differences and cultural and contextual factors, Vol. 2, pp. 471–499). Washington, DC: American Psychological Association.

Goldin, G. A. (2002). Beliefs, meta-affect, and mathematical belief structures. In G. C. Leder, E. Pehkonen, & G. Törner (Eds.), *Beliefs: A hidden variable in mathematics education?* (pp. 59–72). Dordrecht: Kluwer.

Goldin, G. A., Epstein, Y. M., Schorr, R. Y., & Warner, L. B. (2011). Beliefs and engagement structures: Behind the affective dimension of mathematical learning. *ZDM, 43*(4), 547–560.

Green, T. (1971). *The activities of teaching.* New York: McGraw-Hill.

Hodgen, J., & Askew, M. (2007). Emotion, identity and teacher learning: Becoming a primary mathematics teacher. *Oxford Review of Education, 33*(4), 469–487.

Holland, D., Skinner, D., Lachicotte, W., Jr., & Cain, C. (1998). *Identity and agency in cultural worlds*. Cambridge, MA: Harvard University Press.
Horn, I. S. (2007). Fast kids, slow kids, lazy kids: Framing the mismatch problem in mathematics teachers' conversations. *Journal of the Learning Sciences, 16*(1), 37–79.
Horn, I. S., Nolen, S. B., Ward, C., & Campbell, S. S. (2008). Developing practices in multiple worlds: The role of identity in learning to teach. *Teacher Education Quarterly, 35*(3), 61–72.
Hoyles, C. (1992). Mathematics teachers and mathematics teaching: A meta-case study. *For the Learning of Mathematics, 12*(3), 32–44.
Lerman, S. (2001). A review of research perspectives on mathematics teacher education. In F.-L. Lin & T. J. Cooney (Eds.), *Making sense of mathematics teacher education* (pp. 33–52). Dordrecht: Kluwer.
Ma, J. Y., & Singer-Gabella, M. (2011). Learning to teach in the figured world of reform mathematics: Negotiating new models of identity. *Journal of Teacher Education, 62*(1), 8–22.
Malmivuori, M.-L. (2006). Affect and self-regulation. *Educational Studies in Mathematics, 63*(2), 149–164.
Mead, G. H. (1934). *Mind, self, and society from the standpoint of a social behaviorist*. Chicago: University of Chicago.
Nespor, J. (1987). The role of beliefs in the practice of teaching. *Journal of Curriculum Studies, 19*(4), 317–328.
Op't Eynde, P., De Corte, E., & Verschaffel, L. (2006). "Accepting emotional complexity": A socio-constructivist perspective on the role of emotions in the mathematics classroom. *Educational Studies in Mathematics, 63*(2), 193–207.
Palmér, H. (2013). *To become- or not to become – A primary school mathematics teacher*. Ph.D. thesis, Linnaeus University, Växjö, Sweden. Retrieved from: http://lnu.diva-portal.org/smash/get/diva2:613396/FULLTEXT02.pdf
Schoenfeld, A. (1992). Learning to think mathematically: Problem solving, metacognition and sense making in mathematics. In D. A. Grouws (Ed.), *Handbook of research on mathematics teaching and learning* (pp. 334–370). New York: Simon & Schuster Macmillan.
Schoenfeld, A. (2011a). *How we think. A theory of goal-oriented decision making and its educational applications*. New York: Routledge.
Schoenfeld, A. (2011b). Toward professional development for teachers grounded in a theory of decision making. *ZDM – The International Journal on Mathematics Education, 43*(4), 457–469.
Sfard, A. (2007). When the rules of discourse change, but nobody tells you: Making sense of mathematics learning from a commognitive perspective. *Journal of the Learning Sciences, 16*(4), 565–613.
Sfard, A. (2008). *Thinking as communicating. Human development, the growth of discourses, and mathematizing*. Cambridge: Cambridge University Press.
Shott, S. (1979). Emotion and social life: A symbolic interactionist analysis. *The American Journal of Sociology, 84*(6), 1317–1334.
Skott, J. (2001). The emerging practices of a novice teacher: The roles of his school mathematics images. *Journal of Mathematics Teacher Education, 4*(1), 3–28.
Skott, J. (2009). Contextualising the notion of belief enactment. *Journal of Mathematics Teacher Education, 12*(1), 27–46.
Skott, J. (2013). Understanding the role of the teacher in emerging classroom practices: Searching for patterns of participation. *ZDM – The International Journal on Mathematics Education, 45*(4), 547–559.
Skott, J. (2014). The promises, problems, and prospects of research on teachers' beliefs. In H. Fives & M. G. Gill (Eds.), *International handbook of research on teachers' beliefs* (pp. 13–30). New York: Routledge.
Skott, J., Larsen, D. M., & Østergaard, C. H. (2011). From beliefs to patterns of participation: Shifting the research perspective on teachers. *Nordic Studies in Mathematics Education, 16*(1–2), 29–55.

Speer, N. M. (2008). Connecting beliefs and practices: A fine-grained analysis of a college mathematics teacher's collections of beliefs and their relationship to his instructional practices. *Cognition and Instruction, 26*, 218–267.

Stephan, M., Bowers, J., Cobb, P., & Gravemeijer, K. (2003). *Supporting students' development of measurement conceptions: Analyzing students' learning in social context* (Vol. 12). Reston: NCTM.

Sztajn, P. (2003). Adapting reform ideas in different mathematics classrooms: Beliefs beyond mathematics. *Journal of Mathematics Teacher Education, 6*(1), 53–75.

Wagner, D., & Herbel-Eisenmann, B. (2009). Re-mythologizing mathematics through attention to classroom positioning. *Educational Studies in Mathematics, 72*(1), 1–15.

# Of Love, Frustration, and Mathematics: A Cultural-Historical Approach to Emotions in Mathematics Teaching and Learning

**Luis Radford**

*Like all other mental functions, emotions do not remain in the connection in which they are given initially by virtue of the biological organization of the mind. In the process of social life, feelings develop and former connections disintegrate; emotions appear in new relations with other elements of mental life.*

(Vygotsky 1999, p. 244)

**Abstract** Emotions have traditionally been characterized as inner, subjective, and physiological experiences, usually of an irrational nature. Against this subjectivist and physiological position, drawing on cultural psychology and anthropological research, in this article I advocate for a cultural conception of emotions and their role in thinking in general and mathematical thinking in particular. I argue that, rather than momentarily subjective phenomena, emotions (for instance, anger, frustration, love) are historically constituted. Emotions, I contend, are not opposed to thinking, but are an integral part of it. Emotions are as ubiquitous as breathing. I illustrate these ideas through the analysis of Grade 4 students working on a mathematical problem.

**Keywords** Thinking and emotions • Feelings • Cultural historical activity theory • Subjectification • Motives

## Introduction

In his Plenary Lecture at the CERME 7 Conference Hannula (2011) offers a detailed review of the problem of affect in mathematical thinking and learning. In particular, he points out the difficulties that mathematics educators encounter when trying to define the key concepts through which the affective domain can be scrutinized and understood (see also Goldin 2002; Furinghetti and Pehkonen 2002). The result is

L. Radford (✉)
École des sciences de l'éducation, Université Laurentienne, Sudbury, ON, P3E 2C6, Canada

School of Education, The University of Manchester, Manchester, M13 9PL, UK
e-mail: lradford@laurentian.ca



B. Pepin, B. Roesken-Winter (eds.), *From beliefs to dynamic affect systems in mathematics education*, Advances in Mathematics Education,
DOI 10.1007/978-3-319-06808-4_2

obvious: as far as the affective domain remains difficult to understand, its link to mathematics teaching and learning will remain difficult to recognize.

What, indeed, do we mean by affect? And how is it different from emotion and feeling? How do affect, feeling, and emotion relate to motives and motivation? Motivation, Hannula remarks, "is perhaps the most difficult [concept] to define" (2011, p. 44). This is so, I would like to suggest, because motive and motivation require that the manner in which individuals' intentions, needs, and interests relate to the social and cultural context be made unambiguous. Motives are the affective component of projects of life that link the individuals and their contexts, present and future. How to explain this link is not an easy matter. Here resides the central problem of the classical distinction between intrinsic and extrinsic motives, a distinction that remains decidedly dualistic. In dealing with motives, such an account assumes that the individual, while screening his/her sociocultural environment for clues and insights, finds in an allegedly insulated interiority the foundations of what moves him/her towards action. Unavoidably the intrinsic-extrinsic motive account ends up portraying individuals as entities living in solipsistic envelopes. The most profound deficiency of this account is that it assumes a kind of auto-sustained self. Within this model, motives are personal constructs and emotions truly private bodily phenomena.

The point that is missed here is that the affective domain in general and motives and motivation in particular are not only subjective but also sociocultural phenomena. They are subjective and sociocultural in the sense that on the one hand motives are the *motives of a concrete and unique person* but, on the other hand, they relate to a sociocultural and historical world that *transcends the individual.* In its transcendence, the sociocultural historical world indirectly—albeit in a decisive manner—shapes and organizes the individual's motives and emotions. This point, however, is often missed as a result of conceiving the relationship between society and its individuals as a relationship of opposition—society *versus* individuals. Commenting on this oppositional view, A. N. Leont'ev wrote: "the main thing is ignored, that in society man [sic] finds not only his external conditions to which he must adapt his activity, but also that these very social conditions carry in themselves the motives and aims of his activity" (Leont'ev 2009, p. 3).

In the past few years, sociocultural research has made an effort to go beyond the oppositional conception of the individual and the social. Evans and Zan distinguish three trends: (1) a socio-constructivist approach, where "Emotions are seen as social in nature and situated in a specific socio-historical context, because of the social nature of an individual's knowledge and beliefs" (Evans and Zan 2006, p. 44); (2) a discursive approach that considers "emotions as socially organized within a structure of social relations where power is exerted" (p. 43); and (3) an approach based on cultural-historical activity theory where emotions "come from the body… [and are] seen as integral to practical action" (p. 45). Evans and Zan (2006) show clearly how these approaches with their different conceptions of emotions address specific problems through different methodologies (see also Evans 2006).

The conception of emotions that I am about to sketch here draws on cultural psychology and anthropological research. It stresses the role of emotions in thinking.

My goal is to offer evidence of the manner in which thinking and emotions are intertwined in mathematical cognition and to stress some implications for teaching-and-learning. The cultural conception of emotions that I put forward here is located in an important shift that Evans (2006) and Evans and Zan (2006) note. According to these authors, there has been a shift in mathematics education research that goes from the investigation of more or less durable individuals' features (e.g., attitudes and beliefs, in general scrutinized through questionnaires and interviews) to research on emotions considered as volatile and contextual dynamic processes. I argue, however, that the contextual and dynamic nature of emotions cannot be limited to the analysis of their contextual occurrences. My contention is that emotions are dynamic processes, but rather than being singular and momentarily subjective, emotions (for instance, anger, frustration, love), while being socially organized, are historically constituted. The historicity of emotions (despite their formal acknowledgment in the sociocultural and activity theory trends identified by Evans and Zan) has not been a main theme in mathematics education research. The inclusion of the historical dimension into the investigation of emotions in mathematical thinking and learning, I contend, may help us to understand emotions not only as socially organized, or as bodily based, but also as historically structured and produced. The point is not, hence, to assert that we are emotional beings through and through. We are emotional beings, for sure. But the *kind* of emotional beings that we are can only be understood within the scope of cultural forms of subjectification that are available to us. Before going into the subject matter, I start with a brief overview of conceptions of emotions, hoping that the overview may provide a background against which we might better understand the affective domain in mathematics teaching and learning.

## The Naturalistic Approach to Emotions

In his 1932 series of lectures on psychology delivered at the Leningrad Pedagogical Institute, Vygotsky complained that emotions had been conceptualized in biological and naturalist terms only. He lamented that the investigation of emotions was "completely dominated by a pure naturalism of a kind profoundly foreign to other domains of psychological investigation" (Vygotsky 1987, p. 325). Darwin's (1886) famous book *The expression of the emotions in man and animals*, preceded by physiological investigations in France, England, Germany and other countries, paved the way to a conception of human emotions as remnants of our animal nature—vestiges of our irrational forces. Meticulous observations were made to ascertain the bodily modifications that animals and human undergo during emotional experiences. Changes of activity in the autonomic nervous system (e.g., perspiration, pupillary dilation, heart rate) were taken as "expressions" of our emotional life. At the end of the nineteenth century one of the central questions revolved around whether primacy was to be given to the 'bodily disturbances' or to the 'mental states' that occur in an emotional experience. In other words, the question was to determine whether emotion as a psychic state preceded its bodily expression or whether it was the other way around. For the idealist camp, a mental perception of a fact (a dangerous situation, for example) excites a mental affection (considered to be *the* emotion, in this case, fear), which leads to a bodily disturbance (e.g., an increase of the heart rate). For the physio-pragmatist camp, the feeling of the bodily disturbances resulting from an exciting fact (in our case, the feeling of heart rate) *is* the emotion. The latter was William James' (1884) famous position. According to James, "the bodily changes follow directly the *perception* of the exciting fact, and … our feeling of the same changes as they occur *is* the emotion" (James 1884, pp. 189–190; emphasis in the original).

However, several years later some psychologists and physiologists argued that bodily disturbances could not be equated with the sensed emotion. Thus, the works of Sherrington (1900) with dogs and Cannon et al. (1927) with cats showed that the removal of the body parts where bodily changes reside in situations of anger, fear and rage—e.g., the sympathetic channels for nervous discharge in situations of profound excitement—does not affect the expected emotional states. Thus, the chirurgical-altered cats in Cannon et al.'s (1927) experiments behaved emotionally in the same way as intact cats when a dog approached their kittens or when food was taken away. These experiments suggested that emotional states might continue to be present even when the corresponding physiological support is missing.

One of the most significant contributions of Cannon's (1922, 1927), Cannon et al.'s (1927) and Sherrington's (1900) investigations was the distinction between *emotional feeling* (i.e., the uncontrolled and uncontrollably bodily changes, such as adrenaline production occurring during an emotional experience) and *emotion as such*. The psychic aspect of emotion is certainly intertwined with the physiological aspect, but one cannot be reduced to the other. Both together prepare us for action: physio-psychic emotion is not the end of the emotional phenomenon but the beginning of an action—*fight or flight, as* Cannon formulated it. "According to the argument here presented," Cannon wrote, "the strong emotions, as fear and anger, are rightly interpreted as the concomitants of bodily changes which may be of utmost service *in subsequent action*" (1922, p. 212; emphasis added).

Psychologists such as Lewin (1935) moved the conceptualization of emotions to new grounds by showing that human emotional phenomena is not of an instinctual nature, as in the case of animals, but is linked to the meaning of life: "one must not forget that in dealing with psychical processes one is dealing with life processes" (Lewin 1935, p. 63). On a commentary concerning Lewin's view Vygotsky wrote: "the structure of the individual's character is reflected in his emotional life and his character is defined by these emotional experiences" (Vygotsky 1987, p. 333). Emotional phenomena came hence to be seen not as merely transient experiences rooted in our biological apparatus (although without it no emotional life would be possible), but as something entrenched in the manner in which we understand ourselves in the world. This is the view conveyed by Charles Solomon (1978), who suggests that emotions are a "complex system of judgments, about the world, about other people, and about ourselves and our place in our world" (1978, p. 186). Judgments, however, do not refer here to assertive or declarative instances backed up by a logical-deductive apparatus. On the contrary, they mean rather appraising and gauging events involving self and context. As a result, emotions, as systems of judgments or appraisals are not merely declarative or assertive. Through them we do not merely say or state something about the world in a cold, logical way. Through emotions we speak out and relate to events, people, behavior, things, and actions. Emotions do not only drive our affective life; they also shape the manner in which we understand the world and ourselves (Roth 2007). Thus, rather than a crisis or worldly lived incident, emotions are focal points of a whole way of life. They rest on physiological processes, but cannot be reduced to them. They entail a range of cultural conceptual categories that are instantiated differently by different people (e.g. moral and ethical categories; notions of privacy, responsibility, autonomy, etc.).

The picture that emerges from the previous account can be summarized as follows. Emotions are not irrational forces; neither are they momentary incidents or disruptions in our everyday life. Emotions are part of a worldview that, through our participation in cultural and social activities, we come to share. Our emotional life is, in this sense, profoundly shaped by history and culture, although this does not mean that the parameters of what is to come in our emotional life is somehow injected into our being by a kind of mysterious syringe. Like cognition, emotions can only be understood through the interplay of history and the manner emotions develop in ontogeny. That is, emotions can only be understood through the incessant dialectical relationship of past and present and their projection into the future.

## Emotions as Cultural Constructs

To better describe the sense in which I take emotions as cultural constructs, in this section I would like to make an excursus into one chief category of emotional life: love. Such a move should allow me to make my point clear when it will be time to see emotions in mathematical cognition. I should clarify, however, that this is so not because my forthcoming classroom analysis is going to be about love in mathematics; nor is it because there is a straightforward transfer from love to the joys and frustrations that students experience in learning mathematics. The complexity of human life makes it impossible to express its affective domain in terms of homomorphisms and transpositions. The reason of my excursus is to show that in the same way as love is a historically and culturally constituted emotion, so are the alienating or fulfilling emotion students experience in dealing with mathematics. Yet, it is not randomly that I have chosen love as the terrain of my excursus. Love is usually conceptualized as the most intimate conceivably repository of individuality. And although this might be true in Western cultures, in particular since the Romantic movement, it is so to the extent that the manners in which we consider love and practice it are embedded in a concept of subjectivity and individuality that is cultural through and through. Once this point is realized I hope that there will be room to envision emotions in mathematics teaching-and-learning not as mere idiosyncratic features of individuality but as culturally and historically constituted dimensions of the self. There is no doubt that, in walking along this path, I am diving in controversial waters. When the late anthropologist Clifford Geertz argued that passions of the Bali in Indonesia are culturally shaped, he was received with skepticism. Thus, in his review of Geertz' (1980) book *Negara*, Edmund Leach asserted that

> I can make no sense of a line of thought which claims that "passions" are culturally defined. From my prejudiced position as a social anthropologist this passage reveals with startling clarity the ultimately radical weakness of the basic assumption of cultural anthropology, namely, that … human individuals are products of their culture rather than of their genetic predisposition. (Leach 1981, p. 32)

Let us see, then, in what sense love appears as a cultural construct.

Although it has been argued that love is a part of our instinctual kit and that its function is to ensure the species survival, the manner in which love occurs between two adults and how it is felt is not an invariable concept. This point was already recognized by the Andalusian philosopher Ibn Hazm, author of a treatise on love written around 1022 (*Ring of Dove*) where he admits that love for the Bedouins and for the ancients meant two different things (Hazm 1022; Preface). The same can be said of love in the Western Middle Ages and today.

The Medieval concept of love and the manner in which it was felt was mediated by the social-economic structures of the time. These structures, along with cultural aesthetic concepts (such as "beautiful figure"), social ideas of good human personality (like "excellence of character"), and a praised role of language (referred to as "extreme readiness of speech") structured the space within which love was sought, practiced, and felt. In a famous book written ca. 1184 Andreas Capellanus explains how the aesthetic elements, the worthiness of character, and mastering of speech should be advantageously put in motion in obtaining love. These components were articulated differently depending on the social range of the individuals. Thus, Capellanus gives a series of examples: one deals with the case where the man and the woman are both plebeians (or commoners); another example deals with the case where the man is a plebeian and the woman is part of the nobility, etc. Love in each case was conveyed as an intense feeling (an "inborn suffering," as Capellanus put it 1960, p. 28) modulated by aesthetic and ethical concepts such as modesty, loyalty, commitment, and generosity. Intimacy and fulfillment as we know them now were not part of love in the Middle Ages. As Ratner puts it (2000, p. 12), "Personal idiosyncrasies were not cultivated during feudalism and they played no part in evoking romantic love." Instead of personal idiosyncrasies, lovers attended to questions of family social position, along with humility, beauty, and dedication. Love consisted in the contemplation of the soul, and the sentiments of the heart. And if it involved a kiss on the mouth and physical contact, it was in a very chaste and modest manner (de la Croix 2013). This is why "Love," Ratner says, "was a spiritual, almost religious, sentiment that sublimated the base instincts and elevated the soul through dedication to one's loved one. One was a better person through caring for (serving) another. Love was thus a moral act" (Ratner 2000, p. 12).

At the dawn of the twentieth century, the manufacturing forms of production that emerged progressively since the Renaissance reached an unprecedented level of industrial sophistication. This evolution of the forms of production came hand in hand with a range of new divisions of labour out of which new understandings of the self and concepts of others as well as how individuals relate to each other came to be envisioned. Within these societal transformations in the forms of production and modes of interaction, love, as a specific form of human relationship, found itself transformed.

The modern concept of love required indeed a specific concept of self—one that was defined in individualistic and private terms and which came in tandem with a new ethics of consumerism. Sociologist Eva Illouz notes that "The rise of consumerism coincided with the period between the two world wars when the self became both locus and focus of culture… In the new ethos, individuals were encouraged to express themselves 'creatively' and 'authentically'" (Illouz 1997, p. 35). In his studies about love, William Leach connects the emergence of modern romantic love to "the emergence of economic individualism" and goes on to say that the "romantic lover resembled his economic counterpart the risk-taking entrepreneur" (1980, p. 106).

According to Illouz, the transformation of the concept of love was characterized, among other things, by

> the increasing prominence of the theme of love in mass culture, especially in film and advertising; the glorification of the theme of love as a supreme value and the equation of love with happiness; the association of love and consumption, more specifically, the romanticization of commodities; the inclusion of "intensity" and "fun" in the new definitions of romance, marriage, and domesticity. (Illouz 1997, p. 28)

As in the Middle Ages, loved was shaped by the economical dimension of society. But rather than being refracted through a social hierarchy of church officials, nobles, bourgeois, and commoners, love was refracted along the lines of advanced capitalism and its ethos. Thus, instead of being mediated by "excellence of character" and an "extreme readiness of speech" and the ethical categories that made it a moral act in the Middle Ages, love came to be mediated by the expanding industry of commodities. This is what Illouz calls the "commodification of romance" (1997, p. 11). Some signs of consumerist love are: dancing, eating dinner and drinking cocktails at expensive and luxurious places, travelling, vacations, and movie-going. The movie theater, the dance, and the candle-lit dinner became signs of a new intimacy that was made possible by the circulation of capital and the expansion of the working class. While the Middle Ages' love themes revolved around the value and practice of humility, commitment, and praise, the seductive themes of love during the first quarter of the twentieth century evolved from the Victorian morality of domesticity to a plethora of consumerist notions such as exoticism, expenditure, speed, adventure, intensity and the physical care of the self. Instead of the intense ecstatic longing feeling of the Middle Ages courtly love, modern love, in short, appears stimulated by spontaneous and hedonistic desire for commodities (Ratner 2000) and based on an "experience of intense feeling, uninhibited sensuality, instant gratification, spontaneous pleasure, [and] fun" (Illouz 1997, p. 88). And as in the case of all emotions, love is learned through socialization. In the case of contemporary love, much of its socialization is done through mass culture, which provides adolescents with cognitive responses of romantic mannerism, behavior and skills. In a study conducted in the early 1930s by Herbert Blumer on what adolescents learn from movies, one of the respondents—a 21 year-old male—answered:

> The technique of making love to a girl received a considerable amount of my attention, and it was directly through the movies that I learned to kiss a girl on her ears, and cheeks, as well as on the mouth (Blumer 1933, p. 47).

Love as cultural phenomena means hence that the biological arousal that is at emotion's origin evolves into a psychic emotion that goes beyond the biological realm. Elicited by a concrete element in the world (the direct sight of beauty or its mental evocation), the physiological phenomenon—i.e., the *emotional feeling* or uncontrollably bodily changes that Capellanus (1960, p. 28) candidly referred to as the "inborn suffering derived from the sight of and excessive meditation upon the beauty of the opposite sex"—occurs in a world of cultural significations where it comes to be appraised, labeled, and sensed variously ("longing" and "devotion" as in the Middle Ages, "rational friendship" in the Victorian era, the passionate and sensual in the capitalistic culture).

By unavoidably occurring in the world of sociocultural significations in the form of judgments and appraisals (Solomon 1978), emotions entail a moral and ethical dimension. Anger, for instance, involves more than the production of adrenaline, or a neuronal circuitry in the lateral orbitofrontal cortex; it involves moral categories (e.g., offense and transgression), and concepts of the self.[1]

## Emotions and Thinking

In the previous sections I have advocated for a cultural-historical concept of emotions according to which emotions are historically constituted. They are part of the forms of subjectivity that cultures foster. This is why emotions cannot be understood without taking into account the processes of subjectification through which we enter cultural life and come to instantiate the raw forms of being that are culturally available to us at a certain point of culture's development (Radford 2013a). It is indeed within the scope of the various Medieval, Victorian, and capitalistic cultural forms of subjectivity that, in my examples, love is practiced and felt.

Since the self is emotional through and through, it is not surprising that thinking is rooted in emotions too. Yet, a precise functional description of the relationship between thinking and emotions has proved difficult to articulate. Ratner (2000, p. 6) suggests that "Emotions are feelings that accompany thinking. They are the feeling

[1] The same can be said about guilt. Murphy's studies suggest that spread of guilt in Africa during the first half of twentieth century was often associated with a new concept of self as promoted by Protestantism and proto-capitalist forms of production. Within this societal transformation led by new entrepreneurial activities, individuals came to conceive of themselves as planners and masters of their own actions. Unfolding under the presence of an "omniscient God who can read one's thoughts" (Murphy 1978, p. 237), individuals conceptualized themselves as responsible for their actions, as opposed to a former worldview where actions were understood more in collective terms and events attributed to the collective, chance, bad luck, or witchcraft.

side of thoughts; thought-filled feelings; thoughtful feelings." In a commentary on Ratner's position, however, Menon complains that thoughtful feelings may fail to recognize the embodied dimension of affective life:

> While Ratner succeeds in emphasizing the irrelevance of biological processes to emotional experiences, he goes a little too far, perhaps, when he ignores the body and somatic experiences in his discussion. There is passing mention about 'bodily concomitants' (p. 19) but little more. In my view, it is very necessary to explicitly recognize the body in emotional experiencing, because such experiences are grounded in the reality of the bodily self—although I would not go so far as to claim that emotions can be identified with particular feelings. (Menon 2000, p. 43)

Menon finds missing a clear reference to the somatic correlates of emotions and turns to the work of Rosaldo, who has suggested that emotions are "embodied thoughts" (Rosaldo 1983, p. 143) an idea that conveys unequivocally the fact that "emotions are grounded and experienced in our bodily selves" (Menon 2000, p. 44).

Now, if thought is inherently embodied (Radford 2013b; Shusterman 2012; Varela et al. 1991) and emotions are more than physiological processes, what is then the difference between emotion and thought? Perhaps what we should bear in mind is the fact that there is no dividing line between thought, body, and emotion. To refer to emotions as embodied thought is redundant. Our thinking is *necessarily* embodied *and* emotional. During a match, chess players may seem to be exclusively cogitating before a move. Yet, the cogitation is highly emotional. The tensed and sustained gaze at the chessboard and the muscular tension in the otherwise immobile sitting body are two expressions of the ongoing intense emotional and bodily phenomena. Intermingled with rational calculations and logical thinking are the emotions that underpin chess players' activity. Only computers can "think" without feeling anything. They do not even feel the heat of their chips. They feel nothing. They display pure mechanical calculations of which humans are definitely incapable. We can make some calculations, and we can do it while feeling boredom, thrill, excitement, challenge or something else; what we cannot do is simply feel nothing.

## Emotions and Motives

The brief overview of emotions carried out in the first part of this chapter shows that emotions were initially investigated through their *expressive form*. The problem was to understand what happens when we feel something—e.g., anger, fear, or rage. Leont'ev suggested that emotions should rather be investigated in terms of the psychological organizing role that they play in activity, a role that he conceived in terms of "inner signals," and their relationship with the individuals' motives:

> Emotions have the function of inner signals; that is, they do not directly represent the psychological reflection of object-oriented activity. The special feature of emotions is that they reflect relationships between motives (needs) and success, or the possibility of success, of realizing the action of the subject that responds to these motives. (Leont'ev, cited in Holodynski 2013, p. 8)

Emotions are hence related to motives in a time-projection manner: they relate to the *possibility* to succeed (or to fail) to reach the object of the activity. We have to bear in mind here that, for Leont'ev, what characterizes an activity is its object- and motive-orientation (Roth and Radford 2011). Working within the more general 'production paradigm' (Markus 1982) of his time, he conceived of activity as something that is driven towards a result—an outcome. This outcome has to produce something objective, tangible: the product of activity (which can be material or ideal). But activity is not merely a mechanical or technical production of things. Activity has to include the human dimension that Leont'ev captures through the concept of *motive*. This is why "The concept of activity is necessarily bound up with the concept of motive. There is no such thing as activity without a motive" (Leont'ev 2009, p. 6).

However, the concept of motive as theorized by Leont'ev is not easy to formulate. The concept appears at two different levels: the level of activity (where it appears as the activity's motive) and the level of the individual (where it appears as the individual's various motives). To make the distinction, Leont'ev presents the example of hunters, who labour together in order to satisfy their common needs. In this example, there is a perfect match between the motive of the activity and the individuals' motives. However, this coincidence of subjective motives (the individuals' motives) and the motives of activity is rather the exception:

> At the early stages, when people participating in collective labour still have common motives, meanings as phenomena of social consciousness and as phenomena of individual consciousness directly correspond to one another. But this relationship does not endure in further development. (Leont'ev 2009, p. 20)

This non-coincidence between the individuals' motives and the activity's motive is, however, often the rule in classroom activity. What we have there is indeed often a plethora of different motives that may seem to threaten even the possibility of joint activity to occur.

As mentioned previously, Leont'ev resorted to motive and object as the two main vectors of activity. What I want to propose is to see them not as fixed entities but as dynamic and evolving ones. Thus, instead of considering activity as something that has to end up with the materialization of the object in the activity's outcome (which is what Engeström (1987) emphasizes, ending up in a functionalist conception of activity), I suggest that we see activity as an *open system*, driven by an *evolving object* and a *developing web* of interconnected and sometimes *contradictory motives*.

The couple object-motive thus becomes the drive that moves activity and its sentient individuals not towards something to be attained, but rather towards a participation in a cultural way of life and the fulfillment of material and spiritual needs.

Leont'ev did not theorize activity in exactly the way I am suggesting. However, my proposal is not alien to Leont'ev's perspective, as it can be seen in the following passage, where Leont'ev talks about activity's general structure. He says:

> Historically, man's activity does not change its general structure, its "macrostructure". At every stage of historical development it is realised by conscious actions in which goals become objective products, and obeys the motives by which it was stimulated. What does change radically is the character of the relationships that connect the goals and motives of activity. These relationships are psychologically decisive. The point is that for the subject

> himself the comprehension and achievement of concrete goals, his mastering of certain modes and operations of action is a way of asserting, fulfilling his life, satisfying and developing his material and spiritual needs, which are reified and transformed in the motives of his activity. It makes no difference whether the subject is conscious or unconscious of his motives, whether they declare their existence in the form of interest, desire or passion. (Leont'ev 2009, pp. 21–22)

To recap, from a cultural-historical perspective, emotions are both subjective and cultural phenomena simultaneously; they are entrenched in physiological processes and conceptual and ethical categories through which individuals perceive, understand, reflect, and act in the world. Their subjective-social link is to be found in the double-faced nature of motives, which are always personal and cultural.

Let me turn now to my classroom example to see how emotions unfold in activity.

## I Hate to Give My Answers: Frustration, Exasperation and Disappointment

In the rest of the article, I would like to discuss some passages from a lesson in a Grade 4 class (9–10-year-old students). The class is part of a 3-year longitudinal study. The lesson reported here takes place during the third year. I focus on the work of a group of three students: Jay, Thom, and Laura. Jay and Thom have been involved since year 1 in the study. Laura, by contrast, joined the class the third year and was hence new.

The lesson was about sequence generalization and started with a general discussion of how to continue a sequence of numbers. Before working on the problem, the teacher discussed with the students the meaning of group work. To understand the importance of the teacher's emphasis on group work the reader needs to bear in mind that, within the theory of objectification, learning is not conceptualized as a mere acquaintance with cultural forms of thinking (in this case, algebraic thinking). Learning is not only about knowing but also about becoming (Radford 2008a). As a result, the design of the classroom activities (which is made by the teachers and our research team) involves both a thorough design of problems of increasing difficulty whose organization requires the mobilization of the target mathematical concepts in depth, as well as the constitution of meaningful spaces of social interactions where students are encouraged to attend to other voices and ideas, to collaborate with others, and to show support and solidarity (Radford 2012, 2013b).

These mathematical and ethical dimensions provide distinctive basic elements for particular forms of subjectification to occur (e.g., forms based on responsible understanding and solidarity). Although these processes of subjectification cannot be anticipated or predicted beforehand, the elements highlight conceptual and ethical features out of which kinds of plausible intersubjective theoretical-emotional experiences may occur.

The aforementioned mathematical and ethical dimensions have been the driving vectors of our 3-year program. The third year, the teacher summarized with the students what they have been practicing the years before. Collectively, the students and the teacher discussed the introductory problem on the white board; they talked about the meaning of mathematical concepts required in the task (e.g., the regularity in a sequence), and the meaning of group work. The teacher wrote on the white board the students' responses, which included: "ask for help," "listen to the others' ideas," "encourage others," "do not get frustrated."

After collectively solving the introductory problem, the students worked on other generalization problems. The third problem revolved around the sequence of numbers indicated in dark in the following table:

| | | | | | | | | | |
|---|---|---|---|---|---|---|---|---|---|
| 1 | 2 | 3 | 4 | 5 | 6 | 7 | 8 | 9 | 10 |
| 11 | 12 | 13 | 14 | 15 | 16 | 17 | 18 | 19 | 20 |
| 21 | 22 | 23 | 24 | 25 | 26 | 27 | 28 | 29 | 30 |
| 31 | 32 | 33 | 34 | 35 | 36 | 37 | 38 | 39 | 40 |
| 41 | 42 | 43 | 44 | 45 | 46 | 47 | 48 | 49 | 50 |
| 51 | 52 | 53 | 54 | 55 | 56 | 57 | 58 | 59 | 60 |
| 61 | 62 | 63 | 64 | 65 | 66 | 67 | 68 | 69 | 70 |
| 71 | 72 | 73 | 74 | 75 | 76 | 77 | 78 | 79 | 80 |
| 81 | 82 | 83 | 84 | 85 | 86 | 87 | 88 | 89 | 90 |
| 91 | 92 | 93 | 94 | 95 | 96 | 97 | 98 | 99 | 100 |

The students were invited to find out the next three terms, and then the following three terms.

Jay and Thom engaged in an exploration of the sequence, counting on Jay's page the spaces between the dark cells, and exchanging ideas. Laura worked on her own:

1. Jay: (*Starting from cell 1, he points rhythmically to the following cells with his pen*) 1, 2, 3, 4.
2. Thom: (*Who is following Jay's utterances and gestures says at the same time as Jay*) 4 (*short pause*).
3. Jay: (*Pointing to cell 5*) 1.
4. Thom: (*At the same time as Jay and with the same intonation as Jay's says*) 1.
5. Jay: Wait, 1, 2, (*Thom starts counting with him*) 3, 4, 5 (arriving at cell 9).
6. Laura: (*Who has made no eye or other contact with her teammates says lowly without leaving the eyes from her page and if talking to herself*) yeah, 1, 2, 3.
7. Jay: (*Pointing at cell 10 and continuing uttering in a synchronized manner with Thom*) 1, 2, 3, 4, 5, 6, 7 (*arriving at cell 16*).

During their work Thom and Jay show an emotional tension that results from the search for a regularity that they cannot yet grasp. This tension is reflected in the sensuous counting of squares and the tremendous attention that they have to pay to carry out their actions. Jay's utterance "wait" (line 5) interrupts the flow of the counting process: it marks a moment of hesitation that is overcome and reassumed right after with some assurance. The synchronic work of Thom and Jay creates a feeling of closeness and unity that may colour the tension positively. This feeling of closeness in which utterance and gesture are coordinated is highlighted even further by the aural proximity of voice tonality. Laura remains outside of the synergy that is created between her teammates (see Fig. 1). She looks determined and focused.

**Fig. 1** Thom (to the *left*) moves towards Jay (*middle*) and accompanies Jay's counting visually and verbally in a synchrony that relieves the tension of solving a problem with an uncertain outcome. Laura (to the *right*) works on her own, without making contact with her teammates

The teacher came to see the students work. Thom and Jay engaged with the teacher in a process of objectification out of which they started noticing that they had to move 3 cells to the next dark cell, then they had to move 5 cells, 7 cells, etc. Laura continued to work alone:

8. Teacher: Can you start seeing the sequence?
9. Jay: They add 2? (*With some uncertainty*).
10. Teacher: Ah! Two are added each time!
11. Thom: (*Thrilled*) Oh!
12. Teacher: Can you explain it to your teammates?
13. Laura: (*As if referring to something trivial*) No, I know already.

The teacher went to see another group. The students continued their work. Adding successively 15, 17 and 19, Jay and Thomas found that the next terms of the sequence were 64, 81 and 100. Adding successively 15, 18 and 19, Laura found that the next terms were 64, 82 and 101. Jay and Thomas were surprised by the difference:

14. Thom: Laura, can you explain to us what is your idea?
15. Jay: Why is it 101?
16. Laura: (*Referring to the difference between terms, she says*) Well it's because right now I calculated 11 and when I continued I found … um… 13 and then 15 and then 18.
17. Jay: Yes, but we don't understand…
18. Laura: Look, (*pointing to the cells with the pen, confidently she starts counting from cell 82; she makes a mistake and counts 82; Thom counts with her*) (*pointing to cell 82*) 1, 2 3, 4, 5, 6, 7, 8, 9, 10, 11, 12, 13, 14, 15, 16, 17, 18, 19 (*She doesn't arrive to 101, but to 100; she seems hesitant*) (pause) I mean 18 (pointing to cell 99).
19. Thom: 19!
20. Laura: (*Makes a gesture in the air*) Oh! (*She rotates the sheet to put it in front of her and tries to understand*) Wait, I think I put…
21. Jay: It's 19
22. Laura: (*Looking attentively to her page*) Wait, wait.
23. Thom: It's 19, because it went…

**Fig. 2** In pic 1 (*top left*) Laura asks Thom to wait. In pic 2 (*top right*) she scratches number 82 during 4 s. Pic 3 (*bottom*) shows her activity sheet

24. Laura: (*She is scrutinizing the first cells and doesn't want to listen; the tension increases; she makes a "waiting" gesture and says*) Wait, wait (see Fig. 2, Pic 1).
25. Jay: You put 18.
26. Laura: (*In an apologizing tone and passing her pen over cell 100*) I think I forgot by accident to write 100. (*She starts counting 15 from cell 49; then, she counts 17 from cell 64; Thom follows the counting without talking; he replaces utterances with a sequence of short noddings; Jay follows the counting from his post*) 1, 2, 3, 4, 5, 6, 7, 8, 9, 10, 11, 1 2, 3, 4, 5, 6, 7, 8, 9, 10, 11, 12, 13, 14, 15, 16, 17 (*arriving at cell 81 and not to the expected 82 that she had marked on her sheet. She tries to make sense of the unexpected outcome*).
27. Thom: (*Wanting to help, he says*) Plus, 17 plus 2 equals
28. Laura: (*Pointing to cell 82*) 18 (*pause*). (*Talking rather to herself*) I think I made a mistake in my own work maybe (*she crosses out cell 81, still thinking that the right answer is 82. Disappointed, she hits the table with her pen*).
29. Thom: (*Noticing that she has crossed out 81*) No, it's 81!
30. Jay: Yes, its 81.
31. Thom: Yeah, it's 81.
32. Laura: (*With emphasis and dismay*) Oh greeeeat! (*long exhalation; she holds her head with one hand for a while; then with the arms extended in front of her,*

*she says*) This is why I hate to say what I do … Ugh!! (*During 4 s she crosses out with intense circular motion square 82*; see Fig. 2, pics 2 and 3) Mmmgh! Mmmgh! (pause) (*Talking to her teammates*) That's why I HATE (she pronounces the word slowly and louder) to give my answers (*she corrects her mistake on her page*) (pause) (*with frustration*) Mmgh… Mmgh… See? (*With great disappointment*) Cause I get [it] wrong…

33. Thom: (*Talking to Laura in an encouraging tone, and pointing to the 100th cell*) The answer is 100.
34. Laura: [One hundred] and one (*she insists she has the right answer although having noticed she made a mistake in her calculations*)

The episode starts with Thom and Jay inviting Laura to explain her result. The boys confess to not having understood her short explanation. Laura slightly rotates the page towards Jay and starts counting from cell 82, although she makes a mistake and includes 82 in the counting process; Thom and Jay watch her count attentively. Thom joins her while she is counting 13 and both count together the rest of the cells. To her dismay, she does not arrive at the expected cell 101. She hesitates and, in turn 20, with irritation, she moves the arms in front of her. Something went wrong and she still does not understand what or why. She has two options. She may try to get some feedback from Thom and Jay, who have proven collaborative and willing to help, or she can try to sort out the problem by herself. She opts for the second option. When Thom volunteers an explanation, she asks him to wait. She is thinking in a very effortful way. Her body becomes rigid and tense (see Fig. 2, pic 1). In turn 26, she concedes that she might have forgotten to pen cell 100. It is not a mistake. It is an accident, she says. She might not believe the reason she has offered to her teammates, as she starts counting again. The fact that she starts counting from cells 49 and 64 may suggest that she is now unsure of the correctness of her procedure. She needs to check it. Thom, who has been asked to wait, decides not to count aloud with her, but follows her counting with a series of rhythmic short nods of his head. Things become even more complicated as she lands on cell 81, and not 82 that she has marked on her sheet. She shows her disappointment by hitting the desk with the pen. In line 28, although she acknowledges the possibility that she has made an error, she eliminates 81, to the dismay of Jay and Thom, who hurry up to exclaim that 81 is good. She loses control and things escalate. She utters "great" in a heavily pronounced manner, showing confusion and deep frustration. In general, frustration refers to a sense of dissatisfaction associated with difficulties of encounter. In this case, frustration appears around the conceptual dilemma of whether the good cell is 81 or 82. Laura spends 4 s (which is a huge amount of time in the context) scratching cell 82 and voicing her frustration through a sequence of verbal "Mmgh" lamenting sounds. Thom tries to alleviate the tension, talking no longer about cell 81, and says "The answer is 100," while she still insists that the answer is 101.

Laura's and the other students' unfolding emotional experience is a key component of the process of subjectification they are all immersed in. A process of subjectification refers to the always evolving sense of the self that results from the manner we and others recognize and position ourselves socially. Students' sense of the self are to a large extent related to the manners in which they engage in activity and come to position themselves in cultural practices in the public space. Laura's positioning is mined by difficulties that she senses and interprets in ways that colour it rather negatively. She overcomes partially the frustration and starts counting again, although with less confidence. She starts counting with annoyance from cell 81 and, counting 18, she arrives at cell 99 and not to the expected cell 101. After reflecting for a moment, she restarts counting again, but instead of starting from cell 64, she starts from cell 65. This time she arrives at cell 81 and looks puzzled. The dialogue continues as follows:

35. Thom: The answer is 100, the three following numbers are 64…
36. Laura: (*Without listening to Thom, interrupting, she says with great distress*) Oh my god! (*She lifts her arms up*) I am, (*holding her head with her left hand and looking at the numbers*) ugh…!
37. Thom: Because she has 64 right there.
38. Laura: (*With frustration*) I mixed myself up now!
39. Thom: (*Intervenes to try and help*) After
40. Laura: I mixed myself up (*still holding her head with her left hand, she hits repeatedly the desk with the pen in her right hand*).
41. Thom: (*Trying to help, he points to the numbers on Laura's sheet*) 81, and after 100.
42. Laura: I'm all messed up now!
43. Thom: Therefore
44. Laura: Maybe I made a mistake…
45. Jay: It's alright Laura, everyone makes mistakes
46. Laura: I'm all messed up now! (*She still holds her head with her left hand*; see Fig. 3).
47. Thom: It's true.
48. Laura: I always lose my memory. What if I say more than 10?
49. Jay: (*Trying to help*) Laura, just, just do that, scratch that out (*suggesting to cross out 101*).

**Fig. 3** Laura deeply discouraged

## Madam, Now I'm Too Mixed Up...!

A few minutes later the teacher came to see the students work. The teacher hears the students' explanation and engages in a counting process with them. She counts the cells between the numbers of the sequence, starting from the first two terms (i.e., 1 and 4). She notices that Laura is working on her own and wants to include her in the discussion:

50. Teacher: (*Counting on Jay's page*) 1, 2, 3 (*Talking to Laura*) Laura do you agree with that?
51. Laura: 12, 13, 14, 15 (*Continues counting on her page; she lifts her finger up to signify "wait" as in* Fig. 2, pic 1).
52. Teacher: (*In an inviting tone*) Count with me...
53. Jay: (*Laura continues working on her own; Jay and Thom count at the same time, while the teacher points to the cells on Jay's page*) 1, 2, 3, 4, 5
54. Teacher: And after that?
55. Jay: (*At the same time as Thom*) 1, 2, 3, 4, 5, 6, 7...
56. Laura: All my work is mixed up! [...]
57. Teacher: Ok (*She use Laura's page to point to the cells*) So, Laura, did you arrive at 9 here?
58. Laura: Now I'm mixed up...!
59. Teacher: (*In an inviting tone*) Continue to count.
60. Laura: Madam, now I'm too mixed up!
61. Teacher: How many, what number should we have afterwards? After 9 we should have how many? We count up to what?
62. Laura: Madam.
63. Thom: 11 (*Laura makes a gesture of discouragement as she does not understand*).
64. Teacher: Ah! 1, 2, 3, 4, 5, 6, 7, 8, 9, 10, 11 (*the teacher points to the cells on Laura's page; Laura watches her point and count*) Did we arrive at the correct number?
65. Thom: Yes.
66. Teacher: Ah, the next number that I have to count is how much?
67. Laura: Agh... (*Her body falls to the back of the chair, demonstrating a great confusion and frustration*)
68. Thom: 13!
69. Teacher: Why 13?
70. Thom: Because plus 2 [...]
71. Teacher: Do you see Laura?
72. Laura: (*She makes a gesture of discouragement*; see Fig. 4, pic 1) Madam now (*the upper part of her body falls down slowly towards the desk, pics 2–4 and, crying, she says*) I scribbled on my page!
73. Teacher: Ok Laura, can we go out [of the classroom] for a minute?

The efforts that the teacher made to include Laura in the discussion did not pay off. It would be a mistake, however, to think that Laura's emotional dimension has clouded her judgment and impeded her from thinking rationally. From the

**Fig. 4** Laura shows frustration. At the end she cries

cultural-historical perspective here sketched, emotions, as we pointed out in previous sections, are always intertwined with thinking. Emotions, I suggested, are rather entrenched in physiological processes and historical conceptual and ethical categories through which individuals perceive, understand, reflect, and act in the world. In other words, it is not because Laura became emotional that she failed to think and calculate in an appropriate way.[2] Although unpredictable in its details, the emotional-cognitive process that she underwent unfolded shaped by the manner in which she perceives herself in her relationship to knowledge and to others. In the same way as love is practiced and felt culturally, so is the manner in which we experience and practice learning. And in the same way that love is differently instantiated by different lovers from the same culture, so is learning. What the previous excerpts intimate through our interpretative stance is that, drawing on cultural models of being (here modes of learning and learners), Jay and Thom position themselves differently from the manner in which does Laura. This general positioning affords specific ways

[2] This doesn't mean, however, as one of my reviewers notes, "that emotions have to be seen only as the end of a (cultural) process." Emotions are already there, with us; they evolve as we evolve into cultural subjects through subjectification processes, appearing—as Vygotsky suggested—in new relations with other elements of our whole life.

through which to emotionally interpret the world and our actions within it. As the next episode shows, such a positioning is underpinned by what the students understand of what is ethically expected from them.

## The Ethical Dimension of Emotions: Cheating

The students moved to the next part of the task: they were finding the three next terms of the sequence after term 100. The table stopped at cell 100 (see Fig. 2, pic 3), so now the students were expected to move into more abstract mathematical relationships. Instead of counting, they were supposed to add numbers or to come up with a functional relationship: $n \rightarrow n^2$ or $n \rightarrow n \times n$ (although not necessarily expressing the relationship in the alphanumeric symbolism).

74. Thom: (*Laura is mentally calculating; she is whispering numbers to herself*) Laura would you like to see our copy, (*Laura lifts her finger in her classical "wait" gesture*) we're going to explain to you what we have to do now. Jay, are we going to explain it to her now?
75. Jay: Yes!
76. Thom: Ok.
77. Jay: Laura?
78. Laura: (*Thinking with great concentration*) Wait! I'm right there, you're gonna mess up my calculations!
79. Thom: Oh I am going to do the same thing as you.
80. Laura: (*Loosing her train of thought*) Ok! you just messed them up! Ughhh! I was *so* close! I keep messing my stuff up, I'm trying to think, but you messed it up! (*Looks at Thom accusingly*). Ughh! (*She sighs and whispers something*) I tried to do my copy, I know, I know the answer; it's just I'm trying to get there, I know the answer.
81. Thom: Would you like us to explain to you how?
82. Laura: I don't want to... like cheat, Ok, I don't want to cheat.
83. Jay: That's not, that's not cheating! We're a group!
84. Laura: I know but still...
85. Thom: Yeah we work together.
86. Laura: I just want to, I just want to
87. Thom: If they... (*Points to another group*), that would be cheating from them, but if you look at our copy that would not be cheating because we're working together.
88. Jay: Because that's together.
89. Laura: I know, but I don't know.
90. Jay: Would you like some help?
91. Laura: I was so close to knowing what the answer was!
92. Jay: Would you like some help though?

Laura conceives of learning as something that she has to do by herself, despite the fact that the class discussed the meaning of group work before embarking in the mathematical content. She certainly heard the ideas about collaboration, understanding others, etc. But that does not mean that she felt concerned. In fact, to reach the cultural forms of being that we have been promoting in this class (see Radford 2012) and that Jay and Thom instantiated in the previous excerpts, requires a long and sustained endeavour. It is not sufficient to hear words or to utter them. The students have to engage in and practice intersubjective understanding, openness towards others, etc. These fostered cultural forms of being at the heart of our didactic designs bring with them forms of ethical relationships and concomitant forms of sensing. They make students prone to *move* towards certain forms of actions rather than others. This is what emotion means etymologically. Emotion comes from the Latin *emovere*, that is "to move" or "to move out."

Within her conception of learning, Laura has tried systematically and honestly to answer the questions by herself. Her frustration, disappointment and other expressions of the emotional phenomena involved in the episodes appear now clearly comprehensible. So is her tireless refusal to get help from the students. To do otherwise would amount, according to her, to cheating.

The *Encyclopedia of Applied Psychology*, defines cheating as "any intentional action or behavior that violates the established rules governing the administration of a test or the completion of an assignment, [and] gives one student an unfair advantage over other students on a test or an assignment" (Cizek 2004, p. 308). The definition stresses the cultural censuring dimension through the legal governing apparatus of conduct and behaviour. Those regulatory devices frame Laura's motives, which clash however with those of Jay and Thom, moved (or "emotionned" if we continue using emotions in their etymological sense) by a Bakhtinian ethics of solidarity and intersubjective understanding (Radford 2008a, 2012).[3] As Jay argues in turn 83, "that's not cheating! We're a group!" To explain the idea, in turn 87 Thom refers to another group. Cheating would be to look at the work of another group. But cheating cannot occur within the group as long as the group works together: "if you look at our copy that would not be cheating because we're working together." Emotions, as we can see, always reflect "phenomena, perceived and understood from the special point of view of the perspective of a person who is interested in them" (Zaporozhets 2002, p. 61). But emotions cannot be reduced to the panoramic view of the subject, as our analysis intimates. They are rather entrenched in ethical and other cultural categories through which emotions become personal and cultural at the same time.

[3] The Bakhtinian character of the ethics that we foster rests indeed in the primacy of the Other (or Otherness or alterity) in our ways of being. This is why, for Bakhtin as for us, consciousness is always dialogical and intersubjective (see, e.g., Bakhtin 1981, 1990; Radford 2008b.

## Summary and Concluding Remarks

In mathematics education, McLeod's pioneering work has been very important to move the study of the affective domain from stable features of individuals (as in the case of beliefs and attitudes) to dynamic, contextual processes. McLeod's (1989, 1992) tremendous insight, however, remains bounded by the inherent limitations of Mandler's (1984, 1989) cognitive conception of emotions that influenced his views. According to Mandler emotions arise out of interruptions of plans that we carry out. Mandler's view is based on the idea that emotional behavior rests on two systems: arousal and meaning analysis. While the first is cast in behaviorist terms and the idea of stimulus, the second is formulated within the traditional rationalist framework that assumes a lonely individual coping with an ahistorical surrounding through schemas and representations. If Mandler's subject is formulated as *emotional + cognitive*, emotion is formulated as *arousal + meaning*. In the end, the account remains quite behaviorist. It is not surprising that, in drawing from Mandelr's work, McLeod (1989, 1992) ended up picturing the affective domain as *repeated* experiences that depend on the magnitude (or intensity), direction (positive or negative), duration, and control of emotions.

The cultural-historical conceptualization of emotions that I have sketched here draws on previous research, but departs from it in several aspects. Within the cultural-historical conception of emotions that I articulated, emotions are not considered irrational forces or mere disruptions in our everyday life. Emotions are part of a worldview that, through our participation in cultural and social activities, we come to share. Emotions comprise a physiological component but cannot be reduced to it. They are shaped by conceptual and ethical cultural categories out of which we define our stance towards the world, and how we relate to people and events. To illustrate this idea, drawing on the work of Illouz (1997), Ratner (2000), and W. Leach (1980), I discussed the example of love (allegedly the most intimate and personal of our emotional life) and attempted to show that what is expressed through the term 'love' is culturally situated and produced. I contrasted the medieval ideas and feelings about love to the modern consumerist counterpart and tried to show that love is mediated by cultural conceptual and ethical categories.

The second part of the article was an effort to show how emotions are implicated in mathematical thinking. My argument is that it is misleading to believe that emotions obstruct thinking. Emotions and thinking are not separate entities. They are fused together. We cannot think without emotions. Emotions and thought come to constitute a unity in ontogenetic development. In the course of social life, emotions develop and "appear in new relations with other elements of mental life" (Vygotsky 1999, p. 244). They become related in particular to the students' motives, regardless of how they are expressed—e.g., "in the form of interest, desire or passion" (Leont'ev 2009).

In the classroom episodes here discussed, two contrasting forms of motives drove the students' actions. In the case of Laura, motives were cast in terms of an ethics of auto-sufficiency, where individuals come to conceive of themselves in terms of the origin of meaning, cognition, and intentionality. This conception of ethics is not spontaneous: it is cultural and has its specific history (Radford 2012). It paves the way to emotionally engage in activity in certain ways. In this case, Laura felt compelled to work alone. She considered that attending to what the other students are doing is cheating. All that she felt during the episodes—the irritations, disappointments, frustrations, vexations—was felt in tune with her understanding of her own role and her role vis-à-vis others. Thom and Jay's motives, by contrast, were cast in terms of a "communal ethics" that promotes participation in the public space, openness, solidarity, a sense of belonging, and critical awareness (Radford 2012). Like the auto-sufficient ethics that underpins Laura's actions, this concept of ethics is not spontaneous. It has also its own history. Thom's and Jay's continuous attempts to connect with Laura were bounded by such an ethical project. Differences in the cultural ethical stances and the ensuing outlook of the world, people, and events, offer the raw material out of which thinking and their concomitant emotions unfold in activity-bound processes of subjectification. It is in this sense that I hope to have shown that cultures fill, infuse, and permeate our emotional life.

Taken together, the historical example and the classroom episodes remind us that in the same way as lovers and love are socio-cultural constructs, so are the students and what they feel and sense when learning. In the Middle Ages, cultural ideas of love and lovers were conveyed by the songs of troubadours, by literature (written and oral), and by other media. Contemporary cultural ideas of learning and learners are conveyed by schools and other social institutions, family, and mass culture. They provide the elements out of which conceptual-emotional experience unfolds within processes of subjectification. I am not intimating, however, that love and learning are produced in some causal manner. Causality has been the paradigm of the natural sciences since Aristotle and Galileo. Yet, the human psyche seems to escape to mechanical explanations epitomized by causal relations. The relationship between culture and their individuals is one of mutual constitution in a complex dialectical way. They are not separated entities glued together by a third term. They co-evolve together: they mediate each other. Yet, with its persistent emphasis on the pole of the individual and the concomitant subjective outlook of psychological and conceptual phenomena, our longstanding Western philosophical and psychological traditions have enduringly posited the individual as the source of intellectual and emotional life—even if from time to time acknowledgment of the cultural dimension is made, as in the case of McLeod, who suggests that "The role of the culture that shapes [our] beliefs would seem to be particularly important" (1992, p. 578). By sticking to the view of the subject as 'cognitive' *plus* 'emotional,' it becomes practically difficult to understand the formation and transformation of motives and emotions in its relationship to culture and history. The cultural-historical perspective that I have presented here tries to avoid this pitfall. It sees emotions as part of the processes of subjectification, processes out of which we position ourselves as cultural subjects in social and political practices. As Menon contends,

> To me, this appears to be the distinguishing feature of cultural psychology—the idea that culture and psyche cannot be smoothly and easily disentangled one from the other, and it is this premise that gives cultural psychology the theoretical power to achieve a dense understanding of a people's emotional reality. (Menon 2000, p. 45)

The historical and classroom episodes also suggest some elements that might be useful to take into account in the teaching and learning of mathematics. The first insight points to the theoretical-methodological premise that the study of individuals—what they do, how they think and feel—cannot be divorced from the sociocultural contexts in which they live and grow. In other terms, the individuals' cognitive, volitional, and emotional dimensions cannot be disentangled from these contexts, for these contexts are not merely "backgrounds" but rather constitutive elements of the human psyche. Emotions in particular cannot be understood if they are abstracted from these historical, cultural, contexts that shape the individuals' motives. Second, emotions are not natural kinds; emotions are historically and culturally constituted. What people sense about guilt, anger or love is not something invariable in time (chromos) and space (topos). Emotions are chronotopical. Cultures offer a range of emotional possibilities of action and reaction that individuals dialectically actualize or instantiate as they learn, since birth, to interact with others and to engage in material and embodied activity.

**Acknowledgments** This article is a result of a research programs funded by the Social Sciences and Humanities Research Council of Canada (SSHRC/CRSH). I wish to thank the reviewers of this article for their generous comments, help, and advice.

## References

Bakhtin, M. (1981). *The dialogical imagination*. Austin: University of Texas Press.

Bakhtin, M. (1990). *Art and answerability*. Austin: University of Texas Press.

Blumer, H. (1933). *Movies and conduct*. New York: Arno Press/The New York Times.

Cannon, W. (1922). *Bodily changes in pain, hunger, fear and rage*. New York: D. Appleton and Company.

Cannon, W. (1927). The James-Lange theory of emotions: A critical examination and an alternative theory. *The American Journal of Psychology, 39*(1/4), 106–124.

Cannon, W., Lewis, J., & Britton, S. (1927). The dispensability of the sympathetic division of the autonomic nervous system. *The Boston Medical and Surgery Journal, 197*, 514–522.

Capellanus, A. (1960). *The art of courtly love*. New York: Columbia University Press (Original work published ca. 1184).

Cizek, G. (2004). Cheating in academia. In C. Spielberger (Ed.), *The encyclopedia of applied psychology* (Vol. 1, pp. 307–312). Oxford: Elsevier.

Darwin, C. (1886). *The expression of the emotions in man and animals*. New York: D. Appleton and Company.

de la Croix, A. (2013). *L'érotisme au moyen âge. Le corps, le désir, l'amour* [Erotism in the middle ages. Body, desire, love]. Paris: Tallandier.

Engeström, Y. (1987). *Learning by expanding: An activity-theoretical approach to developmental research*. Helsinki: Orienta-Konsultit Oy.

Evans, J. (2006). Affect and emotion in mathematical thinking and learning. In *New mathematics education research and practice* (pp. 233–256). Rotterdam: Sense Publishers.

Evans, J., & Zan, R. (2006). Sociocultural approaches to emotions in mathematics education: Initial comparisons. In J. Novotná, J. Moraová, M. Hrátká, & N. Stehlíková (Eds.), *Proceedings of the 30th conference of the international group for the psychology of mathematics education* (Vol. 3, pp. 41–48). Prague: PME.

Furinghetti, F., & Pehkonen, E. (2002). Rethinking characterizations of beliefs. In G. C. Leder, E. Pehkonen, & G. Tôrner (Eds.), *Beliefs: A hidden variable in mathematics education* (pp. 39–58). Dordrecht: Kluwer.
Geertz, C. (1980). *Negara: The theatre state in 19th century Bali*. Princeton: Princeton University Press.
Goldin, G. (2002). Affect, meta-affect, and mathematical belief structures. In G. Leder, E. Pehkonen, & G. Tôrner (Eds.), *Beliefs: A hidden variable in mathematics education?* (pp. 59–72). Dordrecht: Kluwer.
Hannula, M. (2011). The structure and dynamics of affect in mathematical thinking and learning. In M. Pytlak, T. Rowland, & E. Swoboda (Eds.), *Proceedings of the seventh congress of the European society for research in mathematics education* (pp. 34–60). Rzeszów: Erme.
Hazm, Ibn. (ca. 1022). *The ring of dove. A treatise on the art and practice of Arab love*. Downloaded from: http://www.muslimphilosophy.com/hazm/dove/index.html
Holodynski, M. (2013). The internalization theory of emotions: A cultural historical approach to the development of emotions. *Mind, Culture, and Activity, 20*(1), 4–38.
Illouz, E. (1997). *Consuming the romantic utopia: Love and the cultural contradictions of capitalism*. London: The University of California Press.
James, W. (1884). What is an emotion. *Mind, 9*(34), 188–205.
Leach, W. (1980). *True love and perfect union*. New York: Basic Books.
Leach, E. (1981, April 4). The poetics of power. *The New Republic*, 30–33.
Leont'ev [or Lentyev], A. N. (2009). *Activity and consciousness*. Pacifica: MIA. Retrieved August 29, 2009, from http://www.marxists.org/archive/leontev/works/activity-consciousness.pdf
Lewin, K. (1935). *Dynamic theory of personality*. New York: McGraw-Hill.
Mandler, G. (1984). *Mind and body: Psychology of emotions and stress*. New York: Norton.
Mandler, G. (1989). Affect and learning: Causes and consequences of emotional interactions. In D. Mc Leod & V. Adams (Eds.), *Affect and mathematical problem solving* (pp. 3–19). New York: Springer.
Markus, G. (1982). *Langage et production* [Language and production]. Paris: Éditions Denoël/Gonthier.
McLeod, D. (1989). The role of affect in mathematical problem solving. In D. McLeod & V. M. Adams (Eds.), *Affect and mathematical problem solving* (pp. 21–36). New York: Springer.
McLeod, D. (1992). Research on affect in mathematics education: A reconceptualisation. In D. Grouws (Ed.), *Handbook of research on mathematics teaching and learning* (pp. 575–596). New York: Macmillan.
Menon, U. (2000). Analyzing emotions as culturally constructed scripts. *Culture & Psychology, 6*(1), 40–50.
Murphy, H. (1978). The advent of guilt feelings as a common depressive symptom: A historical comparison of two continents. *Psychiatry, 41*, 229–242.
Radford, L. (2008a). The ethics of being and knowing: Towards a cultural theory of learning. In L. Radford, G. Schubring, & F. Seeger (Eds.), *Semiotics in mathematics education: Epistemology, history, classroom, and culture* (pp. 215–234). Rotterdam: Sense Publishers.
Radford, L. (2008b). Di Sé e degli Altri: Riflessioni su un problema fondamentale dell'educazione [The self and the other: Reflections on a fundamental problem in education]. *La Matematica e la sua didattica, 22*(2),185–205.
Radford, L. (2012). Education and the illusions of emancipation. *Educational Studies in Mathematics, 80*(1), 101–118.
Radford, L. (2013a). Three key concepts of the theory of objectification: Knowledge, knowing, and learning. *Journal of Research in Mathematics Education, 2*(1), 7–44.
Radford, L. (2013b). Sensuous cognition. In D. Martinovic, V. Freiman, & Z. Karadag (Eds.), *Visual mathematics and cyberlearning* (pp. 141–162). New York: Springer.
Ratner, C. (2000). A cultural-psychological analysis of emotions. *Culture & Psychology, 6*(1), 5–39.
Rosaldo, M. (1983). The shame of headhunters and the autonomy of the self. *Ethos, 11*(3), 135–151.
Roth, M. W. (2007). Motive, emotion, and identity at work: A contribution to third-generation cultural historical activity theory. *Mind, Culture, and Activity, 14*(1–2), 40–63.

Roth, W. M., & Radford, L. (2011). *A cultural historical perspective on teaching and learning*. Rotterdam: Sense Publishers.

Sherrington, C. S. (1900). Experiments on the value of vascular and visceral factors for the genesis of emotions. *Proceeding of the Royal Society, 66*, 3–97.

Shusterman, R. (2012). *Thinking through the body. Essays in somaesthetics*. Cambridge: Cambridge University Press.

Solomon, R. (1978). Emotions and anthropology: The logic of emotional world views. *Inquiry, 2*, 189–199.

Varela, F., Thompson, E., & Rosch, E. (1991). *The embodied mind*. Cambridge: MIT.

Vygotsky, L. S. (1987). *Collected works* (vol. 1, R. W. Rieber and A. S. Carton, Eds.). New York: Plenum.

Vygotsky, L. S. (1999). *Collected works* (vol. 6, R. W. Rieber, Ed.). New York: Plenum.

Zaporozhets, A. V. (2002). Towards the question of the genesis, function, and structure of emotional processes in the child. *Journal of Russian and East European Psychology, 40*(2), 45–66.

# The Construct of Attitude in Mathematics Education

**Pietro Di Martino and Rosetta Zan**

**Abstract** This chapter addresses a number of crucial theoretical issues about research on attitude towards mathematics, a field that has a very long tradition in mathematics education, with early studies on attitude being published more than 60 years ago. Over time, research on attitude in mathematics education has developed a range of perspectives and methodologies, dealing with a variety of questions concerning the construct of attitude: discussion and development of tools for measuring/assessing/observing it; analysis of the relationship with other affective constructs and with cognition; investigation of the relationship with achievement; critique of the lack of a suitable theoretical framework. The chapter traces the 'story' of the construct of attitude, providing a theoretical discussion of the issues mentioned above that are crucial to understanding the mosaic of relationships and interactions within the affect field. Through the theoretical debate, the aim of the chapter is to highlight new directions for research on attitude in mathematics education.

**Keywords** Attitude towards mathematics • Affect in mathematics education • Emotions • Beliefs

## Introduction

Research on attitudes towards mathematics can be viewed as paradigmatic of research in mathematics education. This research field lies at the intellectual crossroads of many different domains (e.g. mathematics, psychology, cognitive science, epistemology, semiotics, anthropology), and often deals with constructs that have been developed in those domains to face (new) emerging issues in mathematics education (Sierpinska et al. 1993). The construct of *attitude* was introduced in the first decades of the nineteenth century in the context of social psychology in order

P. Di Martino (✉) • R. Zan
Dipartimento di Matematica, Università di Pisa, Largo Pontecorvo 5, 56127, Pisa, Italy
e-mail: dimartin@dm.unipi.it; zan@dm.unipi.it



B. Pepin, B. Roesken-Winter (eds.), *From beliefs to dynamic affect systems in mathematics education*, Advances in Mathematics Education,
DOI 10.1007/978-3-319-06808-4_3

to foresee individuals' choices in contexts such as voting or buying goods. *Attitude* is seen as a trait of an individual that has a direct influence upon his/her behaviour:

> An attitude is a mental and neural state of readiness, organized through experience, exerting a directive and dynamic influence upon the individual's response to all objects and situations with which it is related. (Allport 1935, p. 810)

In mathematics education, early studies about attitude appear in the middle of the twentieth century. These pioneering studies were deeply affected by the field (social psychology) in which the construct was born, both regarding the characterization of attitude, seen as an individual's trait capable of influencing his/her own behaviour (Aiken 1970), and the methods used to assess and measure it.

In this context, the main goal was the search for a measurement of attitude: Dutton (1951), in one of the first studies concerning attitude and mathematics, stated his aim to measure pupils' and teachers' attitude towards arithmetic using Thurstone scales. As a matter of fact, following the trend in social psychology, the measurement of attitude was mainly carried out by the means of unidimensional ad hoc scaling methods, such as Thurstone and Likert scales.

Many things have changed in the field of research on attitude from those early studies up to now; some of those changes have been deeply influenced by a change of perspectives in mathematics education. At present, attitude is considered (together with beliefs, emotions and values) as one of the constructs that characterize a new field of research: that of affect.

Research on attitude, as often happens, has not followed a linear path. Over the years, the researchers' position on basic issues such as the definition itself of attitude and the instruments used to assess the construct has dramatically changed and new issues and goals have been identified.

This feature of research on attitude has increased the need for a clear theoretical systematization of research results, which has also emerged as a priority in the whole mathematics education field in the last two decades. As a matter of fact, this need has now become a necessity in mathematics education, due to the considerable development of the research field in the last few years and, in particular, to the identification of its *cumulative* and *universal* characters (Boero and Szendrei 1998). This view of the field is strictly linked with the characterization of the nature of research findings:

> Researchers in education have an intellectual obligation to push for greater clarity and specificity (…) [in mathematics education] findings are rarely definitive; they are usually suggestive. Evidence is not on the order of proof, but is cumulative. (Schoenfeld 2000, pp. 647–648)

Therefore, coherently with the *cumulative* characterization of research in mathematics education, we believe that tracing, with critical eyes, the history of research on attitude may bring forward an understanding – through a theoretical lens – of the mosaic of the relationships and interactions between definitions of attitude and instruments to measure it, and of the influence the shift from a normative to an interpretive paradigm had on both these issues.

Moreover, this systematization is necessary to map out the future of research on attitude, including the identification of new issues, the development of suitable methods, and a warning against repeating the same old mistakes.

## Early Studies of Attitude in Mathematics Education: The Problematic Relationship Between Attitude and Achievement

During its early period (ranging from the first half of the twentieth century to the end of the 1980s), research on attitude within mathematics education followed the trend of research in social psychology. The definition of attitude was rarely made explicit, and implicitly it seemed to refer to the tendency to behave in a certain way. A central research topic was the development or refinement of measuring instruments and sampling methods:

> The search for more adequate questionnaire and sampling techniques and factors underlying attitudes toward these subjects [arithmetic and mathematics] continues to be an important area for research. (Dutton 1951, p. 418)

In this period, the predominant methodology was quantitative and statistical: as a matter of fact the quantitative and statistical approach seems to have been considered a sort of warrant for the scientific nature of the discipline.

Research on attitude at this stage reflects the evolution of the field of mathematics education: an in-depth discussion about the very *nature* of this emerging field had not yet been developed. According to Kilpatrick (1992, p. 15), in that period "the measurement movement begins". The quantitative primacy in the methods used had its roots in the search for scientific acceptance of a young discipline that was just beginning to take its first steps:

> From the beginnings of the century through its three-quarter point, such inquiry [inquiry in math education] becomes increasingly "scientific", that is, ostensibly objective and rigorously quantified. (Schoenfeld 1994, p. 698)

On the other hand, the attention paid to measurement instruments was also linked to the main goal of early studies on attitude, which was the identification of causal correlations between attitude and other significant factors.

In the first review of the construct of attitude within mathematics education, Feierabend (1960) highlighted two main reasons for the increasing academic interest in this construct. Drawing on the development of such construct in social psychology, the first reason was related to the view of attitude as a *selective factor* because of its correlation with the choice of enrolling/not enrolling in advanced mathematical courses:

> Mathematics, geometry, and algebra are the courses which, when disliked in high school, have the highest percentage of students who never take a course in this area again. This implies

> the operation of such a strong selective factor that by the time students reach college, only the students with a strong positive attitude will still be taking mathematics; the rest have negative attitudes which may increase in strength with the operation of time and the lack of counteracting influences. (Feierabend 1960, p. 19)

The second reason concerned the relationship between attitude and mathematical achievement:

> A series of recent investigations have attempted to explain differences in school performance among students of equal abilities on the basis of their attitudes. (Feierabend 1960, p. 11)

This point also implied taking into account gender differences in mathematics achievement and in problem-solving ability:

> There are sex differences in problem-solving ability unrelated to general mental ability, special abilities, or specific knowledge (...) [he] attempted to show that the differential performance of the two sexes was due to a difference in attitude toward problem-solving. (Feierabend 1960, p. 17)

In his review Feierabend advanced some criticism towards research on attitude, but his criticism was limited to some aspects related to the development of instruments and to the statistical analysis. There was no reference to the lack of theoretical clarity and in particular no explicit definition of attitude was provided: a naïve view of the construct emerges. The term 'attitude' was used to address different constructs, such as preference, interest, motivation.

Ten years later Aiken (1970) summarized early research on attitude as follows:

> The major topics covered were: methods of measuring attitudes towards arithmetic and mathematics; the distribution and stability of mathematics attitudes; the effects of attitudes on achievement in mathematics; the relationship of mathematics attitudes to ability and personal factors. (Aiken 1970, p. 592)

It is interesting to notice that Aiken's list also does not include reference to the topic 'nature of the construct of *attitude*' (that would become a major topic in research on attitude in the early 1990s).

What emerges from the reviews carried out by Feierabend and Aiken and from the analysis of other literature of that period (Reyes 1984) is that most studies were focused on the search for evidence of a causal relationship between "something called *attitude*" (Neale 1969, p. 631) and other variables, in particular mathematical achievement. This causal relationship is even seen as a hypothesis of the aetiology of attitude towards mathematics (Aiken and Drager 1961). The search for a causal relationship reveals a normative approach, that seems to drive research on attitude and provide a justification, and in some way a reinforcement, for the great attention paid to measurement instruments, rather than to the theoretical clarification of the construct.

Despite its theoretical limitations, this first phase of research on attitude was fruitful and produced meaningful results that, coherently with a *cumulative view* of research, contributed significantly to the new research era that would follow. The most significant contribution was what became the initial assumption of this kind of research, that is, that non-cognitive factors strictly interact with cognitive factors and have a crucial role in the learning of mathematics. This assumption is a sort of *break in the wall* of the purely cognitive approach to mathematics education, and was to be decisive in the development of the specific field of affect in mathematics education: not purely cognitive factors – and in particular attitude – would become a relevant topic in the study of mathematical learning:

> The attitudes of students toward mathematics play a vital part in their learning (...) Important for the study of attitudes toward mathematics is the idea that an attitude involves both cognitive and non-cognitive aspects. (Corcoran and Gibb 1961, p. 105)

In addition to this, the great emphasis placed on methods caused a refinement of many observational instruments. This brings to light important issues related to the observation of attitude (and more in general of affective constructs), as for example the tendency of individuals to reply to questionnaires according to what is socially accepted and valued, rather than expressing their own thoughts – the so-called social desiderability phenomenon (Kloosterman and Stage 1992).

Moreover, in this early period, research on attitude consolidated two significant findings. A first result – confirmed by many studies – was the relationship between attitude towards mathematics and the choice of mathematics courses. For instance, in his literature review of research on attitude, Aiken (1970) stated that there is a good body of evidence showing that the choice of enrolling in advanced mathematics courses is significantly affected by attitude towards mathematics. A second important finding refers to gender differences in mathematical achievements. In particular, the valuable work of Elizabeth Fennema and Julia Sherman highlighted the differences in attitude towards mathematics between males and females, offering a new and important key for the interpretation of gender differences in mathematics achievements:

> Since the study of mathematics appears not to be sex-neutral, attitudes toward mathematics may reflect cultural proscriptions and prescriptions (...) These data certainly indicate that many females have as much mathematical potential as do many males. The generalized belief that females cannot do well in mathematics is not supported. (Fennema and Sherman 1977, p. 69)

This result, that may now appear unquestionable, was not so obvious before the work of Fennema and Sherman.

Even if the first period of research on attitude provides several important findings and suggests a number of research hypotheses, the above discussion has also shown its strong limitations from the very beginning. The identification and analysis of these limitations has been crucial for the development of research on attitude in the following years.

According to Bishop (1992), carrying out a research study in mathematics education requires taking into account three components: enquiry (which concerns the reason for the research activity), evidence and theory. The initial studies on attitude are motivated by the assumption of the existence of a causal relationship between attitude and achievement in mathematics, and seem to be focused on searching for evidence of this relationship rather than developing a theoretical framework or clarifying the nature of the construct. But in spite of the efforts devoted to developing measuring instruments, research fails to show a causal relationship in the direction attitude → achievement, or a clear correlation between them.

Aiken (1970) reported the results of several studies in which this correlation is far from being clear, highlighting the need for clarifying its very nature. Almost 30 years later, Ma and Kishor (1997), analysing the results of 113 different studies, conclude that this correlation is not statistically significant. Assuming that this correlation does exist, Ma and Kishor identify the cause of the failure to prove it in the inappropriateness of the observing instruments used in the research on attitude towards mathematics up to that point. At that stage, the instruments used to measure attitude towards mathematics have been criticized by many researchers, because their nature is considered "exceptionally primitive" (Leder 1985).

However, starting from the 1980s, researchers increasingly acknowledge that the major weakness of this kind of research lies in the lack of clarity at the theoretical level and in the definition of the construct itself. Kulm (1980) suggests the existence of a trend that tends to avoid an explicit definition of attitude towards mathematics and instead adopts operational definitions determined by the types of instruments used to measure attitude. This lack of interest in characterizing the construct produces a gap between the definition of attitude and its measurement (Leder 1985), and results in the lack of reliability of the observational instruments.

Germann's words below summarize the criticism towards the first phase of research about attitude:

> First, the construct of attitude has been vague, inconsistent, and ambiguous. Second, research has often been conducted without a theoretical model of the relationship of attitude with other variables. Third, the attitude instruments themselves are judged to be immature and inadequate. (Germann 1988, p. 689)

In other words, the naïve theoretical approach that characterizes early studies on attitude appears to be inadequate within the normative-positivistic paradigm in which those studies were conducted. As a matter of fact, this paradigm demands isolating and clearly identifying variables in order to interpret statistical results and to be able to compare them across studies:

> Sometimes no description or definition of what is meant by a particular variable is even included in the research report. This makes interpretation of results difficult and detracts from efforts to compare results across studies. (Hart 1984, p. 573)

For this reason, the process of re-thinking research on attitude began at the end of the 1980s, addressing many aspects such as: the paradigm in which it is framed, the goals that it pursues, the construct definition, the relationship between the construct and other (affective and cognitive) factors, the development of observational tools and the discussion about methods for analysing data.

## The Theoretical Debate About *Attitude* in Mathematics Education

In 1992, in the well-known *Handbook of research on mathematics teaching and learning*, McLeod traces the way for a reconceptualization of research on affect in mathematics education. He identifies three different constructs – beliefs, attitudes and emotions[1] – that, in his view, vary in stability and differ in the degree of the role played by cognition. McLeod's work starts with a crucial premise:

> Affective issues play a central role in mathematics learning and instruction (…) If research on learning and instruction is to maximize its impact on students and teachers, affective issues need to occupy a more central position in the minds of researchers. (McLeod 1992, p. 575)

He emphasizes the relationship between the newly acknowledged role assigned to affective factors and the constructivist view of mathematics learning:

> If we believe that the learner is someone who only receives knowledge rather than someone who is actively involved in constructing knowledge, our research program could be entirely different in terms of both the affective and the cognitive domain. (McLeod 1992, p. 576)

The need for reconceptualization is strictly connected with the criticism of the previous research on attitude:

> Research on affect has been voluminous, but not particularly powerful in influencing the field of mathematics education. It seems that research on instruction in most cases goes on without any particular attention to the affective issues (…) A major difficulty is that research on affect has not usually been grounded in a strong theoretical foundation. (McLeod 1992, p. 590)

Therefore, McLeod highlights that research on affect has to pay particular attention to three strictly intertwined aspects: the discussion of theoretical issues, the development of a wider variety of methods, and the analysis of the relationships among affective constructs and between affect and cognition.

Once again, the development of research on attitude is deeply influenced by the simultaneous development of the field of mathematics education at the end of the 1980s. In this period, many scholars debate on the nature of mathematics education and on the criteria for establishing quality of research in this field. In particular, consistently with the goal of universalization of research results, the request for a theoretical clarification of the constructs used in research is emphasized:

> A community of scholars engaged in the research of common areas with common themes, however, has responsibility to communicate ideas and results as clearly as possible using common terms. For these reasons, it is important to use the terms consistently, accurately, and appropriately once their definitions have been agreed on. (Pajares 1992, p. 315)

## *What Is Attitude Towards Mathematics?*

The discussion about the theoretical aspects of research on attitude starts with a 'definition problem': what is attitude towards mathematics?

[1] Later, De Bellis and Goldin (1999) propose 'values' as the fourth construct of the affective domain.

A first critical issue relates to the object that attitude is oriented towards – that is, mathematics. Some researchers refer to a 'unique' attitude toward mathematics (Haladyna et al. 1983), while others claim that many different attitudes exist according to the different topics and activities that are considered (Tirosh 1993); still other scholars propose to distinguish between attitude towards mathematics seen as a branch of scientific knowledge and towards mathematics as school subject (Schoenfeld 1989), or even that attitude can refer to different objects and situations, such as mathematical content, characteristics of mathematics, kind of teaching, mathematical activities in the classroom and mathematics teacher (Kulm 1980).

Moreover, this complexity grows when, in addition to the variety of objects attitude is oriented towards, we also consider the variety of subjects: whose attitude? Research on attitude has dealt with a large variety of individuals: students, prospective and in-service teachers, students' parents, and, more in general, adults.

But the most significant aspect of the complexity regarding the 'definition problem' is that it involves not only the characterization of the construct 'attitude', but also that of positive/negative attitude, a dichotomy that pervades research, both implicitly and explicitly. Classic studies regarding the relationship between attitude and achievement in practice investigate the correlation between *positive* attitude and success. In the same way, studies aiming to change attitude actually end up in setting the objective of transforming a *negative* attitude into a *positive* one.

As already mentioned, a large portion of studies show the lack of a clear definition of the construct: attitude tends to be defined implicitly and a posteriori through the instruments used to measure it (Kulm 1980; Leder 1985; Daskalogianni and Simpson 2000).

In social psychology, the most recent theories agree on the multidimensionality of the construct, and make reference to a *tripartite model*, according to which attitude has a cognitive, an affective, and a behavioural component (Eagly and Chaiken 1998). Within the field of mathematics education many explicit definitions of attitude refer to this tripartite model, describing attitude by means of three components: the emotional disposition towards mathematics, the set of beliefs regarding mathematics, and the behaviour related to mathematics (Hart 1989; Leder 1992; Ruffel et al. 1998). However, some studies – generally in the earliest period of research – adopt a 'simple' characterization, seeing attitude as a general emotional disposition (Haladyna et al. 1983).

Both definitions show their theoretical but also operational and didactical limitations (Di Martino and Zan 2001). The *simple* definition does not make explicit reference to cognitive aspects, although many researchers who subscribe to this definition use models (see Mandler 1984; Ortony et al. 1988) that emphasize the relationship between emotion and cognition, describing emotional experience as the result of a combination of cognitive analyses and physiological responses. In this framework, it is the interpretation given by an individual to an experience that elicits the emotion, and not the experience itself:

> First, the meaning comes out of the cognitive interpretation of the arousal. This meaning will be dependent on what the individual knows or assumes to be true. In other words, the

> individual's knowledge and beliefs play a significant role in the interpretation of the interruption. (McLeod 1992, p. 578)

According to the simple definition, the characterization of positive/negative attitude is clear: a positive (negative) attitude is a positive (negative) emotional disposition towards the subject.

This characterization can be useful when dealing with issues such as the choice of mathematics courses or the comparison between different groups of individuals, but it seems inadequate to deal with complex issues such as success in mathematics. In this context, the idea of positive attitude that emerges from the simple definition is not considered very significant by many mathematics education researchers, who underline the importance of linking a positive emotional disposition with an epistemologically correct view of the discipline (Ernest 1988). In the same vein, the crucial issue of promoting a positive attitude risks losing its significance if the goal of developing a positive emotional disposition toward mathematics is not associated to the goal of promoting a *positive* view of the discipline. Considering only the emotional aspects poses an even greater didactical threat, since teachers may choose to avoid complex tasks in order to prevent producing negative emotions.

Kulm (1980) discusses similar issues about the attitude definition in the early research period and concludes:

> It is probably not possible to offer a definition of attitude towards mathematics that would be suitable for all situations, and even if one were agreed on, it would probably be too general to be useful. (Kulm 1980, p. 358)

The awareness that the *appropriateness* of the construct depends on the studied issues will lead to the idea of a 'working definition' (Daskalogianni and Simpson 2000).

As regards the *tripartite* model, the main critical aspect is that the implicit assumption of a link between attitude and behaviour becomes part of the construct definition itself. This theoretical choice exposes research to the risk of circular reasoning, as eloquently described by Lester (2002) in relation to the belief-construct:

> A central difficulty is that the fundamental assumption undergirding much of this research rests on a shaky logical foundation. Specifically, a basic assumption is that beliefs influence peoples' thinking and actions. However, it is also often assumed that beliefs lie hidden and so can be studied only by inferring them from how people think and act. For researchers to claim that students behave in a particular manner because of their beliefs and then infer the students' beliefs from how they behave involves circular reasoning. (Lester 2002, p. 346)

In the light of these critical aspects, a third definition of attitude emerges in which behaviours are not explicitly mentioned: attitude towards mathematics is described as the pattern of beliefs and emotions associated with mathematics (Daskalogianni and Simpson 2000).

This choice overcomes the risk of circularity, but the theoretical problem of identifying a positive/negative attitude according to a multidimensional definition still remains (Di Martino and Zan 2003). As a matter of fact, there is not only a need for characterization of the positive/negative dichotomy for each dimension (emotions, beliefs, possibly behaviour), but it is also essential to identify if and how the dichotomies related to the single components can result in a unique characterization of positive/negative attitude. This issue is strictly connected with the choice of the instruments used to measure attitude.

## *Instruments Used to Measure Attitude*

As Leder (1985) claims, the lack of interest in characterizing the construct produces a gap between the definition of attitude and its measurement: as a matter of fact the instruments traditionally used to assess and measure attitudes are not consistent with the different definitions and with whether an explicit definition of attitude is given or not.

The instruments used are almost exclusively self-report scales (Kulm 1980; Leder 1985; McLeod 1987) such as Thurstone or Likert scales. These instruments propose items that take into consideration beliefs and behaviours as well as emotions: for example 'Mathematics is useful', 'I think about arithmetic problems outside school', 'I like problem solving'. Therefore, they make implicit reference to the tripartite model, regardless of whether this definition is explicitly selected as a starting point or not. Even if the instruments used appear to be increasingly sophisticated, the measurement generally results in a reduction to the positive/negative bipolarity, which is obtained by summing up the scores related to each of the three dimensions: cognitive, affective and behavioural.

While some scholars underplay this operation by observing that 'the correlation among measures of the three components, although leaving room for some unique variance, are typically of considerable magnitude' (Ajzen 1988, p. 22), others consider this reduction as contradicting the recognized complexity of the tripartite model (Eagly and Chaiken 1998). Reducing the description of attitude to a single score is also in contrast with the original idea of Thurstone and Chave (1929) who claim that attitude is a complex construct that cannot be measured by a single score, but requires several indices. Thurstone and Chave underline that the choice of the characteristics (indices) to be measured depends on the context – in the same way as when measuring a physical object like a table one can decide whether to measure length, width or height.

But the theoretical debate about research on attitude highlights other critical issues in the *measurement process*. First, the separate measurement of each component presents significant problems, due to the limitations of questionnaires. As far as beliefs are concerned, the mismatch between exposed beliefs and beliefs-in-action is well known (Schoenfeld 1989), just like the already mentioned *social desiderability* phenomenon (Kloosterman and Stage 1992). Regarding emotions, researchers have discussed the difference between an *opinion* about an emotion and the *emotion* itself (Ruffel et al. 1998) and the limitations of instruments such as questionnaires and interviews in capturing emotional reactions that are not conscious (Schlöglmann 2002).

A second critical point concerns the choice of items that, in the case of questionnaires, is fully determined by the researchers, while respondents are only asked to express their agreement/disagreement with these items: how can we be sure that the topic of the items is relevant to the respondent? In other words, using the terminology introduced by Green (1971), how can we be sure that the corresponding beliefs are psychologically central in the respondent's belief system?

A third critical aspect concerns the choice of the score to be attributed to each of the items, requiring identification of what a *positive* emotion/belief/behaviour is (this shows the strong relationship that exists between the definition and the measurement problem). Essentially:

- When *positive* refers to an emotion, it normally means 'perceived as pleasurable'. So anxiety when confronting a problem is seen as negative, while pleasure in doing mathematics is evaluated as positive.
- When *positive* refers to beliefs, it is generally used with the meaning 'shared by experts'. The first limitation of this approach is brought to light by a number of studies which highlight that there is no single pattern of beliefs shared by experts in mathematics (Mura 1993, 1995; Grigutsch and Törner 1998). In light of this, identifying several different typical patterns of beliefs towards mathematics shared by experts becomes necessary. At present, this still remains an issue for discussion that could lead to the definition of a number of different patterns to act as models of 'successful views of mathematics'.
- When it refers to a specific behaviour, *positive* generally means 'successful'. In the school context, a successful behaviour is generally identified with high achievement. This characterization leads to the problem of how to assess achievement (Middleton and Spanias 1999).

A further problem is that the differences between the various meanings of positive attitude are rarely made explicit. If the researcher does not declare his/her initial choices, interpreting the results of a study and comparing different studies becomes problematic.

Even if this ambiguity is overcome by making explicit the initial choices and assumptions, in our opinion other problems remain. In some studies the three meanings for 'positive' (related to emotion, belief and behaviour) overlap thanks to implicit assumptions: for example, that a 'positive' belief (i.e. shared by experts) is associated with a successful behaviour and elicits a pleasurable emotion; or that a pleasurable emotion is necessarily associated with a positive behaviour in mathematics, and vice versa for negative emotion.

Evaluating a belief (or an emotion) as 'positive' or 'negative' according to the emotion and behaviour related to it implies using a cause/effect model, according to which the same belief *causes* the same emotion or the same behaviour in all individuals. Moreover, this evaluation not only assumes that a certain belief has an emotional component, but also looks at the significance of that emotional component, that is, not just that it is linked to a behaviour, but also which type of behaviour.

In this case, the cure seems worse than the disease, since this approach does not take into account the very complex nature of the relationship among beliefs, emotions and behaviour.

As a matter of fact, a number of studies about emotions (Evans 2000) suggest the possibility that for certain subjects, an optimal level of anxiety exists, above which, but also below which, performance is reduced.

The relationship between beliefs and emotions was investigated in a study with 211 high school students aged between 14 and 18 (Di Martino and Zan 2002) in the case of the belief 'In mathematics there is a reason for everything', which is an item that is also used in many scales for measuring attitude towards mathematics. Students are asked to fill in a questionnaire including the following item:

> Choose the option you most agree with:
> □ In mathematics there is always a reason for everything (B)
> □ It is not true that in mathematics there is always a reason for everything (not B)
> And: □ I like □ I don't like □ I am indifferent to this characteristic of mathematics

The findings show that only 51.7 % of the sample fell in the two *expected* groups (i.e. 'B – I like' and 'not B – I don't like'). But overall there was no difference in the percentage of belief B-holders between the groups of high achievers and low achievers. The distinction between these two groups is related to the emotion associated with this belief: 76 % of the high-achievers who are belief B-holders liked this characteristic of mathematics, while this percentage dramatically decreased to 28 % within the low-achievers group.

About the combination 'epistemological correct belief – negative emotion', we suggest two possible interpretations. The first interpretation is that the negative emotion is *directly* related to the belief. On the other hand, we also need to consider the possibility that the emotional disposition is not directly linked to that single belief, but to its interaction with other beliefs. This remark questions the possibility of characterizing a single belief as positive or negative, without considering its connection with other beliefs an individual may have (belief system):

> Because they [single beliefs] offer a limited glimpse into a much broader system and because understanding their connections and centrality is essential to understanding the nature of their effect, researchers must study the context-specific effects of beliefs in terms of these connections. (Pajares 1992, p. 326)

More specifically, when describing belief systems Rokeach (1968) recognizes the dimension of *centrality* for a particular belief, highlighting that not all beliefs have the same importance for an individual. Central beliefs play a prominent role in people's belief systems, and consequently in influencing their behaviour. As Eagly (1967) observes, beliefs about self are generally considered more central than other ones.

Consider for example the relationship between belief B used in our study ('In mathematics there is a reason for everything') and the (likely) central self-belief 'I am not able to understand these reasons': the presence of such relationship may result in unproductive behaviours such as avoiding answering a question or giving random answers (Di Martino 2004).

The discussion above highlights that the assumption of the existence of a cause-effect relationship between a specific belief and emotion or behaviour is inadequate. The interaction is more complex, since it involves an individual's belief system (and not only the single belief) and is strongly dependent on the individual.

Following the results of this debate, a movement towards the overcoming of the normative approach and the use of an interpretive approach for research on attitude emerges with the aim of attending to the complexity of the issues at stake.

## The Attitude Construct in the Reconceptualization of the Affective Domain in Mathematics Education

Once again, the history of research on attitude reflects the evolution of the mathematics education field: the theoretical debate about attitude develops in parallel with a new interpretive perspective that begins to emerge within the field of mathematics education. This perspective, in contrast with the normative-positivistic one, significantly affects the discussion about the theoretical characterization of constructs.

The gradual affirmation of the interpretive paradigm in the social sciences, including a greater attention paid to the complexity of human behaviour, leads researchers in mathematics education to abandon the attempt of explaining behaviour through measurements or general rules based on a cause-effect model, and to search instead for new interpretive tools (once again drawing on other domains):

> The purpose of doing interpretivist research (…) is to provide information that will allow the investigator to "make sense" of the world from the perspective of participants. (Eisenhart 1988, p. 103)

This implies a significant shift in focus: an emerging attention to the understanding of a phenomenon ('making sense of the world') replaces the description of the phenomenon itself, which could be seen as a shift from product to process (Schoenfeld 1994).

The interpretive approach has a direct influence on the process of re-thinking research methods since the limitations of the statistical methods become evident:

> Through the 1980s and into the 1990s (…) with a shift in focus there was a concomitant shift in methods (including the reporting of clinical interviews, process and simulation models, field observations and participant observations), because a new class of phenomena required a new set of explanations a new set of tools to uncover them. (Schoenfeld 1994, p. 703)

This shift of perspectives gives new strength to research on attitude that until this point had remained stuck in the causal-relationship paradigm. In particular, attitude gains renewed popularity in the studies about problem-solving activities aimed at interpreting the failure of students who seem to have the required cognitive resources.

The book *Affect and mathematical problem solving* (McLeod and Adams 1989), collecting contributions by several authors, represents the turning point for research on affective constructs, and in particular on attitude. For the first time, affective constructs are used not only to prove the existence of a numerical correlation with an outcome (mathematical achievement), but also to interpret a process (the interactions between affective and cognitive aspects in problem-solving activities). Therefore, the need for a theoretical clarification in mathematics education (that is also related to the possibility and the intention for a cumulative development in the field) appears to become a fundamental issue also specifically for research on affect:

> There was a lack of definition, lack of clarity, and lack of connections to mathematics. It is possible to avoid making the same mistakes again as new ideas and research methodologies are employed. It is hoped that new researchers on affect will be clear about what is being studied, precise in definition, and respectful of what has been learned previously. (Fennema 1989, p. 209)

The double occurrence of the adjective 'new' in Fennema's words is not casual: it shows awareness of the fact that new perspectives and new more complex issues force a rethinking of the affective constructs. In particular, the shift from a normative paradigm to an interpretive one provokes a discussion (re-definition) of goals, definitions and methods.

The belief that research on attitude towards mathematics may offer interpretive instruments to understand the reasons for an individual's intentional actions in the mathematical context grows (Zan et al. 2006). This *belief* is supported by the initial evidence coming from related research in the field of neuroscience:

> There is apparently some neurological basis for asserting a link between affective and cognitive aspects of human functioning. (Silver 1985, p. 253)

More recently, Damasio (1996) highlighted the close relationship between affect and decision-making processes.

The theoretical construct of 'attitude towards mathematics' is no longer a construct aimed at explaining causes of behaviour, thus enabling researchers to predict it, but instead it becomes a flexible and multidimensional interpretive tool, aimed at describing the interactions between affective and cognitive aspects in mathematical activity. In particular, attitude becomes a tool to interpret people's decisions in mathematical activities, and, if necessary, suggest strategies to modify them. In this context, particularly significant is Ruffel, Mason and Allen's position about the definition of the construct of attitude itself:

> Reflecting on them [some previous studies about attitude] led us to challenge the very construct of attitude. We are also led to challenge the cause-and-effect model underlying much attitudinal research. We now see *attitude* as at best a complex notion, and we conjecture that perhaps it is not a quality of an individual but rather a construct of an observer's desire to formulate a story to account for observation. (Ruffel et al. 1998, p. 1)

It could be argued that the same thing can be said about every theoretical construct, not only in mathematics education. In fact, in our opinion, this position shows awareness of the fact that any phenomenon can only be observed from a particular point of view and thus highlights the role of the researcher/observer, who cannot be a mere measurer. This position represents the overcoming of a naïve approach, in which attitude is seen as an objectively measurable quality of an individual, and the transition to a theoretical approach.

In line with this perspective, Daskalogianni and Simpson (2000) assume that the definition of attitude becomes a *working definition*, which is functional to the research questions that researchers pose in each study. Therefore, having different definitions of the construct appears natural, and a definition is no longer evaluated in terms of *correctness* (is it the *right* definition?) but in terms of *suitability* to address a specific research problem in mathematics education (Di Martino and Zan 2010). According to the classification of research proposed by Bishop (1992), this kind of approach characterizes the new trend of research on attitude as *problem-led*.

The theoretical re-thinking of research on attitude leads to the exploration of new methods of inquiry in the field. Coherently with their position, Ruffel et al. (1998) emphasize the inadequacy of the *measurement approach* by replacing the verb 'measuring' with the verb 'probing'.

At the beginning of the new millennium, a strong criticism of the use of quantitative methods in the research on attitude emerged, and a movement towards the use of qualitative approaches has begun. It is understood that qualitative methods, and in particular the use of narratives, enable researchers to take into account those beliefs and emotions which are psychologically central for the respondents. A number of studies using essays, diaries, interviews and also the observation of behaviour in natural settings or in structured situations appear (Karsenty and Vinner 2000; Hannula 2002; Zan and Di Martino 2007; Kaasila 2007).

Differently from what happens with the traditional attitude scales, the respondents are not requested to express agreement/disagreement with respect to items chosen by others, but are asked to tell their mathematical 'stories', through which they can recount all the aspects that they consider relevant in their relationship with mathematics. As a matter of fact, the pivotal motivation for using narratives in educational research is the following:

> Humans are storytelling organisms who, individually and socially, lead storied lives. The study of narrative, therefore, is the study of the ways humans experience the world. (Connelly and Clandinin 1990, p. 2)

As a consequence, almost 10 years after McLeod's *manifesto*, the shift in focus in research on attitude provokes a shift in perspectives and methods: a real *revolution*.

## The TMA Model: A Definition of Attitude Grounded in Students' Narratives

Within the described framework, we have investigated how mathematics teachers use the diagnosis 'this student has a negative attitude' to interpret students' mathematical difficulties in the context of an Italian National Project. The results of the study (Polo and Zan 2006) show that this diagnosis is frequently used (at all school levels) by mathematics teachers to explain students' difficulties, and, above all, that in the majority of cases it represents a case of surrender instead of being used as an interpretive step capable of steering future action.

These findings persuade us that, in order to turn the 'negative attitude diagnosis' into a useful instrument for both practitioners (teachers) and researchers, it is necessary to link the theoretical construct of attitude to practice. This fits with the strong incentive put forward by Phillip "to develop constructs that might be applied to help make sense of teaching and learning environments" (Philipp 2007, p. 264).

Therefore we have designed a study based on the collection and analysis of students' autobiographical narratives and aimed at constructing a characterization of students' attitude towards mathematics in relation to their experience (Di Martino and Zan 2010).

Our reason for choosing to use autobiographical essays is that through this format pupils tend to explicitly evoke events about their past that they deem important and also to paste fragments by introducing causal links, not in a logical perspective but rather in a social, ethical and psychological one (Bruner 1990). We believe that in order to describe the kind of relationship an individual has with mathematics, and consequently to suggest a characterization of attitude towards mathematics strictly linked to experience, this pasting process is more important than an *objective* report of one's experience with the discipline at school. As Bruner claims:

> It does not matter whether the account conforms to what others might say who were witnesses, nor are we in pursuit of such ontologically obscure issues as whether the account is 'self-deceptive' or 'true'. Our interest, rather, is only in what the person thought he did, what he thought he was in, and so on. (Bruner 1990, pp. 119–120)

In doing so, a theoretical model for attitude emerges from the data collected through a cyclical analytical process, that is, through what Glaser and Strauss (1967) call *grounded theory*. In this kind of process, the autobiographical texts are analysed in order to systematically make meaning out of the individuals' narrations: the final outcome is the identification of a set of categories and relationships aimed at understanding and interpreting different behaviours (Demazière and Dubar 1997).

We have collected and analysed 1662 anonymous essays entitled "Maths and me: my relationship with maths up to now", written by students whose school levels ranged from grade 1 to grade 13.[2] The results of our study show that when students describe their relationship with mathematics, almost all of them refer to one (or more) of the following three dimensions:

- emotional disposition towards mathematics,
- vision of mathematics,
- perceived competence in mathematics.

This result suggests the Three-dimensional Model for Attitude (TMA) represented in Fig. 1.

[2] The sample of the study was not chosen on a statistical basis, but we relied on the collaboration of teachers who voluntarily agreed to participate in our research.

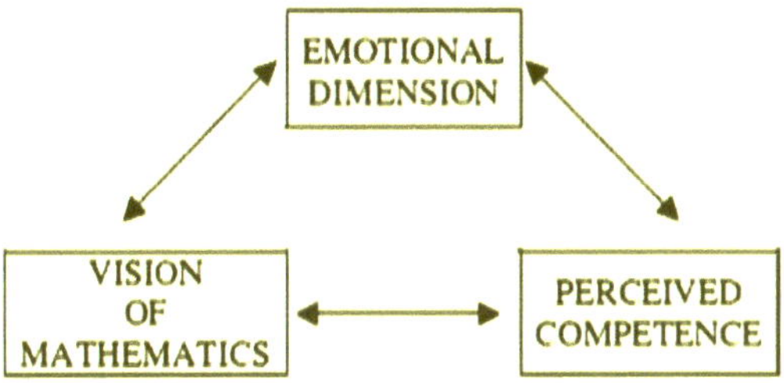


**Fig. 1** The three-dimensional model for attitude

TMA takes explicitly into account the close relationship amongst the three dimensions. The research study also highlights the subjectivity of these interactions, confirming the need for designing suitable observational tools to track it:

> The proposed model of attitude acts as a *bridge* between beliefs and emotions, in that it explicitly takes into account beliefs (about self and mathematics) and emotions, and also the interplay between them. However, in order for it to become effective theoretical and didactical instruments, the construction and use of consistent instruments for observation, capable of taking into account its complexity, is needed. (Di Martino and Zan 2011, p. 479)

The analysis of the students' autobiographical essays also suggests the need for the development of a new approach to the positive/negative characterization of attitude, confirming that the reduction of the dichotomy positive/negative attitude to the emotional dimension is questionable. As a matter of fact, we find that negative emotional dispositions towards mathematics may be associated with different patterns of attitude, depending on the student's perceived competence and vision of mathematics as well as on the relationships amongst the three dimensions. Coherently with this observation, and with the multidimensional characterization of the construct in TMA, we have developed a definition of 'negative attitude' that explicitly makes reference to the negativity of at least one of the three dimensions:

> The multidimensionality of the model underlines the inadequacy of the positive/negative dichotomy for attitude referred only to the emotional dimension (like/dislike), and rather suggests considering an attitude as *negative*, when at least one of the dimensions is *negative*. In this way, we can outline *profiles* of negative attitude, depending on the dimension that appears to be *negative*. (Di Martino and Zan 2010, p. 44)

We identify two polarities for each dimension, and define as negative an emotional disposition resulting in a dislike for mathematics, a low perceived competence, and – according to the characterization of Skemp (1976) – an instrumental vision of mathematics. This definition of negative profiles of attitude within TMA suggests two new interrelated research avenues. One the one hand, the development of observational tools aimed at identifying a student's profile of attitude towards mathematics, in particular at recognizing a possible negative component in this profile. On the other hand, the theoretical construction and implementation of didactical interventions, aimed at preventing or overcoming a negative attitude towards mathematics and differentiated according to the different profiles of negative attitude identified in TMA.

The TMA model, originally created as a model for students' attitudes towards mathematics, also appears suitable for characterizing attitudes towards specific mathematics topics (geometry, algebra, etc.) and for investigating the attitudes towards mathematics held by different groups of people (teachers, adults, etc.). For this reason, the TMA model has recently been used to study and analyse in-service and pre-service primary teachers' attitude towards mathematics and its teaching (Coppola et al. 2012).

## Summing Up and Looking Ahead

In mathematics education, research on attitude has a very long tradition, based on the interest, shared by mathematicians, teachers and mathematics educators, in identifying a causal relationship between something called 'positive attitude' and achievement. In the first period of the research most studies aimed at refining or developing measuring instruments, rather than at clarifying theoretical aspects.

With the evolution of mathematics education as a research field, and with the development of a specific research field on affect, research on attitude towards mathematics has evolved by identifying critical points in the previous phase and setting the need for a theoretical framework as a crucial item in the research agenda. This change has also provoked a shift from a normative paradigm to an interpretive one. Attitude is no longer seen as an individual's trait, useful for predicting his/her behaviour, but as an observer's construct, capable of suggesting an understanding of the individual's intentional actions in a complex context, as is the learning of mathematics: a multidimensional construct that involves beliefs and emotions and acts as a bridge between them (Di Martino and Zan 2011).

The development of research on attitude also suggests new issues to be explored, such as: constructing new observation tools that are consistent with the interpretive approach and the multidimensional characterization of attitude; investigating attitude toward mathematics of different groups of individuals; identifying possible motives underlying a change of attitude; designing and testing didactical paths to prevent or modify attitude.

But the theoretical debate about the quality of research about attitude persists. As a matter of fact, the need for comparing results from different studies and different theoretical frameworks is still a crucial issue, even when studies using questionnaires and statistical analysis have been replaced by qualitative case studies. New paradigms and new methods require the identification of new criteria for research quality: this is an important topic for future research in the affect field.

Despite the fact that many studies on attitude 'look ahead', drawing on the most important findings produced so far, in our opinion some critical issues still remain. The gap between the definition of the construct of attitude and the methods used to assess it is far from being bridged: many studies still use the term 'attitude' without defining it, or propose questionnaires that are not consistent with the chosen characterization of attitude, and, in particular, without clarifying the theoretical choices underlying the studies themselves. Moreover, although the normative approach in the research on attitude has showed all its theoretical limitations, many recent studies place themselves in a normative paradigm, even if, perhaps, this is not a conscious choice made by the researcher.

This lack of a cumulative character in research on attitude is, in our opinion, one of its main weaknesses, a historical weakness that has not yet been overcome. In 1976, in his update on research on attitude, Aiken wrote:

> Regardless of the efforts of this writer and others to bring to the educational research community periodic reviews of studies concerned with attitudes and anxiety toward mathematics, many investigators in this area continue to be unaware or unappreciative of previous research on the topic (…) This oversight is almost certainly due to a failure to search the relevant literature, the first step in any scientific inquiry (Aiken 1976, p. 293)

More than 30 years later, we notice exactly the same phenomenon, in a sort of theoretical and meta-theoretical déjà vu that, we are sure, has limited the development of stronger results in the field. For that reason, we believe that tracing the 'story' of the construct of attitude and discussing the results obtained so far is a very significant step in the development of research in this field.

## References


Aiken, L. (1970). Attitudes toward mathematics. *Review of Educational Research, 40*, 551–596.

Aiken, L. (1976). Update on attitudes and other affective variables in learning mathematics. *Review of Educational Research, 46*(2), 293–311.

Aiken, L., & Drager, R. (1961). The effect of attitudes on performance in mathematics. *Journal of Educational Psychology, 52*, 19–24.

Ajzen, I. (1988). *Attitudes, personality, and behavior*. Chicago: Dorsey Press.

Allport, W. (1935). Attitudes. In C. A. Murchison (Ed.), *A handbook of social psychology*. Worcester: Clark University Press.

Bishop, A. (1992). International perspectives on research in mathematics education. In D. A. Grouws (Ed.), *Handbook of research on mathematics teaching and learning* (pp. 710–723). New York: Macmillan.

Boero, P., & Szendrei, J. (1998). Research and results in mathematics education: Some contradictory aspects. In J. Kilpatrick & A. Sierpinska (Eds.), *Mathematics education as a research domain: A search for identity* (pp. 197–212). Dordrecht: Kluwer.

Bruner, J. (1990). *Acts of meaning*. Cambridge, MA: Harvard University Press.

Connelly, F. M., & Clandinin, D. J. (1990). Stories of experience and narrative inquiry. *Educational Researcher, 19*(5), 2–14.

Coppola, C., Di Martino, P., Pacelli, T., & Sabena, C. (2012). Primary teachers' affect: A crucial variable in the teaching of mathematics. *Nordic Studies in Mathematics Education, 17*(3–4), 101–118.

Corcoran, M., & Gibb, G. (1961). Appraising attitudes in the learning of mathematics. In *Evaluation in mathematics: Twenty-sixth yearbook of the NCTM*. Reston: NCTM.

Damasio, A. (1996). *Descartes' error: Emotion, reason and the human brain*. Basingstoke: Papermac.

Daskalogianni, K., & Simpson, A. (2000). Towards a definition of attitude: The relationship between the affective and the cognitive in pre-university students. In T. Nakahara & M. Koyama (Eds.), *Proceedings of the 24th conference of the IGPME* (Vol. 2, pp. 217–224). Hiroshima, Japan.

De Bellis, V., & Goldin, G. (1999). Aspect of affect: Mathematical intimacy, mathematical integrity. In O. Zaslavsky (Ed.), *Proceedings of the 23rd conference of the IGPME* (Vol. 2, pp. 249–256). Haifa, Israel.

Demazière, D., & Dubar, C. (1997). *Analyser les entretiens biographiques*. Paris: Édition Nathan.

Di Martino, P. (2004). From single beliefs to belief systems: A new observational tool. In M. J. Hoines & A. B. Fuglestad (Eds.), *Proceedings of the 28th conference of the IGPME* (Vol. 2, pp. 271–278). Bergen, Norway.

Di Martino, P., & Zan, R. (2001). Attitude toward mathematics: Some theoretical issues. In M. van den Heuvel-Panhuizen (Ed.), *Proceedings of the 25th conference of the IGPME* (Vol. 3, pp. 351–358). Utrecht, The Netherlands.

Di Martino, P., & Zan, R. (2002). An attempt to describe a 'negative' attitude toward mathematics. In P. Di Martino (Ed.), *Proceedings of the MAVI XI European workshop* (pp. 22–29). Pisa: Università di Pisa Press.

Di Martino, P., & Zan, R. (2003). What does 'positive' attitude really mean? In N. Pateman, B. Doherty, & J. Zilliox (Eds.), *Proceedings of the 27th conference of the IGPME* (Vol. 4, pp. 451–458). Honolulu, Hawaii.

Di Martino, P., & Zan, R. (2010). 'Me and maths': Towards a definition of attitude grounded on students' narratives. *Journal of Mathematics Teacher Education, 13*(1), 27–48.

Di Martino, P., & Zan, R. (2011). Attitude towards mathematics: A bridge between beliefs and emotions. *ZDM – The International Journal on Mathematics Education, 43*(4), 471–482.

Dutton, W. (1951). Attitudes of prospective teachers toward arithmetic. *The Elementary School Journal, 42*, 84–90.

Eagly, A. (1967). Involvement as a determinant of response to favorable and unfavourable information. *Journal of Personality and Social Psychology, 7*, 1–15.

Eagly, A. H., & Chaiken, S. (1998). Attitude structure and function. In D. T. Gilbert, S. T. Fiske, & G. Lindzey (Eds.), *The handbook of social psychology* (4th ed., Vol. 1, pp. 269–322). New York: McGraw-Hill.

Eisenhart, M. (1988). The ethnographic research tradition and mathematics education research. *Journal for Research in Mathematics Education, 19*(2), 99–114.

Ernest, P. (1988). The attitudes and practices of student teachers of primary school mathematics. In A. Barbos (Ed.), *Proceedings of the 12th conference of the IGPME* (Vol. 1, pp. 288–295). Veszprém, Hungary.

Evans, J. (2000). *Adults' mathematical thinking and emotions: A study of numerate practices*. London: Routledge Falmer.

Feierabend, R. (1960). Review of research on psychological problems in mathematics education. *Research Problems in Mathematics Education, 3*, 3–46.

Fennema, E. (1989). The study of affect and mathematics: A proposed generic model for research. In D. McLeod & V. Adams (Eds.), *Affect and mathematical problem solving: A new perspective* (pp. 205–219). New York: Springer.

Fennema, E., & Sherman, J. (1977). Sex-related differences in mathematics achievement, spatial visualization and affective factors. *American Educational Research Journal, 14*(1), 51–71.

Germann, P. (1988). Development of the attitude toward science in school assessment and its use to investigate the relationship between science achievement and attitude toward science in school. *Journal of Research in Science Teaching, 25*(8), 689–703.

Glaser, B., & Strauss, A. (1967). *The discovery of grounded theory: Strategies for qualitative research*. Chicago: Aldine.

Green, T. (1971). *The activities of teaching*. New York: McGraw-Hill.

Grigutsch, S., & Törner, G. (1998). *World views of mathematics held by university teachers of mathematical science*. Duisburg: Gerhard Mercator University, Preprint 420.

Haladyna, T., Shaughnessy, J., & Shaughnessy, M. (1983). A causal analysis of attitude toward mathematics. *Journal for Research in Mathematics Education, 14*, 19–29.

Hannula, M. (2002). Attitude toward mathematics: Emotions, expectations and values. *Educational Studies in Mathematics, 49*, 25–46.

Hart, L. (1984). Affective variables and mathematics education. *The Elementary School Journal, 84*(5), 558–581.

Hart, L. (1989). Describing the affective domain: Saying what we mean. In D. McLeod & V. Adams (Eds.), *Affect and mathematical problem solving* (pp. 37–45). New York: Springer.

Kaasila, R. (2007). Using narrative inquiry for investigating the becoming of a mathematics teacher. *ZDM – The International Journal of Mathematics Education, 39*(3), 205–213.

Karsenty, R., & Vinner, S. (2000). What do we remember when it's over? Adults' recollections of their mathematical experience. In T. Nakahara & M. Koyama (Eds.), *Proceedings of the 24th conference of the IGPME* (Vol. 4, pp. 451–458). Hiroshima, Japan.

Kilpatrick, J. (1992). A history of research in mathematics education. In D. A. Grouws (Ed.), *Handbook of research on mathematics teaching and learning* (pp. 3–38). New York: Macmillan.

Kloosterman, P., & Stage, F. (1992). Measuring beliefs about mathematical problem solving. *School Science and Mathematics, 92*(3), 109–115.

Kulm, G. (1980). Research on mathematics attitude. In R. J. Shumway (Ed.), *Research in mathematics education*. Reston: NCTM.

Leder, G. (1985). Measurement of attitude to mathematics. *For the Learning of Mathematics, 5*(3), 18–22.

Leder, G. (1992). Measuring attitudes to mathematics. In W. Geeslin & K. Graham (Eds.), *Proceedings of the 16th conference of the IGPME* (Vol. 2, pp. 33–39). Durham, USA.

Lester, F. (2002). Implications of research on students' beliefs for classroom practice. In G. Leder, E. Pehkonen, & G. Törner (Eds.), *Beliefs: A hidden variable in mathematics education?* Dordrecht: Kluwer.

Ma, X., & Kishor, N. (1997). Assessing the relationship between attitude toward mathematics and achievement in mathematics: A meta-analysis. *Journal for Research in Mathematics Education, 30*, 65–88.

Mandler, G. (1984). *Mind and body: Psychology of emotion and stress*. New York: Norton.

McLeod, D. (1987). A constructivist approach to research on attitude toward mathematics. In L. Streefland (Ed.), *Proceedings of the 9th conference of the IGPME* (Vol. 1, pp. 133–140). Montreal, Canada.

McLeod, D. (1992). Research on affect in mathematics education: A reconceptualization. In D. A. Grouws (Ed.), *Handbook of research on mathematics teaching and learning* (pp. 575–596). New York: Macmillan.

McLeod, D., & Adams, V. (Eds.). (1989). *Affect and mathematical problem solving: A new perspective*. New York: Springer.

Middleton, J. A., & Spanias, P. A. (1999). Motivation for achievement in mathematics: Findings, generalizations, and criticism of the research. *Journal for Research in Mathematics Education, 30*, 65–88.

Mura, R. (1993). Images of mathematics held by university teachers of mathematical sciences. *Educational Studies in Mathematics, 25*(4), 375–385.

Mura, R. (1995). Images of mathematics held by university teachers of mathematics education. *Educational Studies in Mathematics, 28*(4), 385–399.

Neale, D. (1969). The role of attitudes in learning mathematics. *The Arithmetic Teacher, 16*, 631–641.

Ortony, A., Clore, G., & Collins, A. (1988). *The cognitive structure of emotions*. Cambridge: Cambridge University Press.

Pajares, F. (1992). Teachers' beliefs and educational research: Cleaning up a messy construct. *Review of Educational Research, 62*(3), 307–332.

Philipp, R. A. (2007). Mathematics teachers' beliefs and affect. In F. K. Lester (Ed.), *Second handbook of research on mathematics teaching and learning* (pp. 257–315). Reston: NCTM.

Polo, M., & Zan, R. (2006). Teachers' use of the construct 'attitude'. Preliminary research findings. In M. Bosch (Ed.), *Proceedings of the 4th CERME*. Barcelona: FundEmi.

Reyes, L. (1984). Affective variables and mathematics education. *The Elementary School Journal, 84*(5), 558–581.

Rokeach, M. (1968). *Beliefs, attitudes and values*. San Francisco: Jossey-Bass.

Ruffel, M., Mason, J., & Allen, B. (1998). Studying attitude to mathematics. *Educational Studies in Mathematics, 35*, 1–18.

Schlöglmann, W. (2002). Affect and mathematics learning. In A. D. Cockburn & E. Nardi (Eds.), *Proceedings of the 26th conference of the IGPME* (Vol. 4, pp. 185–192). Norwich, UK.

Schoenfeld, A. (1989). Exploration of students' mathematical beliefs and behavior. *Journal for Research in Mathematics Education, 20*, 338–355.

Schoenfeld, A. (1994). A discourse on methods. *Journal for Research in Mathematics Education, 25*(6), 697–710.

Schoenfeld, A. (2000). Purposes and methods of research in mathematics education. *Notices of the American Mathematical Society, 47*(6), 641–649.

Sierpinska, A., Kilpatrick, J., Balacheff, N., Howson, G., Sfard, A., & Steinbring, H. (1993). What is research in mathematics education, and what are its results? *Journal for Research in Mathematics Education, 24*(2), 274–278.

Silver, E. (1985). Research on teaching mathematical problem solving: Some underrepresented themes and needed directions. In E. Silver (Ed.), *Teaching and learning mathematical problem solving: Multiple research perspectives* (pp. 247–266). Hillsdale: Lawrence Erlbaum.

Skemp, R. (1976). Relational understanding and instrumental understanding. *Mathematics Teaching, 77*, 20–26.

Thurstone, L., & Chave, E. (1929). *The measurement of attitude: A psychophysical method and some experiments with a scale for measuring attitude toward the Church.* Chicago: University of Chicago Press.

Tirosh, D. (1993). Affective and cognitive variables: Students' perspectives. In I. Hirabayashi, N. Nohda, K. Shigematsu, & F. L. Lin (Eds.), *Proceedings of the 17th conference of the IGPME* (Vol. 1, pp. 11–13). Tsukuba, Ibaraki, Japan.

Zan, R., & Di Martino, P. (2007). Attitude toward mathematics: Overcoming the positive/negative dichotomy. *The Montana Mathematics Enthusiast, Monograph 3*, 157–168.

Zan, R., Brown, L., Evans, J., & Hannula, M. (2006). Affect in mathematics education: An introduction. *Educational Studies in Mathematics, 63*(2), 113–121.

# Networking Theories to Understand Beliefs and Their Crucial Role in Mathematics Education

**Katrin Rolka and Bettina Roesken-Winter**

**Abstract** Many publications present research on teacher beliefs, whether concretized for pre-service or in- service teachers. Most of them have in common that they highlight the crucial role that beliefs play in the classroom. In this chapter, we explore more deeply what those crucial aspects are, what they consist of, and how they interact with other variables. Different theoretical lenses will be brought together to underline different perspectives and to gain explanatory power going beyond the single approaches. For the case of practising teachers, we will discuss some thoughts on the classical contributions by Shulman (Educ Res 15(2):4–14, 1986) and Schoenfeld (Issues Educ 4(1):1–94, 1998). On the one hand, we extend the knowledge categorization provided by Shulman to the fields of beliefs and goals. On the other hand, we elaborate on Schoenfeld's theory of Teaching-In-Context. For the case of pre-service teachers, we combine the classification of mathematical beliefs based on the work of Ernest with ideas of conceptual change originally conceived in the context of knowledge (cf. Ernest, J Educ Teach 15:13–33, 1989; Vosniadou and Verschaffel, Learn Instr 14(5):445–451, 2004).

**Keywords** Pedagogical content beliefs • Conceptual change • Networking of theories

## Introduction

In recent years the networking of theories has received much attention in mathematics education, as can be seen in the overview articles by Artigue et al. (2006), Bikner-Ahsbahs and Prediger (2006) and Arzarello et al. (2007). Some authors particularly focus on using theoretical diversity to strengthen theory development, and make suggestions on strategies and methods for networking (cf.

K. Rolka (✉)
Faculty of Mathematics, Ruhr-Universität Bochum, Bochum, Germany
e-mail: katrin.rolka@rub.de

B. Roesken-Winter
Professional School of Education, Department of Mathematics,
Humboldt-Universität zu Berlin, Berlin, Germany
e-mail: bettina.roesken-winter@hu-berlin.de



B. Pepin, B. Roesken-Winter (eds.), *From beliefs to dynamic affect systems in mathematics education*, Advances in Mathematics Education,
DOI 10.1007/978-3-319-06808-4_4

Bikner-Ahsbahs and Prediger 2006), while others discuss whether theoretical plurality and diversity hinder moving forward as a field (Dahl 2006; Lerman 2006). In particular, at CERME 4 and 5 a group of researchers dealt with challenges induced by theoretical diversity (Arzarello et al. 2007) and they remind us that "researchers with different theoretical perspectives consider empirical phenomena [...] from different perspectives and hence come to very different results in their empirical studies" (p. 1618). The authors continue by asking, "How can the results from different studies be integrated or at least understood in their difference?" (p. 1618). That is, results of empirical studies might be incompatible or even contradictory, a fact that, in the long run, can impede progress in the field of mathematics education. In this regard, Artigue et al. (2006) consider it as an important task of the community to pay attention to the different theoretical frameworks in terms of networking:

> If we can develop and maintain a certain degree of networking between some of the advocates of the different theoretical stances that are currently evident within mathematics education, this will constitute an important step on the path towards establishing mathematics education as a scientific discipline. (p. 1242)

As an essential endeavour, the authors identify integrating or synthesizing theoretical approaches into a new framework. Connecting theoretical approaches can then follow a bottom-up development while using a concrete empirical phenomenon as starting point; or a top-down development while using different theories from the beginning and focusing on the relationship of theories (cf. Arzarello et al. 2007). In this chapter, we elaborate on the latter aspect and pursue a deductive approach to networking theories. We will explore how different theories serve to analyse similar data to gain a more comprehensive understanding of relationships and interdependencies of the underlying frameworks to increase their explanatory power. In particular, regarding beliefs research, we point out how research directions were determined while developments took place in two different research fields, mathematics education and psychology, and which, of course, have influenced methodological approaches and choices. To underline our thoughts we present two examples of networking theories that come from those two research areas.

First, we will discuss some thoughts on the classical contributions by Shulman (1986, 2005a, b) and Schoenfeld (1998). On the one hand, we elaborate on Schoenfeld's theory of Teaching-in-Context[1] which explains teacher behaviour from a local view as a function of knowledge, goals and beliefs, while extending the knowledge categorization provided by Shulman (1986) to the fields of beliefs and goals. On the other hand, we use Shulman's (2005a, b) overarching theory of signature pedagogies to additionally understand the significant role of beliefs from a global view. Second, we develop further ideas of conceptual change originally conceived in the context of knowledge (Ernest 1989; Vosniadou and Verschaffel 2004; Appleton 1997) with regard to belief change and, in addition, use the classification

[1] Our ideas are based on Roesken and Rolka (2011).

of mathematical beliefs based on the work by Dionne (1984) and Ernest (1989) to illustrate belief change in more detail.[2]

## Setting the Frame of Networking Theories: Beliefs Research in Psychology and Mathematics Education

From a historical viewpoint, the mathematics-related research in the field of beliefs has its roots in the "failure of the problem-solving based reforms" of the mathematics curriculum in the United States in the late 1980s (Schoenfeld 2007). More concretely, Roesken et al. (2011) emphasize that "numerous studies detected that one reason for that 'failure' were the 'inappropriate' beliefs of teachers concerning mathematics in general, the process of problem solving and characteristics of doing mathematics in particular, in addition to strong teacher convictions concerning students' apparent lack of ability [...]" (p. 452). From that time on, the discussion on the role and significance of beliefs became more elaborated and led to the seminal papers by Pajares (1992) and Thompson (1992) in the 1990s which indicated a starting point for specific and targeted research on beliefs that has entailed numerous studies (cf. Philipp 2007) and encouraged substantial discussion. The reader will find almost no studies on beliefs that do not refer at least to one of these papers.

While comparing the work by Pajares (1992) and Thompson (1992) more closely, it is evident that the two researchers pursue different approaches since they come from different fields of educational research: educational psychology and mathematics education. These different roots entailed particular emphases which are briefly sketched in the following.

Among other aspects, the psychologists' viewpoint emphasizes the epistemological character of beliefs as those "play a key role on knowledge interpretation and cognitive monitoring" (Pajares 1992, p. 324); an issue that has dominated successive research in educational psychology. Moreover, Pajares (1992) refers to the work by Schommer (1990) who "argued that the study of epistemological beliefs may prove more valuable for understanding comprehension than either metacognition or schema theory, neither of which is able to explain students' failure to integrate information or monitor comprehension" (p. 328). As an overarching concept Pajares (1992) suggests *educational beliefs* as a construct that "is itself broad and encompassing" (p. 316). However, he reminds us that the concept's wide scope is difficult to operationalize and thus educational beliefs need to be specified by using the label *educational beliefs about*:

> Therefore, as with more general beliefs, *educational beliefs about* are required – beliefs about confidence to affect students' performance (teacher efficacy), about the nature of knowledge (epistemological beliefs), about causes of teachers' or students' performance (attributions, locus of control, motivation, writing apprehension, math anxiety), about

[2]These ideas are based on Liljedahl et al. (2007a, b) and Rolka et al. (2006).

> perceptions of self and feelings of self-worth (self-concept, self-esteem), about confidence to perform specific tasks (self-efficacy). There are also educational beliefs about specific subjects or disciplines (reading instruction, the nature of reading, whole language). (p. 316)

Subsequent research accentuated different directions in terms of personal epistemology (Hofer and Pintrich 1997; Hofer 2000), epistemic positions (Perry 1968), epistemic cognition (Kitchener 1983), epistemological beliefs (Schommer 1990), epistemological world views (Schraw and Olafson 2002) and epistemological understanding (Hofer 2004). In sum, research on beliefs in the field of pedagogy or psychology has developed strongly towards investigating epistemological aspects.

In comparison, the focus of Thompson (1992) is domain-specific as she explores explicitly teachers' conceptions of mathematics:

> A teacher's conception of the nature of mathematics may be viewed as that teacher's conscious or subconscious beliefs, concepts, meanings, rules, mental images, and preferences concerning the discipline of mathematics. (p. 132)

In the focus are conceptions about the nature of mathematics indicating a philosophical viewpoint that is also pursued in the work by Ernest (1991, 1994) and Lerman (1983). In the beginning of beliefs research in mathematics education, some researchers applied the term mathematical *world views*; these included Schoenfeld (1985), and later Grigutsch (1996) and Grigutsch et al. (1998). A few papers explicitly address epistemology in mathematics education (Sierpinska and Lerman 1996; Steinbring 1998) while mostly such issues are only implicitly included in discussions with roots going back to philosophical positions on mathematics (Hersh 1991, 1997). That is, those articles address beliefs about the origin and acquisition of knowledge, and how such attitudes affect teaching in general and students' learning of mathematics specifically. Some studies explore epistemological beliefs held by mathematics teacher educators and investigate how those shape and influence their prospective teachers' beliefs and even actions in the classroom (e.g. Carter and Norwood 1997; Schraw and Olafson 2002). Roesken and Törner (2007), in this regard, yielded seven dimensions structuring university professors' beliefs about mathematics that encompass factors including characteristics of mathematics, main features of mathematical learning, philosophical aspects and sophisticated views on mathematics.

A subsequent step in binding together various approaches of beliefs and their influences (see also Ernest 1989) was the issue of *ZDM* in 1996 edited by Pehkonen and Törner. The book edited by Leder et al. (2002) reflected many of the views presented in the Special Issue, and beliefs were referred to as 'hidden variables', with reference to a famous paper by Bauersfeld (1980). Meanwhile, much research has been conducted in the field of mathematics-related beliefs.

In sum, the history of beliefs research particularly indicates that it is worthwhile to explore theoretical diversity as two different research strands, which originated in psychology and mathematics education, have developed almost independently of each other and have influenced subsequent research in both fields substantially. In what follows, we present two examples of how to capture the crucial role of beliefs by networking different theoretical lenses that have their roots either in *psychology* or *mathematics education*.

## Example 1: Understanding Beliefs by Combining a Global and a Local View on Teaching

As regards the teaching of mathematics, many researchers have looked for underlying variables in order to understand and explain teacher behaviour in the classroom (cf. Baumert et al. 2010). Examples for such variables are teacher knowledge, teacher beliefs and teacher instructional goals – typically, researchers focus on one of these variables without considering their relationships or potential overlaps (Rösken et al. 2008). Schoenfeld's (1998) merit lies in providing a theory that accounts for a local view on teaching by modelling teacher behaviour as a function of a teacher's knowledge, goals and beliefs. His theory of Teaching-in-Context does not simply go beyond knowledge and beliefs by assigning an essential role to goals, but also emphasizes strongly that the three variables are pieces of a puzzle and the challenge is to explore how these fit together. In sum, the theory explains developments in teaching from a multi-faceted perspective and allows the didactical analysis for focusing on understanding, and explaining rich and complex teaching coherences. A teacher's spontaneous decision-making is characterized in terms of available knowledge, high priority goals, and beliefs. In his latest book, Schoenfeld (2010) modifies his initial theory as follows:

> The main claim in the book is that what people do is a function of their resources (their knowledge, in the context of available material and other resources), goals (the conscious or unconscious aims they are trying to achieve) and orientations (their beliefs, values, biases, dispositions, etc.). (p. viii)

What is new? While attention is still given to goals, Schoenfeld introduces the broader concepts of resources to refer to the category of teacher knowledge and of orientations to encompass the fields of beliefs, values, biases and dispositions. Regarding the former-used category of beliefs, Schoenfeld (2010) explains:

> Beliefs play much the same focal role that they did in my earlier work. Just as students' beliefs about themselves and about mathematics shape what they do while working on mathematics problems, teachers' beliefs about themselves, about mathematics, about teaching, and about their students shape what they do in the classroom. (p. 26)

Still he assigns a major role to beliefs and he gives the following explanation for his shift in terminology:

> The term "beliefs" worked well in characterizing problem solving and teaching (and it fit comfortably with the literature's use of the term), but it seemed less apt when I applied the theoretical ideas to other domains. In cooking, tastes and life style preferences are consequential; in other arenas (e.g., health care) one's values play a major role. For that reason I chose orientations as an all-encompassing term, to play the same role in general as beliefs do in discussions of mathematical and pedagogical behavior. (p. 27)

What is not new? Schoenfeld still aims at explaining comprehensive teaching behaviour:

> I argue that if enough is known, in detail, about a person's orientations, goals, and resources, that person's actions can be explained at both macro and micro levels. That is, they can be explained not only in broad terms, but also on a moment-by-moment basis. (p. viii)

Although Schoenfeld acknowledges a global level for analysing teachers' decision-making processes, we will reveal later in this section that Shulman's theoretical framework of signature pedagogies provides an additional source to understand teacher actions from a more global perspective. But first we elaborate on Shulman's seminal work on teacher knowledge, and his notion of pedagogical content knowledge which "goes beyond knowledge of subject matter per se to the dimension of subject matter knowledge *for teaching*" (Shulman 1986, p. 7). Interestingly, Shulman's starting point for introducing a new category as an additional aspect of teacher knowledge is rooted in his observation that research in teacher cognition so far was either on teacher's subject matter knowledge or teacher's pedagogical knowledge. In a convincing manner, he explains pedagogical content knowledge as an essential link between the two:

> Within the category of pedagogical content knowledge I include, for the most regularly taught topics in one's subject area, the most useful forms of representation of those ideas, the most powerful analogies, illustrations, examples, explanations, and demonstrations – in a word, the ways of representing and formulating the subject that make it comprehensible to others. (p. 7)

Much research followed and led to advances in understanding the knowledge category, but only a few publications applied the typology to explore beliefs in more detail. Törner (2002), for instance, draws on global beliefs as mentioned by Thompson (1992) and mentions subject matter beliefs as those beliefs that relate to aspects of a teacher's subject matter knowledge. Kuntze (2011) brings forward those ideas and investigates local and global components of pedagogical content beliefs. However, he does not distinguish knowledge and beliefs but chooses the pragmatic solution that "beliefs and instruction-related convictions are [...] understood to be contained in the notion of professional knowledge" (Kuntze 2011, p. 2). In what follows, we take up the idea of networking theories in first combining Schoenfeld's theory on teacher behaviour and an adaption of Shulman's categorization that was initially developed for teacher knowledge to understand teachers' actions in the classroom from a local view. That is, we adopt the classification of subject matter and pedagogical content knowledge for the constructs of beliefs and goals, which allows for a more fine-grained analysis of teaching incidents.

Second, we go back to Schoenfeld's theory of Teaching-in-Context, and have a closer look at his aim to explain a teacher's behaviour in the classroom, and the choices that he or she makes in any moment. According to Schoenfeld, teaching can be studied on a fine-grain level and the analysis focuses on the decision-making process, as Schoenfeld (2010) points out in the following:

> Decision making and resource access are largely automatic when people are engaged in well practiced behavior. Mechanisms for routine access to cognitive resources have been extensively studied and have various names in the literature. Depending on the tradition, they may be called *scripts, frames, routines,* or *schemata.* The core idea is the same: when people perceive a situation as being of a familiar type, they have a "default" set of expectations that guide their perceptions and/or actions. (p. 16)

On a more global level, going beyond the single classroom actions, we can find some similar ideas in Shulman's work (2005a, b) on signature pedagogies. While drawing on studies in law, engineering, the clergy, medicine, nursing and teaching, Shulman (2005a, b) describes the signature pedagogy of an entire field. In particular, the construct catches the characteristics of different professions and how those can be described and analysed:

> What I mean by "signature pedagogy" is a mode of teaching that has become inextricably identified with preparing people for a particular profession. This means it has three characteristics: One, it's distinctive in that profession. So you wouldn't expect clinical rounds in a law school. And even though it might be very effective, you wouldn't expect a case dialogue or case method teaching of this sort in a medical school. Second, it is pervasive within the curriculum. So that students learn that as they go from course to course, there are certain continuities that thread through the program that are part of what it means to learn to "think like a lawyer," or "think like a physician," or "think like a priest." There are certain kinds of thinking that are called for in the rules of engagement of each course, even as you go from subject to subject. The third feature is another aspect of pervasiveness, which cuts across institutions and not only courses. Signature pedagogies have become essential to general pedagogy of an entire profession, as elements of instruction and of socialization.

Shulman (2005a, b) also underlines the decisive role of teacher knowledge bundled in routines, but emphasizes additionally that the cultures and characteristics of the professions are transported as signature pedagogies. Signature pedagogy is an emerging concept in teacher education which catches the salient and pervasive teaching practices that characterize an entire field and thus allows for analysing practices from a global viewpoint. The Carnegie Foundation for the Advancement of Teaching has undertaken many studies (cf. Shulman 2005a, b) to describe the signature pedagogy of the different fields. Those studies of education in the professions share that they link the following aspects of a professional role:

> [...] professional education is a synthesis of three apprenticeships – a cognitive apprenticeship wherein one learns to think like a professional, a practical apprenticeship where one learns to perform like a professional, and a moral apprenticeship where one learns to think and act in a responsible and ethical manner that integrates across all three domains. (Shulman 2005b)

Building on knowledge as basis, Shulman (2005a) assigns a fundamental role to signature pedagogies since those "are designed to transform knowledge attained to knowledge-in-use [...]". Moreover, Shulman (2005b) reverts to his knowledge categories and explains that "these forms of knowledge are foundational, necessary but not sufficient". In order to understand more deeply teachers' actions in the classroom, Shulman (2005a) thus refers to the crucial role of signature pedagogies:

> Signature pedagogies are important precisely because they are pervasive. They implicitly define what counts as knowledge in a field and how things become known. They define how knowledge is analyzed, criticized, accepted, or discarded. (p. 54)

Signature pedagogies possess a structure by which a discipline's pedagogies can be examined, elaborated and compared (Shulman 2005a). Following Shulman, we have to distinguish three dimensions: *surface structure*, *deep structure* and *implicit structure*. The surface structure "consists of concrete, operational acts of teaching and learning, of showing and demonstrating, of questioning and answering, of interacting and withholding, of approaching and withdrawing" (Shulman 2005b, p. 54). That is, the surface structure covers overtly social acts associated with teaching and learning the subject. According to Shulman (2005b), "any signature pedagogy also has a deep structure, a set of assumptions about how best to impart a certain body of knowledge and know-how" (p. 55). Thus, the deep structure transports assumptions about the teaching and learning within the field. Finally, Shulman (2005a) points out that the implicit structure addresses a moral dimension comprising a set of beliefs about professional attitudes, values, norms and dispositions.

The dimension especially interesting in our case is the surface structure since it maintains the relevance of the knowledge body of mathematics, a domain with a proud long and international history and many traditions. Roesken and Törner (2010) point out that this surface structure encompasses a specific language and semiotics, as well as particularities regarding the teaching style and the teacher–student relationship. Teaching mathematics has its characteristics that depend strongly on the underlying subject of mathematics and how it is taught at the universities and during teacher education. For instance, the style of speech in the lectures often shows possessive set phrases such as using the plural "we" or an authoritative wording like "let be", so that no one feels invited to say something against. Another interesting example is the notion of w.l.o.g. (without loss of generality), a well-known saying of mathematicians. Who wants to show any weakness by claiming that it is not trivial for him or her? One can imagine that such an education leaves its marks and affects a teacher's later behaviour in the classroom essentially. In the analysis that we present later, we will elaborate on these ideas and show the influence on teacher behaviour in the classroom.

So far, we have briefly sketched the contributions by Schoenfeld and Shulman. In what follows, we elaborate on those theoretical strands and offer some further ideas on Schoenfeld's theory of Teaching-In-Context and Shulman's work on teacher knowledge and signature pedagogies (cf. Rosken and Rolka 2011) by combining both theoretical approaches. In particular, we enrich Schoenfeld's theoretical lens of analysing teacher behaviour locally in terms of knowledge goals, and beliefs by extending Shulman's knowledge categorization to beliefs and goals, and use the notion of signature pedagogies to elaborate on the role of beliefs from a global viewpoint. The ideas emerged throughout our work on a paper that was dedicated to analysing a video-taped school lesson through the lenses of Schoenfeld's approach (Törner et al 2010). We now explicate how the networking approach contributes to a better understanding of the data.

## Illustrating the Networking of the Two Theories with Empirical Data

In the following, we further support our ideas on networking the above- mentioned theories by presenting evidence that we found in an empirical study (Törner et al. 2010). In particular, we show how the networking of theories that tackle aspects of teacher knowledge and beliefs can help to analyse local and global beliefs relevant in the classroom, and how beliefs interfere with teacher goals. This empirical study emerged from a bi-national in-service teacher training[3] that aimed at working out cultural differences and/or similarities in teaching styles. For this purpose, a Dutch and a German lesson on linear functions were videoed, forming the basis of discussions within the teacher training. The German teacher who taught the lesson on linear functions possesses 30 years of professional experiences. She has attended numerous in-service teacher training courses, in particular on using computer algebra systems and open tasks in mathematics teaching. In the lesson, linear relationships as motivation for the treatment of linear functions were embedded in various tasks. Students had to work in small groups of two or three on one of the tasks by using the computer, in particular the software Excel.

The teacher engaged very eagerly to implement newly imparted issues into her teaching on linear functions, a topic that she has taught in rather traditional ways several times previously. Although the teacher planned the lesson thoroughly, its course developed unexpectedly. At the beginning of the lesson, the teacher pursued a rather open and problem-oriented approach where students worked in small groups using the computer. However, as the lesson developed and time seemed to run out, the teacher suddenly changed her teaching style in favour of a more traditional approach. That is, she shifted back to her solid and approved methods in terms of a monologue on definitions in a formalized structure.

Another important data source is an interview that took place several days after the lesson. After watching the video and immediately recognizing the turning point in the course of the lesson, we wanted to find out more about the teacher's goals and beliefs underlying the planning and teaching of that lesson.

On the one hand, we resort to Schoenfeld's theory and identify the teacher's knowledge, goals and beliefs that were observable in the lesson but also expressed by her in the interview. On the other hand, we draw on the work by Shulman (1986) and adapt his categorization for the domain of knowledge to the one of beliefs, and we differentiate between pedagogical content goals and beliefs, and subject matter goals and beliefs. Basically, the knowledge categories can be directly adapted to beliefs. That is, the pedagogical content goals and beliefs concentrate on how to teach the subject of mathematics while the subject matter goals and beliefs are derived from the subject itself. We illustrate the categories by some examples.

In the interview with the teacher after the lesson, we identified statements that can be interpreted as both pedagogical content goals and beliefs. To be more concrete, the expressed goals were strongly rooted in beliefs and the beliefs influenced the goals to be fixed. The conclusion of the duality of the two constructs was even strengthened by the teacher justifying her goals and hence revealing subjective convictions that can be understood as beliefs. For the teacher, the use of the computer plays a central role in her teaching in general, but also in the specific lesson and the interview. She formulates as a goal:

> Whenever possible, I employ the computer in mathematics lessons (pedagogical content goal).

---

[3] Funded by the Robert-Bosch Foundation.

This statement can be interpreted as belief in the sense that employing the computer whenever possible is rooted in the conviction that there is an additional value compared with the abdication of the computer. She complements this goal by a belief that is related to the topic of the lesson:

> The theme linear functions can be mediated by the computer (pedagogical content belief).

This pedagogical content belief was also realized as the teacher actually employed the computer when introducing linear functions. A reformulation of this belief in terms of a goal could have been "Students shall use the computer when dealing with linear functions". Hence, this expressed belief corresponds to an implicit goal.

Although the pedagogical content goals and beliefs were highly relevant during the first 29 min of teaching, the teacher suddenly shifted to her approved and traditional style while the computer lost its central role. Besides articulating frustration about the use of open tasks, she provided some subject matter goals and beliefs that explain the shift in the teaching trajectory from her point of view:

> Linear functions are defined by their slopes. The slope of a linear function is its most important characteristic (subject matter belief).
>
> Functions are important for Calculus in grade 12 (subject matter belief). The central term to be mediated in the context of linear functions is the concept of slope, which prepares students for the concept of derivative (subject matter belief).

From this results the following specific mathematical goal, which can also be identified as a kind of output directive for the lesson:

> The term slope must be mentioned in this lesson (subject matter goal).

This episode underlines that the subject matter beliefs on the relevance of linear functions can be understood as a key prerequisite, which in the last instance characterize unavoidably the subject matter goal that the teacher tried to obtain desperately in the lesson. That is, the moment the teacher realized that she could not achieve her central subject matter goal of introducing the term slope, she let the students simply switch off the computer. From this point onwards, global subject matter goals dominated the lesson activities to reach the one goal: the term slope must be mentioned. In other words, all pedagogical content goals and beliefs lost their rather positive value and stepped aside to make room for subject matter goals and beliefs.

Regarding this teaching episode, the questions arising for us are the following. Why are goals and beliefs so closely connected and attached to the same idea? Does this observation depend on the subject of mathematics and its specific structure? In addition to our analysis, we found some answers on a meta level while drawing on Shulman's work on signature pedagogies that we will discuss in the following.

In a talk at a conference in Germany, Shulman gave some examples for signature pedagogies in the domain of mathematics. For instance, he characterized the domain of teaching mathematics at university as a kind of dorsal teaching while showing a picture of a mathematics lecturer in front of a blackboard, turning his back to the classroom and writing down formulas while the students tried eagerly to copy the text on the boards. That is, all mathematics lectures are given in a specific style and thus elements of a signature pedagogy even permeate the field of teacher education.

Teacher students are confronted with a specific culture that is related to the subject they are studying.

Going deeper into the construct of signature pedagogy, we identify as surface structure influencing the domain of mathematics teaching in school the stable network provided by the discipline in terms of definitions, theorems and examples. In the teaching incident that we observed, the subject structure served as a kind of safety net for the teacher. That is, the subject matter goals and beliefs are rooted robustly in mathematics and dominate the pedagogical content goals and beliefs. The possibility of abandoning the term slope does not occur for the teacher either during the lesson or in the aftermath of the lesson while reflecting on the teaching.

Such a signature, obviously a powerful frame, maybe blurs the differences between goals and beliefs and serves as an overarching theme so that both constructs appear as two sides of the same coin.

## Example 2: Networking a Beliefs Classification and Conceptual Change Approaches to Understand Changes in Beliefs

Whereas conceptual change theories initially focused on knowledge systems, Pintrich et al. (1993) called for also taking into consideration the affective domain. Beliefs are part of the affective domain (McLeod 1992), and can be used to explain why learners who possess the cognitive resources to succeed at mathematical tasks still fail (Di Martino and Zan 2001). In this context, beliefs transport what learners assume to be true about mathematics. Beliefs about mathematics are often based on an individual's own experiences as learner of mathematics. For example, beliefs that mathematics is 'difficult', 'all about one answer', or 'all about memorizing formulas' stem from classroom experiences where these ideas were implicitly conveyed and constantly reinforced. Research has shown that such beliefs are slow to form but, once established, are resistant to change even in the face of intervention (Op't Eynde et al. 2001; Schommer-Aikins 2004). In the context of teaching mathematics, beliefs have been used to explain the discordance between teachers' knowledge of mathematics and their teaching practices. This research has revealed that beliefs about *teaching* mathematics also arise from teachers' experiences as *learners* of mathematics (Calderhead and Robson 1991; Chapman 2002; Feiman-Nemser and Featherstone 1992; Feiman-Nemser et al. 1987; Fosnot 1989; Liljedahl 2006; Lortie 1975; Millsaps 2000; Skott 2001; Uusimaki and Nason 2004). So, a belief that teaching mathematics is 'all about telling how to do it' may come from a belief that learning mathematics is 'all about being told how to do it', which in turn may have come from personal experience as a learner of mathematics. Or it may not have! Implication and causality is difficult to determine in the context of beliefs.

The above-mentioned examples of concrete beliefs about mathematics or its teaching and learning reveal a specific view on mathematics. In accordance with

Dionne (1984) and Ernest (1989, 1991), Törner and Grigutsch (1994) labelled such beliefs as belonging to the "toolbox aspect". Here, mathematics is seen as a set of rules, formulae, skills and procedures; while mathematical activity means calculating as well as using rules or formulae and mastering procedures. Besides this toolbox aspect, one finds in the literature at least two other components, sometimes with varying notions: the "system aspect" and the "process aspect" (Grigutsch et al. 1998). In the system aspect, mathematics is characterized by logic, rigorous proofs, exact definitions and a precise mathematical language, and doing mathematics consists of accurate proofs as well as of the use of a precise and rigorous language. In the process aspect, doing mathematics is considered as a constructivist process where relations between different notions and sentences play an important role. Here the mathematical activity involves creative steps, such as generating rules and formulae, thereby inventing or re- inventing mathematics. In addition to these three perspectives on beliefs about mathematics and its teaching and learning, a further component is the usefulness or utility of mathematics (Grigutsch et al. 1998). Since beliefs are often referred to as a "messy construct" (Furinghetti and Pehkonen 2002; Pajares 1992), considering them as consisting of different components enables some reduction in this "messiness". Besides, this classification allows for identifying changes in beliefs, as will be illustrated later.

However, using this classification in order to code beliefs about mathematics and its teaching and learning, and finally to trace changes in teachers' beliefs, does not explain how and why these changes are occurring. To be more clear, classifying mathematical beliefs as toolbox, system or process aspect at different points in time can show that changes have occurred, but does not give any information about the mechanisms behind this change. For a better understanding of the underlying conditions, two strands of conceptual change approaches will be briefly sketched in the following (Vosniadou and Verschaffel 2004; Appleton 1997).

Worth noting in this regard is that the conditions described for the conceptual change approach by Vosniadou and Verschaffel (2004) refer to the prerequisites that an individual needs to bring before undergoing any change. For conditions on the instructional level that may produce change see Rolka et al. (2006).

The conceptual change approach used in the field of learning and instruction, initially in the domain of science (Posner et al. 1982), is based on the philosophy and history of science (Kuhn 1970), and was afterwards adapted to mathematics learning as well (Vosniadou and Verschaffel 2004). In accordance with Kuhn (1970), Posner et al. (1982) suggest that three conditions must be fulfilled so that a conceptual change can take place: (a) students do not come to instruction as "tabulae rasae" but already possess knowledge about certain phenomena, and, in some cases, this stands in contrast to the accepted scientific theories that explain these phenomena; hence, it is important to note that these "misconceptions" are formed through lived experiences without formal instruction; (b) students must be dissatisfied or feel discomfort with existing conceptions or theories; and (c) there is a phenomenon of theory replacement, initiated by the mechanism of 'cognitive conflict' which basically refers to the assumption that before a new theory can be adopted the current theory needs to be rejected. In the best case, this model can be seen as partial understandings rather than incorrect understandings. The perfection of these models is achieved through further instruction based on constructivist theories of learning.

Although the theory of conceptual change focuses primarily on cognitive aspects of conceptual change, it is equally applicable to metaconceptual, motivational, affective and socio-cultural factors as well (Vosniadou and Verschaffel 2004; Liljedahl et al. 2007b). In the following, we briefly sketch how this approach can be used for describing changes in pre-service teachers' mathematical beliefs by outlining that each of the three criteria (a) to (c) presented above is equally applicable to tracing changes in teachers' mathematical beliefs. In the sense of the criteria "lived experiences", pre-service teachers also do not come to teacher education as "tabulae rasae", as Ball (1988) points out: "Long before they enroll in their first education course or math methods course, they have developed a web of interconnected ideas about mathematics, about teaching and learning mathematics, and about schools". During their time as students of mathematics they first formulated, and then concretized, deep-seated beliefs about mathematics and what it means to learn and teach mathematics. Unfortunately, these deep-seated beliefs often run counter to contemporary research on what constitutes good practice. As such, it is one of the roles of teacher education programmes to reshape these beliefs and extend insufficient beliefs that could impede effective teaching in mathematics (Green 1971). Certainly, one could raise the objection that the formation of pre-service teachers' mathematical beliefs cannot be viewed as being formed outside a context of formal instruction. For sure, their experiences as learners of mathematics are situated within a setting of formal instruction at school, but here the focus of that instruction was on mathematical contents and not on the nature of mathematical knowledge or the question of how mathematics should be taught or learned. Hence, mathematical beliefs are tacitly constructed and, therefore, the condition of "lived experience" is met. In comparison to the original theory of conceptual change in learning and instruction proposed by Posner et al. (1982), we do not aim to judge beliefs as inadequate or inappropriate, as "misbeliefs". Rather, we would like to emphasize that – referring to the above-mentioned classification of mathematical beliefs according to Dionne (1984) and Ernest (1989, 1991) – all three (or four) aspects in some sense do play a valuable role in answering the question "What is mathematics?" In some papers found in the literature, the toolbox aspect is presented as being rather unacceptable. However, which mathematician would, indeed, claim that mathematics has nothing to do with numbers, rules or calculations? It is important to note here that a sole view on mathematics as toolbox is certainly insufficient and it is strongly desirable to enrich this view with ideas from the system and process aspect. In the case of pre-service mathematics teachers, it can be noted that they often come to teacher education courses with a dominance of the toolbox and system aspect (Rolka et al. 2006).

The criteria (b) and (c) are equally given. The teachers must feel some discomfort with their existing beliefs and they must experience that they can benefit from alternative beliefs that are useful, plausible and fruitful for them. However, in line with our remarks above, belief rejection would not be an appropriate term as the notion of incorrect beliefs as such is not appropriate.

Another approach also focusing on conceptual change is given by Appleton (1997), who elaborated a model for describing and analysing students' learning especially during science lessons. This model offers, according to Piaget's terms of assimilation and accommodation, different possibilities of what happens when students are confronted with new information and experiences. When this new information is processed, one of three possibilities is likely to occur:

- **Identical fit**: The new information may form an apparent identical fit with an existing idea. This means that the students are able to make sense of the new information on the basis of their existing knowledge. This does not imply the correctness of the students' explanations.
- **Approximate fit**: The new information forms an approximate fit with an existing idea in which aspects are seen to be related, but details may be unclear. These students encounter new ideas but do not give up old ones. However, even if contradictory, they do not reach a situation where a cognitive conflict could take place. Hence, new information is assimilated but not accommodated.
- **Incomplete fit**: The new information is acknowledged as not being explained by the ideas tried so far. This incomplete fit of information results in a cognitive conflict. When students experience an incomplete fit they try to reduce the conflict by seeking information which might provide a solution.

The main mechanism for change in Appleton's model is *cognitive conflict*. Although it was originally conceived in the context of knowledge change, we explore in the following how the categories can be applied to capture belief changes. The theory of Appleton (1997) enables categorizing the different reactions of students when confronted with new ideas. The main difference between identical, approximate and incomplete fit is the presence of *cognitive conflict*, which proves to be also the decisive tool for change in beliefs.

## Illustrating the Networking of the Two Theories with Empirical Data

The data that are used to illustrate the networking of the beliefs classification and the conceptual change approaches come from a research study that looked more broadly at initiating changes in pre-service teachers' beliefs (Liljedahl et al. 2007a, b). Participants in this study were 39 pre-service elementary school teachers enrolled in a Designs for Learning Elementary Mathematics course which was taught with the implicit goal of teaching for conceptual change in beliefs. The students were immersed into a problem-solving environment for initiating metacognitive discourse about their mathematics-related beliefs. In addition, the students encountered different instructional strategies so that they were encouraged to change their conceptions about the meaning of mathematics teaching and learning. Throughout the course the participants kept a reflective journal in which they documented their beliefs. In the first and final week of the course, they were asked to respond to the following questions:

- What is mathematics?
- What does it mean to learn mathematics?
- What does it mean to teach mathematics?

In this section we start by presenting an example for beliefs change in terms of the classification by Grigutsch, Raatz and Törner, the four criteria (a) to (d) of the theory of conceptual change, and the model of learning by Appleton. In the following, we present excerpts from David's journal entries where he answers the above-mentioned three questions.

At the beginning of the course, David writes the following:

> When first pondering the question, "What is mathematics?" I initially thought that mathematics is about numbers and rules. It is something that you just do and will do well as long as you follow the rules or principles that were created by some magical man thousands of years ago. [...] To be honest, I don't like math. I found it so boring and so robotic. Lessons were set up in a robotic way. The teachers would show us the principles and then we would do the exercises.

David nicely articulates that his view on mathematics is strongly informed by his experiences with mathematics learning. Using the above-mentioned classification of mathematical beliefs, his answer is coded as "toolbox aspect". Using the conceptual change approach, this is an adequate example of illustrating the first condition for a possible change, namely the role of experiences made by an individual. His lived experience as a student of mathematics is informing his current understanding of what mathematics is. It is also informing his understanding of what it means to teach mathematics = robotic.

He continues his remarks by stating:

> I wish my initial definition could be different but this is the kind of math that I was exposed to.

Here it becomes obvious that David is not satisfied with his view on mathematics or – as he calls it = definition of mathematics, which is part of the second condition of the conceptual change approach. Interestingly, David entered the course already expressing a certain discomfort with his beliefs about mathematics and the teaching and learning of mathematics. However, as there exists no alternative for him, he has not yet fully let go these beliefs, as becomes apparent from further analysis of his journal entries. Although not initiated through instruction, it could be said that David has already experienced cognitive conflict with respect to his beliefs.

In his last journal entry, David expresses and explicitly reflects his change:

> However, after experiencing a couple of challenging problems and exciting classes, I have to say that my definition [of mathematics] can be summed up very simply. To me, mathematics is not about answers, it's about process. Mathematics is about exploring, investigating, representing, and explaining problems and solutions. Learning math is about inquiry and the development of strategies. It is about using your intuition, experimenting with strategies and discussing the outcome. It is about risk taking and experimenting. To teach mathematics is to welcome all ideas that are generated and facilitate discussion. It is about letting the students make sense of the math in their own way, not 'my way'. The teacher's role is about guiding the process, but handing the problem over to the students.

Not only can a change be noted, but using the classification introduced above one can say that this answer corresponds to the process aspect. In terms of the conceptual change approach, it becomes clear that he not only distances himself from his original beliefs ("math is not about answers") but also expresses another belief instead ("it's about process") which illustrates the third condition.

Finally, in order to make sense of David's change in terms of the model established by Appleton (1997), his entries are exemplary of an incomplete change.

In the following, we concentrate on providing more evidence that the theory of conceptual change can be adapted to describe changes in beliefs by examples for the two other possibilities of change introduced by Appleton (1997).

Jacqueline is an example where we observed an identical fit. In her first journal entry, Jacqueline writes the following:

> To teach mathematics, is to guide the learner through the process. It is not the job of the teacher to supply the answer, but to scaffold the process in order for the learners to be successful problem solvers. Guiding the students through the process also allows the learners to discover at their own pace and be at the centre of their learning.

Jacqueline focuses on the role of the teacher as a guide. In her last entry she still remains in this position:

> Finally to teach mathematics is to teach through facilitation. The teacher is there to guide students through the process and supply them with the most efficient tools to solve a problem. It is ultimately up to the student to discover for themselves. [...] It is also the role of the teacher not to provide the answer but put this on the students to solve in the way that best suits them.

This example shows that the ideas offered by the course seem to fit perfectly with what Jacqueline has experienced so far. There is no apparent need for her to change her beliefs.

Aleksandra is an example where an approximate fit took place. She writes in her first journal entry the following:

> I think mathematics is something more than just the use of numbers. It is a way of thinking, a way of knowing things and figuring things out. I believe that it is one of the many ways that some people understand life, connected to multiple intelligences. What I mean is that it is beyond just looking at the world "numerically" and calculating things – it is logical reasoning. Mathematics is a belief that everything has a rational explanation. It is an abstract and conceptual way of thinking about the world around us and solving logical problems.

In few words, Aleksandra views mathematics as a way of thinking. In her last entry, she states:

> I now realize that my understanding of what mathematics is has not really changed but expanded through the course of this class. I would add to this definition [that she used in her first entry] that it is also the way we examine information and analyse it. It is the use of mathematical concepts in real life situations and the flexible way of thinking about numbers, algorithms, patterns, etc. that apply to life. It is an abstract way of looking at the world, through the visualization of number and spatial concepts. It is also using logical and deductive reasoning and making inferences, evaluation problems and situations and making judgments and decisions in given situations. It is the ability to predict and plan and visualize things that are not necessarily presented to us visually.

Aleksandra articulates that her understanding of what mathematics is has not really changed but she emphasizes that she added some beliefs to her already existing ones. Hence, the course did not succeed in producing a fundamental change in her beliefs.

The last example was chosen to give evidence for our remark above that beliefs are not simply replaced but how reflection serves as an important catalyst to allow for changes. In her first entry, Nicola answers the question "What is mathematics?" as follows:

> The first thing that comes to my mind is numbers. I think of math as being calculations such as adding and subtracting or dividing and multiplying using numbers. [...]

Concerning the question "Why do we teach mathematics?" she states:

> I believe that we encounter math everyday in our lives. For example, buying groceries, we need to know how to add and the value of coins and dollars. We need to know how to budget our money. We need to measure cups and table spoons when we are cooking and add up the calories we are eating. [...] Therefore I believe we teach mathematics to function in our daily lives.

In her last entry she states:

> I realize that math is more of a process. There often is a right or a wrong answer but we can't focus on that. We need to value the fact that there is a thinking process of how we feel and of what we did to solve the problem. [...] Mathematics is a set of tools. The more we use the tools, the better that we become with them. [...] I still do believe that math involves an element of memorization. What I do think as well is that before memorization happens, comprehension and the "why" needs to happen. There is no sense in memorizing things we don't understand because we will be sure to get stuck later down the road. [...] So if learning math needs a "why" then we must teach to the "why". I think the best way to figure out why is through self- discovery. I think as a teacher it is important to have interactive thought provoking activities that provide a time for students to ask themselves, their classmates and their teacher questions about why and how.

It becomes obvious that Nicola justifies her former beliefs in some sense but also makes clear that new beliefs have been added.

## Conclusions

The field of beliefs research has developed into different directions which can roughly be sketched by the different research paradigms that were developed in psychology and mathematics education. In addition, much discussion on mathematics beliefs research has concentrated on the difference between knowledge and beliefs and has led to cognitive theories that mainly omitted beliefs research. Our approach has been to extend theories initially developed for knowledge categorization and development to the field of beliefs.

We showed two examples for networking prominent theories from psychology and mathematics education that helped to extend the understanding of the role of beliefs. First, we elaborated on Schoenfeld's theory of *Teaching-In-Context* that captures teacher behaviour from a local view as a function of teacher knowledge, goals and beliefs by extending the knowledge categorization provided by Shulman (1986) to the fields of beliefs and goals. In addition, we used Shulman's (2005a, b) overarching theory of signature pedagogies to understand the significant role of beliefs from a global view. By analysing a specific teaching episode we showed that those theoretical lenses helped to clarify the turning point that occurred during the lesson under observation.

Second, we developed further ideas of conceptual change originally conceived in the context of knowledge (Ernest 1989; Vosniadou and Verschaffel 2004; Appleton 1997) to explore belief changes which were analysed by using the classification of mathematical beliefs based on the work by Dionne (1984) and Ernest (1989). Changes in beliefs could then be illustrated in more detail. As suggested in the literature, we used different theoretical lenses to analyse the same data set. We found evidence in our data that broadening the theoretical approach is fruitful for gaining a deeper understanding of the construct.

## References

Appleton, K. (1997). Analysis and description of students' learning during science classes using a constructivist-based model. *Journal of Research in Science Teaching, 34*(3), 303–318.

Artigue, M., Bartolini-Bussi, M., Dreyfus, T., Gray, E., Prediger, S., et al. (2006). Different theoretical perspectives and approaches in research in mathematics education. In M. Bosch et al. (Eds.), *Proceedings of the 4th CERME, Sant Feliu de Guixols* (pp. 1239–1244). Spain: Fundemi IQS – Universitat.

Arzarello, F., Bosch, M., Lenfant, A., & Prediger, S. (2007). Different theoretical perspectives in research. In D. Pitta-Pantazi et al. (Eds.), *Proceedings of the 5th CERME 5, Larnaca* (pp. 1618–1627). Cyprus: ERME.

Ball, D. L. (1988). *Knowledge and reasoning in mathematical pedagogy: Examining what prospective teachers bring to teacher education.* Unpublished doctoral dissertation, Michigan State University, East Lansing. http://www-personal.umich.edu/~dball/books/index.html. Accessed 24 Jan 2014.

Bauersfeld, H. (1980). Hidden dimensions in the so-called reality of a mathematics classroom. *Educational Studies in Mathematics, 11*, 23–41.

Baumert, J., et al. (2010). Teachers' mathematical knowledge, cognitive activation in the classroom, and student progress. *American Educational Research Journal, 47*(1), 133–180.

Bikner-Ahsbahs, A., & Prediger, S. (2006). Diversity of theories in mathematics education: How can we deal with it? *ZDM – The International Journal on Mathematics Education, 38*(1), 52–57.

Calderhead, J., & Robson, M. (1991). Images of teaching: The student teachers' early conception of classroom practice. *Teaching & Teacher Education, 7*(1), 1–8.

Carter, G., & Norwood, K. S. (1997). The relationship between teacher and student beliefs about mathematics. *School Science and Mathematics, 97*(2), 62–67.

Chapman, O. (2002). Belief structures and inservice high school mathematics teacher growth. In G. Leder, E. Pehkonen, & G. Törner (Eds.), *Beliefs: A hidden variable in mathematics education?* (pp. 177–194). Dordrecht: Kluwer.

Dahl, B. (2006). Various theoretical frameworks in concept construction and how to move forward as a field: A commentary to Pegg and Tall. *ZDM – The International Journal on Mathematics Education, 38*(1), 63–69.

Di Martino, P., & Zan, R. (2001). Attitude toward mathematics: some theoretical issues. In *Proceedings of 25th annual conference for the psychology of mathematics education* (Vol. 3, pp. 351–358). Utrecht: PME.

Dionne, J. J. (1984). The perception of mathematics among elementary school teachers. In J. M. Moser (Ed.), *Proceedings of the 6th annual meeting of the North American chapter of the international group for the psychology of mathematics education* (pp. 223–228). Madison: University of Wisconsin, PME-NA.

Ernest, P. (1989). The knowledge, beliefs, and attitudes of the mathematics teacher: A model. *Journal of Education for Teaching, 15*, 13–33.

Ernest, P. (1991). *The philosophy of mathematics education*. Hampshire: Falmer.

Ernest, P. (1994). The philosophy of mathematics and the didactics of mathematics. In R. Biehler, R. W. Scholz, R. Strasser, & B. Winkelmann (Eds.), *Didactics of mathematics as scientific discipline* (pp. 335–349). Dordrecht: Kluwer.

Feiman-Nemser, S., & Featherstone, H. (1992). The student, the teacher, and the moon. In S. Feiman-Nemser & H. Featherstone (Eds.), *Exploring teaching: Reinventing an introductory course*. New York: Teacher College Press.

Feiman-Nemser, S., McDiarmid, G., Melnick, S., & Parker, M. (1987). *Changing beginning teachers' conceptions: A description of an introductory teacher education course*. Paper presented at American Educational Research Association, Washington, DC.

Fosnot, C. (1989). *Enquiring teachers, enquiring learners: A constructivist approach for teaching*. New York: Teachers College Press.

Furinghetti, F., & Pehkonen, E. (2002). Rethinking characterizations of beliefs. In G. Leder, E. Pehkonen, & G. Törner (Eds.), *Beliefs: A hidden variable in mathematics education?* (pp. 39–57). Dordrecht: Kluwer.

Green, T. (1971). *The activities of teaching*. New York: McGraw-Hill.

Grigutsch, S. (1996). *Mathematische Weltbilder von Schülern. Struktur, Entwicklung, Einflußfaktoren* (unveröffentlichte Dissertation).

Grigutsch, S., Raatz, U., & Törner, G. (1998). Mathematische Weltbilder bei Mathematiklehrern. *Journal für Mathematikdidaktik, 19*, 3–45.

Hersh, R. (1991). Mathematics has a front and a back. New directions in the philosophy of mathematics. *Synthese, 88*(2), 127–133.

Hersh, R. (1997). *What is mathematics, really?* New York: Oxford University Press.

Hofer, B. K. (2000). Dimensionality and disciplinary differences in personal epistemology. *Contemporary Educational Psychology, 25*, 378–405.

Hofer, B. K. (2004). Epistemological understanding as a metacognitive process: Thinking aloud during online searching. *Educational Psychologist, 39*(1), 43–55.

Hofer, B. K., & Pintrich, P. R. (1997). The development of epistemological theories: Beliefs about knowledge and knowing and their relation to learning. *Review of Educational Research, 67*, 88–140.

Kitchener, K. S. (1983). Cognition, metacognition, and epistemic cognition. *Human Development, 26*, 222–232.

Kuhn, T. S. (1970). *The structure of scientific revolutions* (2nd enlarged ed.). Chicago: University of Chicago Press.

Kuntze, S. (2011). Pedagogical content beliefs: Global, content domain-related and situation-specific components. *Educational Studies in Mathematics, 79*(2), 273–292.

Leder, G. C., Pehkonen, E., & Törner, G. (2002). *Beliefs: A hidden variable in mathematics education?* Dordrecht: Kluwer.

Lerman, S. (1983). Problem-solving or knowledge centered: The influence of philosophy on mathematics teaching. *International Journal of Mathematical Education in Science and Technology, 14*(1), 59–66.

Lerman, S. (2006). Theories of mathematics education: Is plurality a problem? *ZDM – The International Journal on Mathematics Education, 38*(1), 8–13.

Liljedahl, P. (2006). Pentominoes and AHA!'s. In R. Zazkis & S. Campbell (Eds.), *Number theory in mathematics education: Perspectives and prospects*. Mahwah: Lawrence Erlbaum.

Liljedahl, P., Rolka, K., & Rösken, B. (2007a). Affecting affect: the reeducation of preservice teachers' beliefs about mathematics and mathematics teaching and learning. In W. G. Martin, M. E. Strutchens & P. C. Elliott (Eds.), *The learning of mathematics. Sixty-ninth yearbook of the National Council of Teachers of Mathematics* (pp. 319–330). Reston: NCTM.

Liljedahl, P., Rolka, K., & Rösken, B. (2007b, February 22–26). Belief change as conceptual change. In *Proceedings of CERME 5: Fifth conference of the European society for research in mathematics education*, Larnaca, Cyprus.

Lortie, D. (1975). *Schoolteacher: A sociological study*. Chicago: University of Chicago Press.

McLeod, D. (1992). Research on the affect in mathematics education: A reconceptualization. In D. A. Grouws (Ed.), *Handbook of research on mathematics teaching and learning* (pp. 575–596). New York: Macmillan.

Millsaps, G. (2000). Secondary mathematics teachers' mathematics autobiographies: Definitions of mathematics and beliefs about mathematics instructional practice. *Focus on Learning Problems in Mathematics, 22*(1), 45–67.

Op't Eynde, P., De Corte, E., & Verschaffel, L. (2001). Problem solving in the mathematics classroom: a socio-constructivist account of the role of students' emotions. In *Proceedings of 25th annual conference for the psychology of mathematics education* (Vol. 4, pp. 25–32). Utrecht: PME.

Pajares, M. F. (1992). Teachers' beliefs and educational research: Cleaning up a messy construct. *Review of Educational Research, 62*(3), 307–332.

Perry, W. G. (1968). *Patterns of development in thought and values of students in a liberal arts college: A validation of a scheme* (Final Report Project No. 5–0825, Contract No. SAE-8873). Cambridge, MA: Bureau of Study Counsel, Harvard University.

Philipp, R. A. (2007). Mathematics teachers' beliefs and affect. In F. K. Lester (Ed.), *Second handbook of research on mathematics teaching and learning: A project of the National Council of Teachers of Mathematics* (pp. 157–224). Charlotte: Information Age.

Pintrich, P. R., Marx, R. W., & Boyle, R. A. (1993). Beyond conceptual change: The role of motivational beliefs and classroom contextual factors in the process of conceptual change. *Review of Educational Research, 63*, 167–199.

Posner, G., Strike, K., Hewson, P., & Gertzog, W. (1982). Accommodation of a scientific conception: Towards a theory of conceptual change. *Science Education, 66*, 211–227.

Rolka, K., Rösken, B., & Liljedahl, P. (2006). Challenging the mathematical beliefs of preservice elementary school teachers. In *Proceedings of the 30th international conference for psychology of mathematics education*. Prague: Czech Republic.

Rösken, B., & Rolka, K. (2011). Goals and beliefs – Two sides of the same coin. In K. Kislenko (Ed.), *Proceedings of MAVI 16, The 16th conference on mathematical views* (pp. 1–9), Tallinn, Estonia, June 26–29, 2010.

Rösken, B., & Törner, G. (2007, April 9–13). *Beliefs of university teachers of mathematics – Revisited as epistemological world views.* Paper presented at the 2007 annual meeting of the American Educational Research Association, Chicago, USA.

Rösken, B., & Törner, G. (2010, April 29–May 4). *Aspects of accomplished mathematics teaching reflected as characteristics of a signature pedagogy.* Paper presented to the 2010 annual meeting of the American Educational Research Association (AERA), Denver, USA.

Rösken, B., Höchsmann, K., & Törner, G. (2008, March 5–8). *Pedagogies in action: The role of mathematics teachers' professional routines.* Paper presented to the symposium held on the occasion of the 100th anniversary of ICMI, Rome.

Rösken, B., Pepin, B., & Törner, G. (2011). Editorial. Beliefs and beyond: Affecting the teaching and learning of mathematics. *ZDM – The International Journal on Mathematics Education, 43*(4), 451–455.

Schoenfeld, A. H. (1985). *Mathematical problem solving*. New York: Academic.

Schoenfeld, A. H. (1998). Toward a theory of teaching-in-context. *Issues in Education, 4*(1), 1–94.

Schoenfeld, A. H. (2007). Problem solving in the United States, 1970–2008: Research and theory, practice and politics. *ZDM – The International Journal on Mathematics Education, 39*(5–6), 537–551.

Schoenfeld, A. H. (2010). *How we think. A theory of goal-oriented decision making and its educational applications*. New York: Routledge.

Schommer, M. (1990). The effects of beliefs about the nature of knowledge on comprehension. *Journal of Educational Psychology, 82*, 498–504.

Schommer-Aikins, M. (2004). Explaining the epistemological belief system: Introducing the embedded systemic model and coordinated research approach. *Educational Psychologist, 39*(1), 19–29.

Schraw, G., & Olafson, L. (2002). Teachers' epistemological world views and educational practices. *Issues in Education, 8*(2), 99–148.

Shulman, L. S. (1986). Those who understand: Knowledge growth in teaching. *Educational Researcher, 15*(2), 4–14.

Shulman, L. S. (2005a). Signature pedagogies in the professions. *Daedalus*, Summer Issue, 52–59.

Shulman, L. S. (2005b, February 6–8). *The signature pedagogies of the profession of law, medicine, engineering, and the clergy: Potential lessons for the education of teachers.* Manuscript delivered at the Math Science Partnerships (MSP) Workshop 'Teacher Education for Effective Teaching and Learning' hosted by the National Research Council's Center for Education, Irvine, CA.

Sierpinska, A., & Lerman, S. (1996). Epistemologies of mathematics and mathematics education. In A. J. Bishop et al. (Eds.), *International handbook of mathematics education* (pp. 827–876). Dordrecht: Kluwer.

Skott, J. (2001). The emerging practices of novice teachers: The roles of his school mathematics images. *Journal of Mathematics Teacher Education, 4*(1), 3–28.

Steinbring, H. (1998). Elements of epistemological knowledge for mathematics teachers. *Journal of Mathematics Teacher Education, 1*, 157–189.

Thompson, A. G. (1992). Teachers' beliefs and conceptions: A synthesis of the research. In D. A. Grouws (Ed.), *Handbook of research on mathematics teaching and learning* (pp. 127–146). New York: Macmillan.

Törner, G. (2002). Mathematical beliefs – A search for a common ground: some theoretical considerations on structuring beliefs, some research questions, and some phenomenological observations. In G. Leder, E. Pehkonen, & G. Törner (Eds.), *Beliefs: A hidden variable in mathematics education?* (pp. 73–94). Dordrecht: Kluwer.

Törner, G., & Grigutsch, S. (1994). Mathematische Weltbilder bei Studienanfängern – eine Erhebung. *Journal für Mathematikdidaktik, 15*(3/4), 211–252.

Törner, G., Rolka, K., Rösken, B., & Sriraman, B. (2010). Understanding a teacher's actions in the classroom by applying Schoenfeld's theory Teaching-in-Context: Reflecting on goals and beliefs. In B. Sriraman & L. English (Eds.), *Theories of mathematics education: Seeking new frontiers* (pp. 401–417). Berlin/Heidelberg/New York: Springer.

Uusimaki, L., & Nason, R. (2004). Causes underlying pre-service teachers' negative beliefs and anxieties about mathematics. In *Proceedings of 28th annual conference for the psychology of mathematics education* (Vol. 4, pp. 369–376). Bergen: PME.

Vosniadou, S., & Verschaffel, L. (2004). Extending the conceptual change approach to mathematics learning and teaching. *Learning and Instruction, 14*(5), 445–451.

# Developments in Mathematics Teachers' Efficacy Beliefs

**George N. Philippou and Marilena Pantziara**

> *Promoting teachers' efficacy beliefs within teacher education programs may have the unintended effects of promoting problematic types of teachers' efficacy confidence, suppression of potentially beneficial teacher doubts, and fostering maladaptive motivation patterns.*
>
> (Wheatley 2005, p. 758)

**Abstract** Teacher efficacy beliefs refer to beliefs about one's capacity to organize and execute courses of action to accomplish a specific task. After almost 40 years of intensive research on teacher efficacy beliefs, and despite the impressive support of the claim that teacher efficacy beliefs constitute an important influence on teacher behaviour and student achievement, motivation and beliefs, the construct remains under serious criticism. It is not a surprise that after this huge effort of the international educational community, researchers expected more safe theoretical and practical outcomes.

Specifically, researchers contend that rather than being at the verge of maturity, research on teacher efficacy beliefs is still lacking clarity and demands radical reconceptualization. The weaknesses or objections raised recently include:

*The definition of the construct.* Despite Bandura's demand for specificity of the task of reference, in most studies the actual measure refers to global efficacy, which is ambiguous and hard to prove operationally.

*Scale development.* There are significant limitations to most of the established scales and a need for distinction between personal and collective teaching efficacy.

*The sources of teacher efficacy beliefs.* Apart from Bandura's four sources of efficacy beliefs, there are other contextual sources which deserve more attention and analysis.

G.N. Philippou (✉)
Department of Educational Studies, University of Nicosia, Nicosia, Cyprus
e-mail: edphilip@ucy.ac.cy

M. Pantziara
Department of Teachers' In-service Training, Cyprus Pedagogical Institute
University of Nicosia, Nicosia, Cyprus

 
B. Pepin, B. Roesken-Winter (eds.), *From beliefs to dynamic affect systems in mathematics education*, Advances in Mathematics Education,
DOI 10.1007/978-3-319-06808-4_5

*Methodologies*. The great majority of studies are quantitative with items mostly limited to global efficacy. Doubs are raised about the utility of numerical confidence levels and the related measures; the need for more qualitative and mixed research designs is underlined.

*Complexity and multidimensionality of the construct*. There is an urgent need for studies focusing on specific interpretive meaning of the concept and certain dimensions of efficacy in connection with the outcomes of practice.

In the present chapter we discuss and analyse these and other contentions with respect to current research; we draw on recent research, particularly but not only on review papers, in connection with studies that have examined efficacy beliefs with respect to teaching mathematics.

**Keywords** Teacher • Beliefs • Efficacy • Scales • Sources • Effects

## Introduction

The construct "perceived efficacy beliefs" has been under study for almost 40 years. During this period an increasing volume of research has enriched our understanding of the construct and its role in human behavior. Since beliefs have been thought of as lenses through which one looks in interpreting the world (Philipp 2007, p. 258), it seems plausible to expect beliefs in one's efficacy to be a key personal resource in self-development, successful adaptation, and change. Bandura (2006a) asserted that perceived efficacy beliefs affect people's goals and aspirations, how well they motivate themselves, and their perseverance in the face of adversity. However, at the same time various research findings have been faced by other scholars with skepticism (see e.g. Wheatley 2005).

Bandura (1997, p. 3) defined perceived self-efficacy as "beliefs in one's capabilities to organize and execute the courses of action required to produce given attainments". The construct *teacher efficacy beliefs (*TEBs*)* results from specifying "given attainments" as educational goals; it can be defined as "the teacher's beliefs in his/her capability to organize and execute courses of action required to successfully accomplish a specific teaching task in a particular context" (Tschannen-Moran et al. 1998, p. 233). As self-perception of competence, TEBs do not necessarily reflect an accurate assessment of capabilities (Goddard et al. 2004). The study of TEBs has captured the attention of many researchers, who have studied the meaning of the construct and its relevance to educational practice from a variety of standpoints and in many different contexts and cultures, particularly as a factor in improving teacher education and promoting educational reform (Chan 2008a). Research has demonstrated that TEBs beliefs, have been positively correlated with teaching practices and teacher classroom behaviors and a broad range of positive student outcomes (Betoret 2009; Brown 2005; Nie et al. 2013; Tschannen-Moran and Woolfolk Hoy 2001).

In this chapter we review progress of research on TEBs in two parts. First, we discuss research reported before the end of the past century and the questions that gradually arose, and next we focus on recent developments, with emphasis in teachers' mathematics efficacy research. In conclusion we propose some ideas for future research.

## Early Progress and (Un) Expected Problems

### *Clarifying the Concept*

The construct TEBs is conceptualized as both context and content specific; it differs from the general perception of *confidence* which refers to more generalized conceptions of competence (Bandura et al. 1996). TEBs concern beliefs that vary across contents and teaching tasks, as well as according to the group of students and the environment. Setting the borders of the construct, Bandura (2006b, p. 309) distinguished perceived *self-efficacy* from *self-esteem*, *locus of control*, and *outcome expectancy*. Self-efficacy is a judgment of capability; self-esteem is a judgment of self-worth, and locus of control concerns beliefs about factors influencing an outcome, whether determined by one's actions rather than by forces outside one's control. Outcome expectancies are judgments about the results that are likely to occur from the execution of the task. Bandura conceptualized TEBs as an operative capability, while outcome expectancy refers to the effect of the execution of this teaching. In other words, *efficacy expectancies* refer to perceived ability to execute specific teaching actions, while *outcome expectancies* refer to teachers' beliefs about the effects that specific teaching actions will have on students. A teacher may highly perceive his/her ability to execute a teaching task, but may have doubts about the final outcome. High outcome expectancy reflects the degree to which a teacher or a group of teachers believe that the family background and the wider environment could be controlled.

Three types of TEBs have been identified: *Personal teacher efficacy* refers to an individual trait; *general teacher efficacy* refers to beliefs in the ability of teachers in general to bring about the required learning outcomes (Tschannen-Moran et al. 1998); *collective teacher efficacy beliefs* (CTEBs) refer to judgments of teachers in a school that the faculty as a whole is capable to effectively organize and execute teaching actions (Goddard et al. 2004). CTEBs – the resultant of personal efficacies of a group of teachers – reflects the extent to which the group believes in their skills to promote students' learning, via interactive, coordinative, and synergistic dynamics of their transactions (Bandura 2006b). Although the first two types of TEBs were consistently extracted through factor analysis using the scale by Gibson and Dembo (1984), general teaching efficacy was criticized as similar to locus of control and its frequent failure in terms of validity and reliability (Henson 2002; Tschannen-Moran and Woolfolk Hoy 2001). Some researchers have recently suggested that this concept should be abandoned (Tuchman and Isaacs 2011).

A point of agreement among the scholars (Tschannen-Moran et al. 1998; Wheatley 2005; Wyatt 2014) concerns the specificity of TEBs and the consequent differing levels of TEBs, according to tasks. One's capability to apply instructional methods may differ from one's beliefs about one's ability to keep discipline in the class, or to create a positive climate. Since TEBs are content and task specific, global measures of the concept can hardly help educational practice. Yet, the use of global measures by researchers who claim specificity continues to be one of the biggest anomalies in the field (Wyatt 2014). Specificity, however, comes at the expense of generalizability, so the target is an optimal level of specificity (Tschannen-Moran and Woolfolk Hoy 2001). For instance, scale items asking for "efficacy for teaching" are too global, the same could be argued for items asking for "efficacy for teaching mathematics", but efficacy to teach a certain addition of fractions or to solve a specific quadratic equation may result in diminishing the practical relevance of the findings. Apart from the optimal specificity, a complementary solution would be more use of qualitative or mixed research designs that have the potential to produce insightful findings that can make the study of TEBs of greater use to teacher educators.

## Recognized Importance of Teacher Efficacy Beliefs and Unresolved Problems

Bandura (2006a, p. 10) refers to "three main pathways" through which efficacy beliefs play a key role in cognitive development, namely: students' beliefs in their competence to learn, teachers' beliefs in their personal efficacy to promote students' learning, and faculties' collective sense of efficacy. High efficacy teachers are expected to better influence their students' learning, view difficulties as surmountable, and persist in the face of obstacles, while low efficacy teachers are easily convinced of the futility of their effort, tend to be disappointed in the face of difficulties and may give up trying (Bandura 2006a). Indeed, TEBs have been consistently associated with teacher behavior, student attitudes, and student achievement (Tschannen-Moran and Woolfolk Hoy 2007, p. 954) and with factors of interest to teacher educators and reformers, such as teacher retention, commitment and willingness to experiment (Wheatley 2002). Mathematics self-efficacy beliefs were found to be a better predictor of students' performance than students' mathematics anxiety and their conceptions about the usefulness of mathematics (Pajares and Miller 1994). Furthermore, TEBs were found to be linked to pre-service teachers' ability to construct mathematical problems, their mathematical background and their ability to teach problem posing (Philippou et al. 2001).

Inquiry into CTEBs emphasizes that teachers have not only self-referent efficacy perceptions but also beliefs about the conjoint capability of a school faculty (Goddard et al. 2004, p. 4). CTEBs were empirically found to be linked to differences in students' achievement among schools and to differences among schools with regard to teachers' TEBs; the effect of CTEBs was found to be a significant predictor of between schools differences in students' mathematics achievement (Goddard 2002). CTEBs were also found to have stronger effects on student achievement than student race or socio-economic status, even after controlling for students' prior achievement, race/ethnicity, and gender (Goddard et al. 2004).

The enthusiasm for efficacy research, however, could not hide complaints and uncertainties about the field. Tschannen-Moran et al. (1998) wondered whether research on efficacy beliefs was close to maturity highlighting questions that continued to perplex researchers in the field, such as: Do TEBs constitute a trait that can be captured by a self-report instrument, or they are specific to given contexts? Does the concept need to be refined or expanded to capture more aspects of teachers' self-efficacy? What contributes to the development of positive TEBs, how malleable is a sense of efficacy once it is established, and in what ways do TEBs influence teaching behavior?

Although similar issues have frequently been raised, the end of the confusion continues to be out of sight. Researchers urge for a clearer meaning of the construct, deeper examination of its genesis and development, reconsideration of the measures and methodologies, and more relevance to educational practice (see e.g. Labone 2004). Critics question the assumption of causality from findings of correlational nature, the conclusions drawn on the basis of global measures (Wheatley 2002), and the assumption that high efficacy leads to greater effort and better teaching outcomes. Wheatley (2002) identified several types of TEBs that can obstruct educational reform and analyzed potential benefits of efficacy doubts. A list of these benefits includes *teacher learning and change, fostering reflection and productive collaboration, and supporting motivation to learn.* Overconfident teachers are satisfied from current practices and have no reason for critical reflection and change, running the danger of falling into stagnation. On the contrary, a dose of uncertainty may motivate reexamination of old practices and lead to a state of disequilibrium that constitutes the basis for new knowledge, in an era of reformed curricula moving towards meaning-centered education.

To bring research on efficacy to maturity, Labone (2004) asked for diverse research methodologies and more focus on the interpretivists and the critical theorists, which had been somewhat neglected as well as for a theoretical grounding for the study of the development of TEBs, and for broadening the construct to explore dimensions that facilitate educational reform. On the theoretical side, the multiple meanings of teacher efficacy make it problematic for teacher educators to interpret and use research findings; on the practical level, Wheatley (2005, p. 747) could not "identify any tools from teacher efficacy research that can be consistently useful to teacher education". In addition, Wheatley (2005) warned that efforts to increase TEBs can back fire, because teachers' confidence in their capability may produce more bad outcomes than positive, especially in the context of teacher education.

Progress in the field during the present century has so far failed to silence complaints about global measures, rare use of qualitative and mixed research designs. These designs, including the use of interviews and classroom observations may produce insightful findings useful to teacher educators. For instance, Klassen et al. (2011) refer to unfulfilled promises despite some signs of progress; they found that only 8.7 % of the reviewed studies were qualitative. Wyatt (2014) highlighted the continuing problematic situation in efficacy research right in his title; his objective was *"re-conceptualization" of* TEBs, *"tackling enduring problems with the quantitative research and moving on".*

## Recent Developments

### *The Concept and Its Measurement*

Operationalizing TEBs relies on consistent measurement of the construct; a process traditionally performed using self-report scales comprised of items that address a range of teaching tasks and situations. Most scales, however, were widely recognized as global measures, nonaligned with demands for specificity. As Bandura (2006b, p. 307) argued any *"one measure fits all"* approach has limited explanatory and predictive value due to tenuous relevance to the domain of functioning. Substantiating this argument, Klassen et al. (2011) found that almost one half of the 218 studies reviewed used measures incongruent with efficacy theory; they did not assess teachers' capabilities to carry out a course of action. On the same line Goddard et al. (2004) asserted that the broadening of the scope of the construct by adding new areas of teacher functioning at work has led to a need for developing specific scales for tasks in terms of content and domains, e.g. specific teaching skills, relations with peers, and ability to influence the organization.

In recent studies many researchers have used the Teacher Sense of Efficacy Scale (TSES) by Tschannen-Moran and Woolfolk Hoy (2001) – a long and a short version (with 24 and 12 items, respectively). TSES comprise three factors: efficacy for *instructional strategies*, *classroom management*, and *student engagement*. Most TSES items have the stem "To what extent can you…?" and "How much can you do to…?" TSES has been characterized as "superior to previous measures" because it has a unified and stable factor structure and is closely aligned with self-efficacy theory (Klassen et al. 2009). Klassen et al. (2009) have validated the TSES using theory-testing techniques in Canada, Cyprus, Korea, Singapore, and the United States. To this end, six groups of teachers were chosen to enable tests of validity across levels (elementary, middle and secondary schools) and cultural/geographical settings. The study established the importance of the construct across diverse teaching conditions examining measurement invariance of the scale and the relationship between TEBs and job satisfaction. Internal consistency and the three factor model of the TSES were confirmed as well as its reliability and measurement invariance across the five countries. In addition, the study provided evidence that TEBs is a valid construct across culturally diverse settings and also that, TEBs showed a similar relationship with teachers' job satisfaction in five contrasting settings.

Recognizing the predominance of the TSES over other scales, Duffin et al. (2012) examined its factor structure in an attempt to resolve "discrepancies in the interpretation of Bandura's theory in the process of creating TEBs measures, which led to questioning of the psychometric properties of different measures used" (p. 828). They analyzed the scores of pre-service teachers at their early stage of development to gather evidence of internal structure validity. Two plausible rival models derived from prior research (a single factor and three-factor model) were tested using confirmatory factor analyses. Results showed good fit for both models, while high interfactor correlations indicated strong support for the uni-dimensional model. The findings suggested that pre-service teachers who lack pedagogical knowledge and teaching experience do not differentiate between the different aspects of teaching measured by the TSES.

Extending the three dimensions of the TSES, Chan (2008a) developed a scale focusing on teaching functions in secondary schools in times of education reform. The scale (TSES-18) was designed to assess TEBs in six domains: *teaching high ability learners, classroom management, guidance and counseling, student engagement, teaching to accommodate diversity, and teaching for enriched learning*. The scale and its subscales were found to be valid, internally consistent, and also related positively with an equivalent and convergent measure of TEBs and with a measure of personal accomplishment, while they related negatively- slightly or moderately – with two components of burnout -emotional exhaustion and depersonalization (Chan 2008a, p. 191).

Several approaches to measuring CTEBs have been proposed (Goddard et al. 2004); by taking the average of the measures of individual members in the school about their own personal efficacy; by considering the average on items measuring individual member beliefs about the group's capability, and by asking the members of the group to discuss and collectively respond on the items. Goddard (2002) developed and validated a scale for measuring CTEBs in line with the cyclical model by Tschannen-Moran et al. (1998). This model examines the development of TEBs as the outcome of processing efficacy sources, analyzing the task and assessing personal competence, followed by examining the consequences of actual performance and finally reconsidering the sources in a new cycle. Goddard provided evidence that using the short version (12-item of TSES) could be equally effective as using the long version of the scale (24-item). In an effort to advance awareness about CTEBs Goddard et al. (2004) developed a conceptual model to explain the formation and influence of these beliefs. They argued that the connections between CTEBs and student outcomes partially depend on the reciprocal relationships among teachers' collective efficacy, personal efficacy and their professional practices.

Enochs et al. (2000) developed the Mathematics Teaching Efficacy Belief Instrument (MTEBI). It comprised 21 items in two subscales, measuring Personal Mathematics Teaching Efficacy and Mathematics Teaching Outcome Expectancy, respectively. The scale was subjected to testing for factorial validity and also to confirmatory factor analysis, utilizing the structural equation modeling software EQS. Two indicative items: *Even if I try very hard, I will not teach mathematics as well as I will most subjects (teaching efficacy); the mathematics achievement of some students cannot generally be blamed on their teachers (outcome expectancy).* MTEBI has been widely used in studies focusing on mathematics TEBs (Bates et al. 2011; Evans 2011; Gresham 2008; Tran et al. 2012).

## *The Genesis and Development of Teacher Efficacy Beliefs*

### Sources of Efficacy Beliefs

Identifying potential sources which contribute to the genesis and development of TEBs is of major interest for teacher educators in their effort to facilitate teachers' acquisition of positive efficacy beliefs. Bandura (1997) postulated four cognitive sources: *mastery experience, vicarious experience, social persuasion, and physiological and emotional arousal.* Mastery experience refers to one's sense of competence and is empowered by success. However, not all successful experiences reinforce efficacy. Success in trivial tasks does not influence efficacy beliefs. Vicarious experience i.e., observing other's actions, may enhance one's confidence, particularly if the observed person is perceived as having similar qualities to the observer. Social persuasion refers to feedback provided by significant others, i.e., faith in one's capabilities by teachers, parents and superiors. Finally, relaxation and positive emotions relate to self-assurance and the anticipation of future success, leading to higher self-efficacy.

Bandura's sources of TEBs have been studied through different approaches, with respect to their influence on teachers' efficacy, and in connection to other educational parameters, such as the role of TEBs as coping resources against job stressors (Betoret 2009; Brand and Wilkins 2007; Chang 2009). Much of the relevant research (Charalambous et al. 2008; Tschannen-Moran and Woolfolk Hoy 2007) has focused on the weight of each source in the formation of TEBs and also on their effect on TEBs of teachers being at different career stages. All four Bandura's sources were found to contribute to high teacher efficacy, while teacher efficacy was negatively connected to job stressors (Betoret 2009). Usher and Pajares (2006) examined the influence of Bandura's sources on academic self efficacy and efficacy for self regulation. They found that all four sources predicted academic self efficacy with the effect of mastery experience being stronger that the effect of the other three sources.

Brand and Wilkins (2007) examined elementary pre-service teachers' development as effective teachers of science and mathematics, through a relevant methods course. They used naturalistic inquiry to examine TEBs, drawing on participants' experiences of participation in course activities. Analyzing written reflections, at the end of the semester, with regard to factors that influenced teaching capability, the authors reported that all Bandura's sources influenced TEBs, with mastery experiences being the most influential. A relationship was also found between mastery experiences and the other sources, indicating that this source is a function of the other three sources. The conclusion was that vicarious experiences, social persuasion, and stress reduction influence mastery experiences and thus indirectly influence efficacy beliefs.

Additional sources of TEBs have also been investigated. Tschannen-Moran and Woolfolk Hoy (2007) referred to external and internal efficacy sources. External sources relate to the teaching task, including the resources available, students' factors (e.g. ability, motivation), and contextual factors (school principal, colleagues' support, teaching resources). Internal sources concern teachers' judgments about personal capabilities. Tschannen-Moran and Woolfolk Hoy (2007) examined TEBs sources of 255 novice and experienced elementary, middle and high school teachers. Different contextual variables were found to be related to teachers' TEBs. Teaching resources made a significant independent contribution to explaining variance in novice teachers' TEBs beliefs, whereas the School Level Taught made a significant contribution to explaining the variance in experienced teachers' TEBs, with higher TEBs found among teachers who taught younger students. Verbal persuasion and specifically the support of colleagues and the community made a significant contribution to explaining only novice teachers' TEBs. Mastery experience measured as satisfaction with past professional performance was moderately related to both novice and experienced teachers' efficacy beliefs. Mastery experiences of experienced teachers were related to the support they received in the form of verbal persuasion from administrators, colleagues, parents and the community. Mastery experiences of novice teachers were related to support from parents and the community. Contextual factors and mastery experience explained 49 % of the variance for novice teachers' TEBs compared with 19 % for experienced teachers' TEBs. Mastery experiences were found to make the strongest contribution to TEBs for both samples.

Phelps (2010) examined the sources that pre-service elementary teachers use when they construct TEBs and their learning goals. She used narrative interviews (stories about participants' past experiences and their development as learners of mathematics) with 22 participants with regard to mathematics and its importance, to examine retrospectively factors that they reported as affecting the development of their motivational profiles. Phelps considered self-efficacy as one construct of expectancy theory, and learning goals as one construct of reasons for engagement, jointly providing a "picture of pre-service elementary teachers' motivation to learn mathematics" (p. 294). Results revealed that participants relied on multiple sources to construct their TEBs and goals, including past performance, vicarious experiences, verbal persuasions, career goals, and the fit between their views of mathematics and the nature of mathematics activities, as experienced in their classes.

## The Development of Efficacy Beliefs During Teacher Education

Since Bandura (1997) stated that efficacy beliefs are most at play early in learning and, once constructed, become resistant to change, teacher education programs attracted the interest of researchers, as a means to develop efficacy beliefs in pre-service teachers. During teacher education programs pre-service teachers acquire familiarity with the basics of pedagogy and teaching skills, which may help mold TEBs at the time of genesis. Charalambous et al. (2009) examined the effect of a mathematics program on pre-service primary teachers' attitudes, epistemological beliefs, and TEBs. The program comprised two content courses taught successively during the first and the second year of studies. The TSES scale was administered to 91 students before and after each of the courses and semi structured interviews were conducted. The analyses showed mixed changes in students' attitudes and beliefs depending on their background. A positive change in TEBs was observed for the group with high mathematics ability (as determined by their option to take mathematics in the university entrance exams).

Evans (2011) studied the effect of a mathematics method course on Teaching Fellows in an alternative teacher – recruitment program, with regard to mathematical content knowledge, attitudes toward mathematics and TEBs. He analyzed data collected at the beginning and at the end of the semester, using a mathematics content test, an attitude scale and the MTEBI, as well as participants' reflective journals on their teaching and learning reported during the semester. Findings indicated significant improvements in both mathematical knowledge and attitudes toward mathematics, but no significant increase was found in TEBs (neither for teaching efficacy, nor for expectancy efficacy). Both measures, however, correlated with attitudes and were above the neutral point on a five-point scale.

Field-work provides pre-service teachers with precisely the type of enactive mastery experiences that Bandura suggests as a source of TEBs. Charalambous et al. (2008) investigated pre-service teachers' mathematics PTE beliefs during fieldwork. Exploratory factor analysis of longitudinal data (using TSES at the beginning, middle and end of the field work) resulted in a two-factor model (emerged in all three scale administrations), reflecting TEBs in relation to mathematics instruction and in classroom management. The results indicated that during fieldwork, pre-service teachers' TEBs in mathematics improved but not in uniform ways. The analysis of semi-structured interviews with eight participants, suggested that pre-service teachers' TEBs were mainly informed by enactive mastery experiences, vicarious experiences, and social persuasion (experimentation with teaching, and interaction with mentors, tutors, peers, and pupils).

Chang (2009) used a multi-case study to explore the developmental process in beginning elementary mathematics teachers' efficacy with and without mathematics and science backgrounds. Participants were six teachers, three with and three without mathematics and science background, in both cases one with low, one with medium, and one with high efficacy. Data analyzed included initial and follow up interviews, recordings, observations, and reflection notes. Chang found a cyclical developmental model with five gradations continuous over time. The characteristics of each gradation were identified after being subjected to at least 1 month of continuous observations and also verified through the interview process and participants' reflection notes. Posttest scores revealed that all six participants' efficacy ratings rose during the first year of teaching, while the five-gradation model, showed that beginning mathematics teachers with different levels of efficacy exhibited different characteristics of efficacy development. The qualitative findings showed that during the first year of teaching the two low efficacy teachers reached the first and the second gradations, the two medium-efficacy teachers reached the third gradation, one of the high-efficacy teachers exhibited the characteristics of the fourth gradation and the other one, who possessed a mathematics and science background, even entered the fifth gradation. In conclusion, beginning mathematics teachers who had the same level of TEBs tended to exhibit substantial similarities in their developmental processes, though there were slight differences between two teachers with the same efficacy level and different backgrounds.

## Intervention Studies Aimed at Enhancing Teacher Efficacy Beliefs

Tuchman and Isaacs (2011) asserted that "efforts to increase teacher self-efficacy through in-service and other similar interventions have met with mixed success, and no clear pattern can be concluded from prior studies" (p. 415). In this section we summarize two recent intervention studies; the first examined the impact of peer coaching on TEBs and the other the effect of new technologies on TEBs. The studies provide ideas for teacher training programs focusing in developing TEBs.

In the context of communities of practice, peers can influence each other's practices by jointly attempting specific strategies that help participants experience success (a joint mastery experience) (Bruce and Ross 2008). Peers can also influence each other's efficacy through social persuasion, as one observes a peer implementing successful strategies (vicarious experience), and also through enhancing feelings arising from effective teaching or reducing negative feelings arising from negative teaching experiences (physiological and emotional cues) (Bruce and Ross 2008). Peer coaching in relation to mathematics teaching practices and efficacy beliefs are expected to have an impact on student learning. In Bruce and Ross' study four pairs of Grade 3 teachers and two pairs of Grade 6 teachers participated in a professional development program over 6 months. The program focused on three dimensions of mathematics teaching: facilitating student teacher interaction, supporting student construction of mathematical meaning, and selecting effective mathematics tasks. Data included teacher classroom observation (at the beginning and at the end of the program), self assessment, interviews, and field notes that focused on the above three teaching dimensions. The analyses indicated that all pairs successfully implemented the main steps of peer coaching and key strategies for effective mathematics teaching, especially in facilitating student interaction and improving the quality of tasks assigned. As a result they moved toward a more constructivist approach (student directed, manipulative-based, and conceptually-focused learning) and toward facilitating student-student interaction, assigning open- ended student tasks that encourage multiple solutions. The data indicated that participants TEBs improved presumably as a result of a "nexus of sources of efficacy information" (p. 360). By the end of the program, teachers reported feeling more capable of teaching mathematics with an emphasis on conceptual understanding. They attributed this change to several facets of the program recognizing that some of their existing practices were similar to those modeled by presenters (vicarious experience); receiving positive feedback from their peer coaching partners (social and verbal persuasion, physiological, and emotional cues), and by acquiring and applying new instructional strategies in their own classrooms (mastery experiences).

Though the use of technology has long ago been recognized as essential for teaching and learning mathematics (NCTM 2000), we still know little about the use of technological innovations to facilitate mathematics instructions and the impact they have on teachers' efficacy and classroom practices. To explore these issues, Tran et al. (2012) reported the effects of a computer-based teaching tool known as Spatial Temporal Mathematics (ST Math) on teacher efficacy, outcome expectancy, and instructional practice. This program utilizes images to help students develop spatial–temporal cognition that can improve understanding of mathematical concepts and operations. In an experimental design 339 Grade 2–5 teachers were randomly assigned to a control or treatment group, to examine the effects of ST Math approach on teacher beliefs about mathematics teaching. For the treatment group the program involved a minimum of two 45-min sessions of the ST Math program per week, while the control group experienced the regular mathematics instruction in the same content. Data sources included questionnaires with items asking teachers to describe experiences from the implementation of ST Math in their classrooms, measuring TEBs and teaching outcome expectancy using the scale MTEBI, and their instructional practices as related to mathematics (integration of scientific reasoning in the classroom). After a first year implementation the results indicated that ST Math had a positive impact on student achievement in mathematics. Hierarchical linear modeling showed that students' time on ST Math and the integration of ST Math into daily instruction were positively associated with TEBs, outcome expectance and instructional practices.

## *Effects of Teacher Efficacy Beliefs*

In this section we focus on studies examining the relation of TEBs with and the impact on educational practice, including the provision of ideas for teacher educators to enhance teachers' capability to apply democratic education methods.

### Correlation of Teacher Efficacy Beliefs with Other Parameters

Empirical studies have shown that TEBs relate to students' performance and to teachers' behaviors (Betoret 2009; Tschannen-Moran and Woolfolk Hoy 2001; Tran et al. 2012). Bagaka's (2011) examined whether teacher characteristics and practices can enhance secondary school students' mathematics efficacy beliefs. He analyzed self-report data from 3173 secondary school students and their mathematics teachers (193). Using the principal component factor analysis he identified two dimensions of TEBs and practices: *interest and enjoyment of mathematics*, and *ability and competence in teaching mathematics* (p. 823) and five dimensions of students' mathematics self-efficacy: (a) students' lack of interest in and fear of mathematics; (b) students' competence in mathematics; (c) students' mathematics self-confidence and competence; (d) students' interest in, effort in, and perception of the importance of mathematics; and (e) mathematics anxiety. Teachers' frequent use of mathematics homework, their level of interest and enjoyment of mathematics, and their ability and competence in teaching mathematics were found to play a key role in promoting students' mathematics self-efficacy and in narrowing the gender gap in students' confidence and competence in mathematics. The problem with this study concerns alienation with theory. "Interest and enjoyment of mathematics" is considered as a dimension of TEBs, and "lack of interest", and "mathematics anxiety" as dimensions of students' mathematics efficacy. The concepts of interest and anxiety do not fall under the construct "efficacy".

Tella (2011) examined mathematics teachers' internet self-efficacy and its influence on mathematics instruction. She used self- report data from 90 math teachers and interviews with 15 heads of mathematics, seeking information from the respondents about their internet self-efficacy and on the ways the internet has influenced their teaching of mathematics. The analyses indicated that participants had high internet self-efficacy and that correlations existed between mathematics teachers' age, internet usage and internet self-efficacy. The author concludes that "internet self-efficacy and usage were revealed to improve the way teacher teach mathematics and conduct research" (p. 156). The findings justified increase of internet usage, on the part of mathematics teachers to enhance high internet self-efficacy.

Brown (2005) examined the relationship between early childhood teachers' efficacy beliefs, their beliefs about the importance of early childhood mathematics, and their mathematics instructional practices. She analyzed self- report data from 94 prekindergarten teachers using the TSES scale for TEBs, and the instrument by Kowalski et al. (2001) for beliefs about mathematics, and data from recorded observations of classroom practices of 20 of these teachers. Brown found that high efficacy participants rated the importance of mathematics higher on the belief scale than their lower efficacy colleagues; the findings also confirmed that in assessing their capabilities high efficacy participants rated themselves higher in instructional strategies, classroom management, and student engagement, though they did not rate their efficacy beliefs in doing mathematics as high as their TEBs, meaning that low efficacy in mathematics does not inhibit mathematics TEBs.

Bates et al. (2011) examined pre-service early-childhood teachers' mathematics self-efficacy and mathematics teaching efficacy in connection to their actual mathematical performance. They analyzed self-report data from 89 participants, using a scale measuring mathematics self-efficacy, the MTEBI for mathematics TEBs, and a test of basic mathematics skills. The results indicated that the participants' mathematics self-efficacy was positively correlated with their personal mathematics teaching efficacy, and that their mathematical performance was related to their mathematics self-efficacy and mathematics teaching efficacy. As it was expected, in regard to student outcomes, only those participants with high teaching efficacy were found to believe that they could have an effect on their students.

## Teacher Efficacy Beliefs in Reform Efforts

Educational reforms typically impose new demands on the already complex work of teaching, thus aggravating teachers' concerns (Charalambous and Philippou 2010). Instructional innovation typically requires the adoption of newer evidence-based instruction to replace more traditional teaching. Student-centered and constructivist teaching has been proposed as a means to enhance students' potential to be active, creative, and reflective self-directed learners in a changing world. In this respect, Nie et al. (2013) examined the connection between TEBs and constructivist instruction analyzing data collected from 2,139 primary teachers, using the TSES and a scale measuring constructivist instruction. Structural equation modeling showed that positive TEBs predicted constructivist instruction. Although TEBs also predicted didactic instruction, the strength of the prediction was different in terms of the effect size measures (variance explained 39 % for the constructivist against 4.6 % for the didactic instruction), indicating the connection of constructivist views and high TEBs.

In any attempt to introduce educational reforms teachers' concerns about the reform are a crucial variable, as concerns influence teachers' behavior and may obstruct the whole effort. Charalambous and Philippou (2010) examined the connection among elementary mathematics teachers' concerns about the introduction of a new curriculum reform, regarding problem solving, and their TEBs. Data sources included self-report data from 151 elementary teachers' regarding their concerns and efficacy beliefs, 5 years into the mandated curriculum, and qualitative data from 53 teacher logs. The study provided support for a model suggesting that teacher' concerns in preceding stages inform their concerns of succeeding stages, and that TEBs about using the reform affect their task and impact concerns and are in turn, informed by their self-concerns. TEBs about employing pre-reform instructional approaches were found to influence all types of teacher concerns. Data from 53 teacher logs provided additional support to this model.

Gabriele and Joram (2007) explored the sources of efficacy among first- and second-grade teachers by analyzing the events in a lesson they had just taught. Ten teachers participating in a reform-oriented mathematics teacher development project, voluntarily participated in a talk-aloud process regarding the criteria they used to evaluate their teaching effectiveness, after the lesson. The authors compared veterans and newcomers in reform-based mathematics teaching, in terms of their evaluations of the success of the lesson, with regard to events they attended to and used as evidence to support their evaluations of success. The analysis showed that veterans focused more on student thinking and described it in more precise and specific terms, while newcomers tended to talk more about their curricular goals for the lesson. Even in cases they did talk about students' thinking, they described it in less-precise terms. Newcomers, however, more frequently reported positive affective reactions when describing their progress toward achieving instructional goals and outcomes. The authors concluded that involvement in reform-based programs promotes TEBs.

## Teacher Efficacy Beliefs and Coping with Stress

Teachers' stressors spring from various sources including students' misbehavior and poor motivation for work, heavy workloads and time pressure, improper relationships with school administration, and pressure from parents (Chan 2008b). In confronting these stressors teachers may develop psychological symptoms of varying severity. The question is why some teachers are less vulnerable than other teachers in the face of similar work stress? What kind of relations between which personal variables interfere as coping resources in the context of the stress–illness or stress–distress relationship. Chan (2008b) assessed emotional intelligence – the competence in perceiving emotion, facilitating thought, understanding and monitoring emotions- and TEBs to represent personal resources, facilitating active and passive coping, in a sample of prospective and in-service teachers. Intrapersonal and interpersonal emotional intelligence were found to predict active coping strategies, but TEBs did not contribute independently to predicting of active coping, even though for male teachers there was some evidence of interaction with intrapersonal emotional intelligence in the prediction of active coping.

Betoret (2009) examined the relationship between school resources, potential stressors and coping resources, i.e. physical, psychological, social, or material factors which help teachers overcome job related stressors. The results indicated that external (school support resources) and internal (self-efficacy in classroom management and instruction) have a negative and significant effect on potential job stressors, mainly for secondary school teachers. Job stressors were found to have a positive and significant effect on teachers' emotional exhaustion, depersonalization, reduced job satisfaction, and burnout.

Teacher turnover appears to be a worldwide phenomenon with detrimental educational and economic effects. In the USA about one quarter of novice educators leave the profession within 3 years (Martin et al. 2012). Apart from the consequences for the teachers themselves, intent-to-leave can result in decreased school effectiveness. Identifying the variables that contribute to reducing this phenomenon is a prerequisite to creating effective teacher retention and job satisfaction programs. Martin et al. (2012) examined the TEBs about student engagement and instructional management, in relation to job satisfaction and student behavior stressors. Positing that the teacher's approach to instructional management sets the tone for the overall classroom atmosphere and ultimately student behavior stressors, they analyzed data from 631 elementary and middle school teachers using several scales – including the student engagement subscale of the TSES – and inventories measuring teacher burnout, job satisfaction, intent–to leave, and teacher stressor. The analyses fitted models showing a complex relationship between TEBs in student engagement and intent-to-leave (teacher turnover) mediated by variables related to the classroom context, such as TEBs in relation to instructional management, student behavior stressors, aspects of burnout, and job satisfaction.

Pre-service elementary teachers' mathematics anxiety was examined by Gresham (2008) in relationship to mathematics TEBs. Data sources included the MTEBI, the Mathematics Anxiety Rating Scale (Richardson and Suinn 1972) and interviews. Applying Pearson Product–moment Correlation (the two MTEBI subscales were analyzed both as separate subscales and combined) Gresham found a significant, negative correlation between mathematics anxiety and mathematics TEBs. The pre-service teachers with the lowest degree of mathematics anxiety had the highest levels of mathematics teacher efficacy. The interview data indicated that mathematics anxiety is associated with TEBs and with efficacy in mathematics, and that pre-service teachers' attitudes toward mathematics affect their mathematics TEBs.

## Teacher Efficacy Beliefs in the School Context

Instead of the social cognitive perspective, Takahashi (2011) considered a sociocultural perspective regarding the development of TEBs, on the argument that teachers' meaning-making in their communities of practice shape their efficacy beliefs (p. 733). Viewing the context as both constituting and constituted by individuals, rather than as something separate, the author focused on "communities of practice" as a framework where learning occurs in shared work activities. In such a framework teachers' practices are characterized by evidence-based decision-making, so the focus of this case study was to look for connections between teachers' evidence-based decision-making practices and their TEBs. In other words, Takahashi examined how teachers engaged in shared practices when co-constructing understandings of their teaching responsibilities, describe the relationship between their experiences of evidence-based decision- making and their individual and collective efficacy beliefs. He analyzed data from semi-structured interviews with four Junior High school teachers, who taught in one of two academic areas, English Language Arts and Mathematics, on three separate occasions, spanning a 3 month period: one interview not connected to any observations, one after a day of classroom observation, and one after observation of a teachers' evidence-based decision-making meeting. All interviewed teachers expressed high efficacy beliefs, as evident in their discussions about students who struggle academically. The analysis indicated that teachers co-constructed their TEBs in shared practices, suggesting the usefulness of communities of practice theory to more fully understand teachers' efficacy. The author concluded that through reflections on their evidence-based decision-making practices, the participants appeared to reify the processes of collectively examining data as tools of instructional improvement, and also about student learning as they reflected on their teaching. These constructions were fundamentally connected to an identification of teachers as responsible for student learning and contributed to the improvement of their TEBs.

*Coupling* "refers to an organizational and interpersonal structure that serves to link together selected elements of the environment" (Kurz and Knight 2004, p. 114). Research has suggested that schools are viewed as simultaneously loosely and tightly coupled organizations. This means that schools are coupled along a continuum, with different dimensions of the organization varying in their degree of coupling. One such dimension of organizational coupling that has been strongly linked to school effectiveness is goal consensus/vision, which refers to one coupling dimension linked to school effectiveness. Teachers in high consensus schools hold shared goals, beliefs, and values which emphasize teacher and student success (Creemers and Reezigt 1999). Kurz and Knight (2004) explored the relationship between high school teachers' TEBs and CTEBs, and among these two types of teachers' efficacy and their perceptions of goal consensus/vision. Data were collected from 113 teachers of several subjects including mathematics and science, during teacher in-service meetings, using the scales by Gibson and Dembo (1984) and by Goddard (2002). They found that CTEBs were correlated with all of the other variables examined, and most highly correlated with goal consensus/vision. Individual TEBs were found to be related to CTEBs, but not to goal consensus/vision. These authors realized (p. 125) that this result might be due to shortcomings of the Gibson and Dembo scale; the TSES could have yield different outcomes.

There are multiple ways of considering how learning occurs within any organization, how an organization moves toward change, and the role of collective reflective practice in this movement (Kennedy and Smith 2013). The culture of a school community improves through advancement of individual efficacy and jointly facing difficulties finding collective solutions to challenges. Research has found links between collective efficacy of a school community and students' achievement (Goddard et al. 2000). Kennedy and Smith (2013) studied the role of collective reflective practices that affect sources of teacher efficacy as they occur within the school community. They sought to identify the behaviors within a school organization that affected change by assessing impact on physiological sources of teacher efficacy and specifically to determine if there was a relationship between the reflective practice behavior within a Professional Learning Community model and either internal or external sources of teacher efficacy. Analyzing survey data from 661 teachers from 42 elementary and secondary school as regards the impact of school level organizational behaviors and practices on the individual teacher efficacy, they found a relationship between collaborative organizational culture and physiological efficacy sources. Furthermore, the authors identified efficacy sources that have a positive or negative relationship with organizational behaviors supporting the professional learning community behaviors. Specifically, more collective reflective practice was associated with external input such as administrative observations, student outcome data, and colleague observation, while high involvement in school leadership and vision related to uncomfortable feelings with making comparisons to other teachers or engaging parents.

## Conclusions

At the opening of their paper Klassen et al. (2011) questioned whether the field has made progress, or whether early promises remained unfulfilled. In concluding, they referred to some progress in terms of methodological diversity, domain specificity, and a focus on collective efficacy. On the pessimistic note they mentioned insufficient attention paid to sources of teacher efficacy, a dearth of research showing links between teacher efficacy and student outcomes, and lack of conceptual clarity in measuring the construct. Research published after 2009 has resulted in progress but has not overcome the above problems.

In the next paragraphs we summarize the current state of research as regards the foundations and measurement of the construct, the sources and development, and the relevance of TEBs with educational practice. We also provide ideas for further research.

## *The Construct and It Measurement*

In a theory focused review Wyatt (2014) highlighted the continuing misalignment between theory and method with reference to 'conceptually suspect' studies (Klassen et al. 2011). The classical definition of TEBs refers to capability to undertake a specific task, irrespective of the accuracy of reported ability. With emphasis on specificity, the definition might read as "beliefs in capabilities to perform a specific task at a specified level of quality in a specified situation", that is classified as an agent-means definition (Wyatt 2014, p. 166). Considering teachers' interest about the outcome of their effort, any proper definition should retain specificity and include both the agent-means and the agent-end capacity, incorporating both efficacy expectancy and outcome expectancy. Furthermore, it can be argued that Wheatly's (2005) plea for the potential value of efficacy doubt, involves a broader understanding of PTE beliefs, in connection to other beliefs worth exploring in their own right. Doubt and reflection are central to an understanding of how beliefs change. However, any attempt to incorporate all these meanings would result in an omnibus definition making the operational use of the construct too difficult to manage.

As regards efficacy scales, Bandura (2006b, p. 308) advised that efficacy items should accurately reflect the construct which concerns perceived capability. Recent research shows that the demand for clarity in measurement has to some extend being resolved. Scholars have practically recognized that TEBs refer to a teacher or a group of teachers' sense of capability. The TSES scale (Tschannen-Moran and Woolfolk Hoy 2001) has become popular, as an instrument that can be used for teachers in all subjects in all contexts, though it does not embrace all aspects of a wide definition of the construct. Most recent studies on mathematics TEBs have used either an adapted version of the TSES, validated in the specific context, or the MTEBI, which measures mathematics TEBs and caters for both efficacy expectancy and outcome expectancy. The TSES does not provide for outcome expectancy and is limited to three aspects of teachers' work. Important as they are, these three aspects do not cover all aspects of teachers work. Bandura (2006b) refers to six domains of teachers' work: instruction, discipline, influence decision making, enlist parental involvement, enlists community involvement, and creation of positive school climate. The scale by Chan (2008a) measures TEBs in six domains – some of them different from Bandura's. There is a need for further research in this direction.

The importance of CTEBs has been well recognized; yet we know little about how they are formed in school settings and how they are affected by the context, while the question of measuring this concept is still open. The scale by Goddard (2002), designed to assess a faculty "perceptions of group competence and the level of difficulty inherent in the educational task faced by the school" (p. 97), has seldom being used in empirical research. Furthermore, uncertainty prevails as to whether the hypothesized sources of personal TEBs hold true at the group level (Klassen et al. 2011).

The question of specificity in measuring TEBs versus generalizability remains a challenge, while qualitative research is needed to refine patterns of efficacy sources as teachers are observed and interviewed. At the same time further research on teachers' organizational behavior is needed to increase understanding of reflective practice and its relation to teachers' collective efficacy.

## *The Formation and Development of Teacher Efficacy*

Progress with regard to efficacy sources can be acknowledged. Bandura's sources have been examined both quantitatively (Chang 2009) and qualitatively (Usher and Pajares 2009), and efforts to build a stronger measure of the sources of self-efficacy have been reported (Klassen et al. 2011). Some social and contextual sources of TEBs, including the school climate and characteristics of the group of students, have also been examined (Tschannen-Moran and Woolfolk Hoy 2007). More research into the four sources' relative weight in the formation of mathematics TEBs as well as on other external sources will be of interest to teachers' educators.

Research studies (Evans 2011; Charalambous et al. 2009) seem to indicate that one or two courses of pre-service or in-service courses cannot impact mathematics TEBs. On the other hand the effect of field work on mathematics TEBs was found to be quite significant (Charalambous et al. 2008); a relationship between teachers involvement in communities of practice with TEBs (Bruce and Ross 2008) was found, and a positive effect of a computer based teaching tool on TEBs was also found (Tran et al. 2012). Further longitudinal studies focusing on pre-service teachers' mathematics TEBs throughout their educational program would help teacher educators to better understand the developmental process and find means to enhance TEBs. Longitudinal designs could also focus on periods of instability and stability of mathematics TEBs in various stages of teachers' career.

Comparative studies can also illuminate the role of the context in the development of mathematics TEBs and investigate the impact of the official educational policy (e.g. the role of administrators and inspectors) on teachers' TEBs. Examining similarities and differences among countries related to contextual factors, like school leadership, resources available, and collaboration with the colleagues may enhance understanding of the sources of TEBs. An interesting avenue of research will also be the comparison between TEBs in relation to mathematics of primary and secondary school teachers. This will possibly lead to a discussion about the impact of mathematics content knowledge in the formation of TEBs.

The effect of TEBs on teaching practices and student outcomes has been well documented (Betoret 2009; Brown 2005; Nie et al. 2013; Tschannen-Moran and Woolfolk Hoy 2001) and the significance of TEBs in coping capabilities in stressful situations has also been demonstrated (Chan 2008b; Gresham 2008). Considering the complexity of the construct, the domain will benefit from further qualitative and mixed method studies including observations, involving teachers with the same level of efficacy beliefs in teaching mathematics but different behavior in the classroom.

Teacher educators draw attention to the role of TEBs in pursuing reform-oriented goals, such as critical thinking and reflective self-directed learning and hence the need for teachers to teach in new ways. The study by Charalambous and Philippou (2010), indicating a relationship between TEBs and teachers' concerns about the implementation of an innovation in mathematics curricula could be extended to include several aspects of reformed curricula.

### Some Final Remarks

The role of efficacy beliefs in human behavior and the benefits of high TEBs seem to be taken for granted. Despite the fact that some important ideas for teacher educators have been proposed, the possibility that high efficacy may have negative effects (Wheatley 2005) remains unattended. The voices for a wider and deeper conceptualization of the construct, in clear operative terms, incorporating more components, while retaining specificity and covering all domains of teachers' work add new complexities, that increase the difficulty of distinguishing conditions under which "high efficacy may back fire" (Wyatt 2014). The effects of high teacher efficacy on practice and specifically on teachers' persistence and on the level of their openness to new ideas and change, depends on their personal or group characteristics and on class conditions. It would be of interest to see more interpretive research focusing on the effect of these variables on TEBs and its consequences in relation to efficacy doubt and the need for change. In this direction further research could examine teachers' perceptions of their readiness to change views and approaches and their relationship with TEBs.

TEBs concern capability in a complex activity that involves facilitating access to knowledge, helping learners to develop analytical tools that help them learn on their own, providing a classroom environment conducive to learning, and encouraging the social interactions that support learning goals Wyatt (2014). Therefore, irrespective of the construct's degree of maturation and despite of accumulated findings, the field of teacher efficacy continues to hide a treasure waiting to be unearthed. This treasure would be of major importance in teaching mathematics, a subject in which students face difficulties and develop phobias, while teachers experience stressful situations.

## References

Bagaka's, J. G. (2011). The role of teacher characteristics and practices on upper secondary school students' mathematics self-efficacy in Nyanza province of Kenya: A multilevel analysis. *International Journal of Science and Mathematics Education, 9*, 817–842.

Bandura, A. (1997). *Self-efficacy: The exercise of control*. New York: W. H. Freeman.

Bandura, A. (2006a). Adolescent development from an agentic perspective. In F. Pajares & T. Urdan (Eds.), *Self-efficacy beliefs of adolescents* (Vol. 5, pp. 1–43). Greenwich: Information Age Publishing.

Bandura, A. (2006b). Guide for constructing self-efficacy scales. In F. Pajares & T. Urdan (Eds.), *Self-efficacy beliefs of adolescents* (Vol. 5, pp. 307–337). Greenwich: Information Age Publishing.

Bandura, A., Barbaranelli, C., Caprara, G. V., & Pastorelli, C. (1996). Multifaceted impact of self-efficacy beliefs on academic functioning. *Child Development, 67*, 1206–1222.

Bates, A. B., Latham, A., & Jin-ah Kim, J. (2011). Linking preservice teachers' mathematics self-efficacy and mathematics teaching efficacy to their mathematical performance. *School Science and Mathematics, 111*(7), 325–333.

Betoret, F. D. (2009). Self-efficacy, school resources, job stressors and burnout among Spanish primary and secondary school teachers: A structural equation approach. *Educational Psychology, 29*(1), 45–68.

Brand, B. R., & Wilkins, J. L. M. (2007). Using self-efficacy as a construct for evaluating science and mathematics methods courses. *Journal of Science Teacher Education, 18*, 297–317.

Brown, E. T. (2005). The influence of teachers' efficacy and beliefs regarding mathematics instruction in the early childhood classroom. *Journal of Early Childhood Teacher Education, 26*, 239–257.

Bruce, C. D., & Ross, J. A. (2008). A model for increasing reform implementation and teacher efficacy: Teacher peer coaching in grades 3 and 6 mathematics. *Canadian Journal of Education, 31*(2), 346–370.

Chan, D. W. (2008a). Dimensions of teacher self-efficacy among Chinese secondary school teachers in Hong Kong. *Educational Psychology, 28*(2), 181–194.

Chan, D. W. (2008b). Emotional intelligence, self-efficacy, and coping among Chinese prospective and in-service teachers in Hong Kong. *Educational Psychology, 28*(4), 397–408.

Chang, Y.-L. (2009). A case study of elementary beginning mathematics teachers' efficacy development. *International Journal of Science and Mathematics Education, 8*, 271–297.

Charalambous, C. Y., & Philippou, G. N. (2010). Teachers' concerns and efficacy beliefs about implementing a mathematics curriculum reform: Integrating two lines of inquiry. *Educational Studies in Mathematics, 73*, 1–21.

Charalambous, C. Y., Philippou, G. N., & Kyriakides, L. (2008). Tracing the development of pre-service teachers' efficacy beliefs in teaching mathematics during fieldwork. *Educational Studies in Mathematics, 67*, 125–142.

Charalambous, C. Y., Panaoura, A., & Philippou, G. N. (2009). Using history of mathematics to induce changes in pre-service teachers' beliefs and attitudes: Insights from evaluating a teacher education program. *Educational Studies in Mathematics, 71*, 161–180.

Creemers, B., & Reezigt, G. J. (1999). The concept of vision in educational effectiveness, theory and research. *Leaning Environments Research, 2*, 107–135.

Duffin, L. C., French, B. F., & Patrick, H. (2012). The teachers' sense of efficacy scale: Confirming the factor structure with beginning pre-service teachers. *Teaching and Teacher Education, 28*, 827–834.

Enochs, L. G., Smith, P. L., & Huinker, D. (2000). Establishing factorial validity of the mathematics teaching efficacy beliefs instrument. *School Science and Mathematics, 100*(4), 194–202.

Evans, B. R. (2011). Content knowledge, attitudes, and self-efficacy in the mathematics New York City teaching fellows (NYCTF) program. *School Science and Mathematics, 111*(5), 225–235.

Gabriele, A. J., & Joram, E. (2007). Teachers' reflections on their reform-based teaching in mathematics: Implications for the development of teacher self-efficacy. *Action in Teacher Education, 29*, 60–74.

Gibson, S., & Dembo, M. (1984). Teacher efficacy: Construct validation. *Journal of Educational Psychology, 76*, 569–582.

Goddard, R. (2002). A theoretical and empirical analysis of the measurement of collective efficacy: The development of a short form. *Educational and Psychological Measurement, 62*(1), 97–110.

Goddard, R. D., Hoy, W. K., & Woolfolk Hoy, A. (2000). Collective teacher efficacy: Its meaning, measure, and effect on student achievement. *American Education Research Journal, 37*(2), 479–507.

Goddard, R. G., Hoy, W. K., & Woolfolk Hoy, A. (2004). Collective efficacy beliefs: Theoretical developments, empirical evidence, and future directions. *Educational Researcher, 33*(3), 3–13.

Gresham, G. (2008). Mathematics anxiety and mathematics teacher efficacy in elementary pre-service teachers. *Teaching Education, 19*(3), 171–184.

Henson, R. (2002). From adolescent angst to adulthood: Substantive implications and measurement dilemmas in the development of teacher efficacy research. *Educational Psychologist, 37*, 137–150.
Kennedy, S. Y., & Smith, J. B. (2013). The relationship between school collective reflective practice and teacher physiological efficacy sources. *Teaching and Teacher Education, 29*, 132–143.
Klassen, R. M., Bong, M., Usher, E. L., et al. (2009). Exploring the validity of a teachers' self-efficacy scale in five countries. *Contemporary Educational Psychology, 34*, 67–76.
Klassen, R. M., Tze, V. M. C., Betts, S. M., & Gordon, K. A. (2011). Teacher efficacy research 1998–2009: Signs of progress or unfulfilled promise? *Review of Educational Research, 23*, 21–43.
Kowalski, K., Pretti-Frontczak, K., & Johnson, L. (2001). Preschool teachers' beliefs concerning the importance of various developmental skills and abilities. *Journal of Research in Childhood Education, 16*, 5–14.
Kurz, B. T., & Knight, S. L. (2004). An exploration of the relationship among teacher efficacy, collective teacher efficacy, and goal consensus. *Learning Environments Research, 7*, 111–128.
Labone, E. (2004). Teacher efficacy: Maturing the construct through research in alternative paradigms. *Teaching and Teacher Education, 20*, 341–359.
Martin, N. K., Sass, D. A., & Schmitt, T. A. (2012). Teacher efficacy in student engagement, instructional management, student stressors, and burnout: A theoretical model using in-class variables to predict teachers' intent-to-leave. *Teaching and Teacher Education, 28*, 546–559.
National Council of Teachers of Mathematics. (2000). *Principles and standards for school mathematics*. Reston: The National Council of Teachers of Mathematics.
Nie, Y., Tan, G. H., Liau, A. K., Shun Lau, S., & Chua, B. L. (2013). The roles of teacher efficacy in instructional innovation: Its predictive relations to constructivist and didactic instruction. *Educational Research for Policy and Practice, 12*, 67–77.
Pajares, F., & Miller, M. D. (1994). Role of self-efficacy and self-concept beliefs in mathematical problem solving: A path analysis. *Journal of Educational Psychology, 86*, 193–203.
Phelps, C. M. (2010). Factors that pre-service elementary teachers perceive as affecting their motivational profiles in mathematics. *Educational Studies in Mathematics, 75*, 293–309.
Philipp, R. A. (2007). Mathematics teachers' beliefs and affect. In F. K. Lester (Ed.), *Second handbook of research on mathematics teaching and learning* (pp. 257–315). Charlotte: Information Age Publishing.
Philippou, G. N., Charalambous, C. Y., & Christou, C. (2001). Efficacy in problem posing and teaching problem posing. In M. Heuvel-Panhuizen (Ed.), *Proceedings of the 25th annual PME conference* (Vol. 4, pp. 41–48). Utrecht: Utrecht University.
Richardson, F. C., & Suinn, R. M. (1972). The mathematics anxiety rating scale: Psychometric data. *Journal of Counseling Psychology, 19*, 39–47.
Takahashi, S. (2011). Co-constructing efficacy: A "communities of practice" perspective on teachers' efficacy beliefs. *Teaching and Teacher Education, 27*, 732–741.
Tella, A. (2011). An assessment of mathematics teachers' internet self-efficacy: Implications on teachers' delivery of mathematics instruction. *International Journal of Mathematical Education in Science and Technology, 42*(2), 155–174.
Tran, N. A., Schneider, S., Duran, L., et al. (2012). The effects of mathematics instruction using spatial temporal cognition on teacher efficacy and instructional practices. *Computers in Human Behavior, 28*, 340–349.
Tschannen-Moran, M., & Woolfolk Hoy, A. (2001). Teacher efficacy: Capturing an elusive construct. *Teaching and Teacher Education, 17*, 783–805.
Tschannen-Moran, M., & Woolfolk Hoy, A. (2007). The differential antecedents of self-efficacy beliefs of novice and experienced teachers. *Teaching and Teacher Education, 23*, 944–956.
Tschannen-Moran, M., Woolfolk Hoy, A., & Hoy, W. K. (1998). Teacher efficacy: Its meaning and measure. *Review of Educational Research, 68*(2), 202–248.
Tuchman, E., & Isaacs, J. (2011). The influence of formal and informal formative pre-service experiences on teacher self-efficacy. *Educational Psychology, 31*(4), 413–433.
Usher, E. L., & Pajares, F. (2006). Sources of academic and self-regulatory efficacy beliefs of entering middle school students. *Contemporary Educational Psychology, 31*, 125–141.

Usher, E. L., & Pajares, F. (2009). Sources of self-efficacy in mathematics: A validation study. *Contemporary Educational Psychology, 34*, 89–101.

Wheatley, K. F. (2002). The potential benefits of teacher efficacy doubts for educational reform. *Teaching and Teacher Education, 18*, 5–22.

Wheatley, K. F. (2005). The case for re-conceptualizing teacher efficacy research. *Teaching and Teacher Education, 18*, 747–766.

Wyatt, (2014). Towards a re-conceptualization of teachers' self-efficacy beliefs: tackling enduring problems with the quantitative research and moving on. *International Journal of Research & Method in Education, 37*(2), 166–189.

# Reaction to Section 1: Faith, Hope and Charity: Theoretical Lenses on Affect

**David Clarke**

**Abstract** Within the landscape evoked by the term "affect" are an ecosystem of entities that alternately function as objects and as connections; constituted and constituting. Historically, to invoke affect is to simultaneously invoke cognition in the sense that reference to either one of a dichotomous pair simultaneously calls the other into being. Yet the authors of these chapters contest the simplistic dichotomisation of affect and cognition, and consistently argue for the fundamental, complex and intimate connection of the various facets of affect: belief and emotions, for instance, with aspects of cognition, such as learning and meaning. An additional consistency across the chapters is the commitment to locating affect in social practice, rather than locking it from sight within the individual. Indeed, the argument for the inextricability of the individual and the social seems relatively easy to make in relation to affect, where the social performance of affect is so visibly consequent upon personal history modulo the cultural considerations that frame and shape the social expression of emotion and belief.

**Keyword** Affect

The chapters in this section offer a range of perspectives on theories connecting affect to the learning of mathematics. It has been argued elsewhere (Clarke 2011) that multi-theoretic analyses of particular social situations, held as common referents for the purpose of the analyses, lead not only to more complex, complete and connected insights into those situations, but, through juxtaposition of the accounts generated by the various theories, afford the interrogation of the theories and their respective capacities to accommodate and to explain phenomena in the referent situations. In this collection of chapters, "affect" and its conceptual correlates and/or constituents (depending on your theory) become the (more or less) shared referent/s for each chapter's discussion. This supports the examination of how each theory

D. Clarke (✉)
International Centre for Classroom Research, Melbourne Graduate School of Education
University of Melbourne, Melbourne, Australia
e-mail: d.clarke@unimelb.edu.au

 
B. Pepin, B. Roesken-Winter (eds.), *From beliefs to dynamic affect systems in mathematics education*, Advances in Mathematics Education,
DOI 10.1007/978-3-319-06808-4_6

partitions and portrays the affective landscape and connects the dominant features, so identified, with posited objects, events, and relationships in the coincident universe of cognition. The chapter by Rolka and Roesken-Winter represents an example of the value of exploiting multiple theoretical perspectives in interrogating how particular constructs are fore-grounded, elaborated, or even discarded through the adoption of a particular theoretical perspective (see also Clarke et al. 2012).

In seeking an organisational referent for this discussion, after a careful reading of the chapters, I found value in the triadic conjunction of virtues: faith, hope and charity. These seem to me to combine the necessary elements of enduring and evolving belief, aspiring motive, and the acceptance of both obligation and debt to the other. The same combination of terms simultaneously invokes a sense of location and situation (culturally, historically and socially): an inevitable affiliation with a community and the accompanying rights and responsibilities of that affiliation. With this symbiosis in mind, the following comments represent my reflections on the various chapters. I have labelled each reflection by the dominant construct from the associated chapter.

## Belief

Skott (Chapter "Towards a Participatory Approach to 'Beliefs' in Mathematics Education") characterises traditional "belief research" as predicated on a model in which "students and teachers are expected to come to hold or possess reified mental entities, beliefs, through processes of assimilation and accommodation as they engage in social interaction." Practice is then the enactment of these reifications, modulo certain contextual constraints. This conception leaves the central construct of "belief/s" ill-defined and uncertain with regard to empirical demonstration. Skott's solution is to reconceive beliefs as "shifting modes of participation in different social practices," where the constituent actions of those practices are recognised to be "affectively laden." Such an orientation of beliefs research towards a performative or participatory conception of affect, mirrors similar developments in theories of cognition that seek to extend the conceptualisation of cognition as residing "outside the brain of the learner" (Hutchins 1991) or as enacted through institutionalised forms of practice (Lave and Wenger 1991).

Part of Skott's motivation in the development of a more action-oriented conception of affect is the persistent expectation that research should resolve an anticipated causal connection between beliefs and practice. Given this goal, one is allowed to ask whether the strategy of seeing beliefs as performatively realised in practice actually resolves the "beliefs-practice quandary" or simply removes the necessity to identify a relationship, by transforming connection into identity. Skott's proposal that some of the conceptual and methodological problems of belief research can be addressed by adoption a "Patterns of Participation" framework echoes similar arguments that have been made with respect to research into cognition.

In fact, much of Skott's argument mirrors the parallel situation in cognition research, and the parallel becomes vividly evident if one substitutes "understanding" for "belief" at any point in Skott's argument. The posited correspondence between beliefs and practice (what Skott calls the "congruity thesis") and the performative expression of belief in practice might be usefully illuminated by taking the parallel connection between understanding and practice and posing the question, "If practice is the enactment of both understanding and belief, what inferences about either can be construed from the performances encountered in classroom settings?" By reconstructing the correspondence as a form of identity: such that practices constitute enacted beliefs and understandings, then two problems of interpretation appear to be circumvented. Within the domain of belief (although understanding would serve just as well), we can ask the questions: "What forms of practice might be taken by the enactment of specific beliefs? and its inversion, "What particular beliefs might be signified by which specific forms of practice?" As Skott argues, belief research is largely a consequence of the connection presumed in both of these questions. Reflecting within and about this conception of beliefs, Skott points to the possible filtering role by which beliefs shape an individual's interpretation of experience.

Skott problematizes the claim that "beliefs explain behaviour" by suggesting that the enactment of beliefs is contingent on the context in which that enactment occurs and so neither can the enactment be taken as directly indicative of held belief, nor can the holding of any particular belief be taken as definitively predictive of consequent action.

Instead of relying on a connection of such questionable causality, Skott cites *engagement structures* (as proposed by Goldin et al. 2011) as indicative of attempts to acknowledge contingencies on the performative realisation of beliefs in practice within the traditional paradigm of belief research.

One of the several useful achievements of Skott's paper is the organisation of perspectives on the role of teachers' beliefs for practice into the perspective categories: *enactment*, *activation*, *situatedness* and *emergence*. Without simply restating Skott's description of these perspectives, the classification serves to underline the "traditional" conception of beliefs as existing independent of practice as a separate, relatively stable, personal attribute requiring enactment for social expression, but reflexively connected to practice, albeit mediated through filters contingent on context. Claims as to the situatedness of beliefs ascribe a constitutive role to context (rather than just a qualifying or moderating role) and posits the possibility that contradictory beliefs might be enacted in differing contexts. The emergent perspective sees beliefs and social norms [of practice] as reflexively related, being a joint construction by teachers and students (drawing on Cobb and his co-workers; eg. Cobb and Yackel 1996). Possibly Skott's most useful contribution is the location of the alternatives enactment, activation, situatedness and emergence within the 2×2 grid, where the dimensions are "stability across contexts" and "impact of beliefs on practice." This useful table affords the interrogation of the assumptions implicit in each perspective.

The parallel between Skott's critique of belief research and the established paradigm/s of cognitive research are visible again in the use Skott makes of Sfard's contrasting of "acquisition" and "participation" as offering alternative metaphors for learning and knowing (Sfard 2008). The reification of beliefs "acquired" from experience is usefully contrasted with a "patterns of participation" approach, which sees both beliefs and practice as simultaneously and reflexively emergent and negotiated, rather than either being fixed and causally connected to the other.

## Emotion/Motivation

I want to open discussion of Radford's chapter with his assertion that "it is misleading to believe that emotions obstruct thinking." He further asserts "Emotions and thinking are not separate entities" and suggests that "thinking and … concomitant emotions unfold in activity-bound processes of subjectification." This unification is prefaced in the chapter's introduction by some rhetorical questions that, by way of contrast, appear to be calling for recognition of distinctions between affect, feeling, emotions, motives and motivation. The immediately subsequent discussion outlines the historical bases for such distinctions. This introductory framing establishes a commitment to connection between the affective, the cognitive and the social that is pursued in all five chapters, while simultaneously seeking clarification of the distinctions between those constructs of which the affective domain is constituted. Both demands are fundamentally structural and it is the lack of structure that poses the greatest challenge for research into the relationship between affect and learning.

Importantly, the personal basis of motives and emotions is problematized and the recognition of the socio-historico-cultural bases of emotions is portrayed as a movement away from oppositional conceptions of the social and the individual towards the contemporary recognition of their mutually constitutive nature. The depiction of theoretical progression is achieved very effectively through the focus on the historico-cultural development of notions of love. The discussion of the evolving connection of romantic love to "the emergence of the economic individualism" provides useful support to Radford's general thesis regarding the importance (and relevance) of a cultural-historical perspective for the effective theorising and investigation of emotion in the context of mathematical problem solving and learning.

Radford primarily discusses "emotions" interspersed with repeated reference to "motive." These two constructs serve Radford particularly well in pursuing his thesis of the fundamental necessity to see affect in cultural-historical terms. Emotional response to a situation can be plausibly presented as the culmination of personal histories and evolving participation in cultural practice. The argument that emotions are historically (and culturally) constituted provides a useful and enlightening entry point embodied in a well-chosen illustration involving a group of three students engaged in a common mathematical exploration. Radford's early assertion that emotions entail a moral and ethical dimension is well-illustrated by the discrepancies between the Thom and Jay's conception of legitimate group problem solving behaviour and Laura's more self-critical conception.

Propositions that "Emotions are feelings that accompany learning" and "Emotions are embodied thoughts" represent a fundamental dilemma in the conceptualisation of the role of emotion in cognition. Radford's resolution of this dilemma takes the form of asserting that "Our thinking is *necessarily* embodied *and* emotional" (original emphasis). This embodiment includes the attribution of motive to the activity and the possibility of non-coincidence of the motives of the individual/s and the activity. Leont'ev's identification of motive and object as the two main vectors of activity is extended by Radford to accommodate a fundamental dynamism in their social evolution within the performance of any activity, realised through "the double-faced nature of motives, which are always personal and cultural."

In presenting, at some length, the example of three Grade 4 students' problem solving attempts, Radford illustrates the methodological burden we accept in ascribing motives, thoughts, feelings, emotions to individuals based on our observations of their social behaviour. The issue here is not only methodological. The participants in the depicted setting must make the same ascriptions and their ascriptions are even more central to the situation than are the researcher's. The confidence with which another's emotions can be "read" is a problematic central to our participation in social interactions. These readings of emotion require more careful discussion than is typically provided.

Research reports of social interactions seldom include any acknowledgement of the problematic nature of participant individuals' reading of each other's emotional state, and even less of the researcher's warrant for the attribution of emotion. Intersubjectivity has been variously and widely discussed the domain of cognition (e.g. Clarke 2001; Lerman 1996) in relation to "taken as shared" meanings. The social process whereby individuals attune themselves to the emotional state/s of others should not be taken for granted in affective research. If the intention is that the research community draw upon the characterisation of the social negotiation of meaning in its investigation of affect, then this needs be made explicit and both the warrant for that presumption and the consequent warrant for any interpretive claims be provided. It should be noted that Radford argues that, "the contextual and dynamic nature of emotions cannot be limited to the analysis of their contextual occurrences" and asserts that "emotions *are* dynamic processes" (emphasis mine), which, while being socially organised and enacted are both historically and culturally constituted. The argument for consideration of the historical and cultural origins of emotions is compelling. It should also be acknowledged that the performance of emotional response "in the moment" cannot be treated as unambiguous. Not only are the historical and cultural precedents relevant for the researcher's understanding of the enacted emotion, such precedents also frame and shape the interpretations and reactions of those in the social situation under consideration and these interpretations and reaction will, themselves, be various and reflect the participants' personal histories and cultural heritage.

Radford's contention that "contexts are not "backgrounds" but rather constitutive elements of the human psyche" is an important point, but falls short of invoking the full reflexivity posited by Clarke and Helme (1998), who proposed that context is most usefully seen as a construction of the participating individual/s. The essential point, however, is that Radford's vision is an integrative one, in which, with Menon (2000), he asserts that culture and psyche cannot easily be disentangled one from the other. Shakespeare had a related thought when he wrote, "there is nothing either good or bad, but thinking makes it so" (Hamlet, Act 2, Scene 2). Our participation in any social interaction is constituted by the context as we have come to construct it and by the historical and cultural inclinations that find their expression both in our emotive acts and in our interpretations of the acts of others.

## Attitude

Di Martino and Zan take "attitude" as their central construct and the focus of their chapter. In their useful introduction to the historical evolution of the construct of attitude, they posit attitude as one of a suite of constructs characterising research into affect, and accept the obligation to systematize the research into affect by clarifying both the definitions currently employed and the instruments by which attitude is measured. Of particular interest is their account of the evolution in the reasons for undertaking research into attitude. Motivations such as its role as gatekeeper to further study and/or the possible connection between achievement and attitude were studied historically in an under-theorised fashion. In their account, attitude subsequently became a central construct in studies seeking to explain documented gender differences in mathematics achievement, but research continued to be hampered by lack of theoretical clarity and sophistication.

With improved theoretical sophistication, new distinctions emerged between attitudes to mathematics as a branch of scientific knowledge and mathematics as a school subject, together with the recognition that the referents of attitude were many: "mathematical content, characteristics of mathematics, kind of teaching, mathematical activities in the classroom and mathematics teacher (Kulm 1980)." Most importantly, Di Martino and Zan suggest that "a large portion of studies show the lack of a clear definition of the construct: attitude tends to be defined implicitly and a posteriori through the instruments used to measure it." This *definition through description*, if left unattended, would disadvantage the field immensely. Initial attempts defined attitude as "a general emotional disposition" (Haladyna et al. 1983), or as an amalgam of an emotional disposition towards mathematics, a set of beliefs regarding mathematics, and behaviours related to mathematics. Both attempts had their limitations. In particular, while such simple models seemed adequate for questions related to subject choice, they were unable to address success in mathematics. This left attitude to mathematics as a goal in itself (lacking an empirical or theoretical connection between attitude and achievement), and posed a serious didactical threat, if teachers replaced achievement goals with attitudinal goals.

Di Martino and Zan noted that a central element in the tripartite model in which attitude had a cognitive, an affective, and a behavioural component was that the assumption of a link between attitude and behaviour became part of the construct definition itself. The circularity of this model prompted a further definition of attitude, as "pattern of beliefs and emotions associated with mathematics" (Daskalogianni and Simpson 2000), which excluded behaviours from the definition. Nonetheless, the study of attitude remained tied to simplistic instruments that conflated behavioural aspects with beliefs and emotions and persisted in dichotomising attitude into positive and negative.

Consideration of the instruments used to measure (or portray) attitude, the vast majority depend on self-report mechanisms, ultimately reproducing the positive/negative bipolarity that seems to underlie almost all attitude research. Further difficulties include the questionable legitimacy of measuring each component separately (cognitive, affective and behavioural), particularly where any such measurements (e.g. via a questionnaire) are insensitive to the difference between espoused beliefs and beliefs-in-action. Further concerns occur in relation to the determination of the response frame by the researcher, where the highlighted attitudes may not be personally central to the respondent's value system and some attitudes may not even exist prior the completion of the questionnaire or interview. An additional concern, noted above, is the classification of some responses as reflective of positive attitudes and some of negative. This distinction cannot always be made with confidence and Di Martino and Zan provide useful detail on the measurement implications of this concern.

The movement in the 1990s towards an interest in understanding, rather than just describing, phenomena was usefully accompanied by more sophisticated methods of data generation, including clinical interviews, field observations and participant observations. Adoption of an interpretivist paradigm allowed the field to move beyond its narrow focus on causality. Interest in attitude as providing explanations for individuals' decision-making led to new approaches addressing the contribution of attitudes to mathematics education.

A particularly useful statement, quoted by Di Martino and Zan, was "[attitude] is not a quality of an individual but rather a construct of an observer's desire to formulate a story to account for observation" (Ruffel et al. 1998). If this statement is accepted, then theories of affect are not theories about the individual but theories about the interpretation of particular types of social performance. Much of the chapter by Di Martino and Zan concerns itself with questions of research methodology conducted in a virtual theoretical vacuum. The question of whether attitude exists only as a researcher construct takes the discussion back to the realm of theory, albeit grounded in the very practical consideration of what is actually being "measured" or "described" or "probed" (to quote Ruffel et al. 1998). A consequence of the growing status of interpretivist approaches was the acceptance of multiple definitions of the same construct depending on the purpose and situation of the observer and the observed and the rejection of any notion of a singular consensus definition, universally applicable across all situations. The culmination of this progression from measurement to the elicitation of individuals' mathematical "stories" is described by Di Martino and Zan as a "revolution." If this methodological innovation translates into a reconceptualization of the goal/s of attitude research, then "revolution" might well be an appropriate term.

Di Martino and Zan report the development of the "Three-dimensional Model for Attitude (TMA)", which consists of the three dimensions:

- emotional disposition towards mathematics,
- vision of mathematics,
- perceived competence in mathematics.

The TMA framework was used to analyse students' autobiographical essays, and is described as providing a bridge from beliefs to emotions. The emphasis is on the construction of attitudinal profiles of students.

In their concluding remarks, Di Martino and Zan assert that "Attitude is no longer seen as an individual's trait, useful for predicting his/her behaviour, but as an observer's construct, capable of suggesting an understanding of the individual's intentional actions in a complex context, as is the learning of mathematics: a multi-dimensional construct that involves beliefs and emotions and acts as a bridge between them (Di Martino and Zan 2011)." This is a novel position in the context of this book and provides part of the rationale for the authors' dissatisfaction with research into attitude, which they characterise as lacking a cumulative character. To be cumulative, research on affect must first establish theoretical coherence and consistency in the use of the constituent constructs that populate the pages of publications such as this. Such coherence requires not consensus but clarity.

## Multiple Theories Involving Belief/s

The metaphor of "networking theories" is in increasing use. It has several benefits, one of which is the clear negation of any aspiration to construct grand theory through some form of theoretical synthesis or integration. The other contribution of this approach is the recognition of the situated relevance of particular theories with respect to particular problems or particular settings. This recognition that no one theory can be asked to explain everything and that each theory has its area of optimal applicability leads to some very fruitful implications for research. At another level of grain-size, each theory attends to some constructs and ignores others. A commitment to a multi-theory approach to research design (Clarke et al. 2012) or at least to a tolerant co-existence of theories is a great advance in educational research.

Rolka and Roesken-Winter bring together different theoretical lenses in their investigation of beliefs in mathematics classrooms. They further advocate networking as a community activity among researchers in mathematics education and see this activity as central to the development of a scholarly community and to the status of mathematics education as a scientific discipline. There is some ambiguity over whether the networking proposed is a social activity engaged in by the "advocates of the different theoretical stances" (Artigue et al. 2006, p. 1242) or a theoretical activity by which the constructs and connections that characterise one theory are situated in relation to those of another theory. In relation to the second goal, the authors undertake to "explore how different theories serve to analyse similar data to gain a more comprehensive understanding of relationships and interdependencies of the underlying frameworks to increase their explanatory power." Further, the authors make the historical observation that "research directions were determined while developments took place in two different research fields, mathematics education and psychology, and which, of course, have influenced methodological approaches and choices."

By way of background, useful comparison is made between the work of Pajares (1992) and Thompson (1992), coming as they do from the disciplines of educational psychology and mathematics education, respectively. The two approaches are contrasted through the comparison of Pajares' focus on the epistemological character of beliefs, which posits "educational beliefs" as a potentially useful and extremely inclusive construct, with Thompson's focus, which is much more discipline-specific. The discussion culminates in the point that it is highly worthwhile to explore the mutual influences of two "research strands" which have undergone largely separate development. These posited benefits are then illustrated in two examples in which beliefs were studied by networking different theoretical "lenses" with their roots in either psychology or mathematics education.

In implementing their program of theory networking, Rolka and Roesken-Winter seek to connect Shulman's work on teacher knowledge (Shulman 1986) (and subsequently on signature pedagogies (Shulman 2005)) with Schoenfeld's work on modelling Teaching-in-Context (Schoenfeld 1998). The authors emphasise that what Schoenfeld offers is "a local view on teaching by modelling teacher behaviour as a function of a teacher's knowledge, goals and beliefs." In particular, Schoenfeld characterizes teacher's spontaneous decision-making in terms of available knowledge, high priority goals, and beliefs. In his later writings, Schoenfeld replaced "beliefs" with "orientations," considering orientations to be a more encompassing term (Schoenfeld 2010). Schoenfeld's aspirations are explicit: if you know enough about a person's orientations, goals and resources, then it should be possible to explain that person's actions "at both macro and micro levels." That is, Schoenfeld's theory is explicitly predictive at the level of the individual teacher in the classroom. Shulman proposed pedagogical content knowledge as a distinct teacher knowledge domain and the idea has been widely investigated, particularly in the context of mathematics education. Similarly, Shulman put up the idea of "signature pedagogies" as a global characterisation of the means by which a profession initiated new members into its community. Rolka and Roesken-Winter characterise Schoenfeld's theory as local and Shulman's as global. The theories have different zones of relevance, different goals and invoke different constructs. As such they offer an interesting combination for a networking exercise.

Applying both Schoenfeld's and Shulman's theories to the analysis of a videotaped lesson, supplemented by an interview with the teacher, Rolka and Roesken-Winter "identified statements that can be interpreted as both pedagogical content goals and beliefs. To be more concrete, the expressed goals were strongly rooted in beliefs and the beliefs influenced the goals to be fixed." This attempt to connect data reflective of the construct central to one theory with data reflective of a key construct in another theory represents an empirically-grounded approach to the connection (networking) of two well-articulated theories. The respective analyses using Schoenfeld's and Shulman's theories reward careful reading and provide an excellent illustration of the capacity of multi-theoretic analyses to interrogate the constructs central to each theory.

The second example provided by Rolka and Roesken-Winter contrasts a belief classification system with an analysis focusing on conceptual change to explore the process of change in teacher belief. The research narrative that emerges takes the reader usefully through the logic of classifying teacher beliefs for the purpose of documenting change in belief to the realisation that such an analysis on its own "does not explain how and why these changes are occurring." To address this shortcoming, the work of Vosniadou and Verschaffel (2004) was employed to develop conceptual change explanations for the documented changes in beliefs. In particular, the conceptual change theory identified the conditions that must be fulfilled in order that the documented change should take place. This form of theory networking is highly pragmatic in orientation and analyses are tuned to the zones of relevance of the relevant theories. A further theory by Appleton (1997) provided a contrasting conceptual change approach, centring on cognitive conflict. One approach to networking theories is very nicely illustrated through the selective application of these theories to a common data set. The principal strength and message of this chapter is precisely this narrative of the selective, purposeful and pragmatic utilisation of specific theories and their associated constructs for the purpose of better conceptualising and understanding change in teacher beliefs.

## Teachers' Efficacy Beliefs

Philippou and Pantziara build on 40 years of research into efficacy beliefs; that is, into "beliefs about one's capacity to organise and execute courses of action to accomplish a specific task." They argue that these beliefs affect a wide variety of personal behaviours and underlie individual motivation and decision-making. The chapter addresses past research to the turn of the century and the questions posed by that research and then moves to recent developments, specifically in relation to teachers' mathematics efficacy research.

Clarification of construct is the mantra of all the chapters in this section. In the chapter by Philippou and Pantziara, self-efficacy is distinguished from constructs such as self-esteem, locus of control, and outcome expectancy. Teacher efficacy is further partitioned into personal, general and collective teacher efficacy. After reflecting on the early research in the field, the authors identify problems related to teacher self-efficacy that were left unresolved by the end of the last century. Shortcomings related to the difficulty of measuring self-efficacy, the possibility that the current form of the construct omits relevant aspects, the malleability of self-efficacy once established and the ways in which self-efficacy influenced individual behaviour. The authors repeat

Labone's (2004) call for diverse research methods and more focus on interpretivist and critical theories. They further frame an agenda that seeks to connect self-efficacy research with the facilitation of educational reform.

In reporting recent developments in research into self-efficacy, the authors cite Bandura's argument that a general theory will not have the explanatory and predictive value needed, due to the tenuousness of the connection to the domain or context in which the individual's self-efficacy must ultimately find expression. Philippou and Pantziara provide a detailed discussion of the refinement of instruments designed to measure teacher self-efficacy.

Equipped with new methodologies and new tools, researchers undertook studies examining the genesis and development of teacher efficacy beliefs. Bandura (1997) postulated four cognitive sources: *mastery experience*, *vicarious experience*, *social persuasion* and *physiological and emotional arousal* by which teacher self-efficacy might be facilitated. While all four of Bandura's sources were found to contribute to high teacher efficacy, teacher efficacy was negatively connected to job stressors (Betoret 2009). The discussion slides uneasily between teacher efficacy and teacher efficacy beliefs. Studies of self-perceptions of competence compared with measured competence have been undertaken in relation to many professions (e.g. teaching and nursing). One danger, as with all affective research, is that the affective attribute (e.g. high teacher efficacy belief or student attitude to mathematics) becomes the surrogate goal of an intervention initiative, without sufficient attention to the construct for which the affective response is felt to be the mediating agent. If student or teacher affective response is to be manipulated or nurtured, its development should also be in the service of relevant cognitive or social outcomes.

Useful discussion is provided of the contribution that teacher education might make to the development of teacher efficacy beliefs. In particular, the argument is made that the subsequent resilience of efficacy beliefs, once established, makes their active nurturing during teacher education a priority. However, the results of research into the use of enactive mastery experiences for the development of teacher efficacy beliefs were suggestive rather than decisive. Two points suggested by this research concern the dependence of change in student attitude and beliefs on student background and the correlation of measures of teacher efficacy beliefs with measures of attitude. These are not startling observations, but they serve to emphasise the interconnected nature of constructs within the affective domain and the difficulty of according primacy to particular constructs, such as teacher efficacy beliefs, for the purposes of instructional advocacy in teacher education programs or the development of intervention studies.

In one study, a computer-based program designed to help teacher education students develop spatial-temporal cognition that can improve understanding of mathematical concepts and operations was associated with improvement in measured teacher efficacy beliefs. This raises the important question, however, as to whether such attitudinal gains (teacher efficacy beliefs or any other) occur consequent to the acquisition of particular skills, whether mathematical, pedagogical or organisational. The nature of the connection between teacher beliefs, the skills and situations about which those beliefs are held, and teacher capacity to act effectively in classroom situations remains under-theorised, with consequences for our capacity to design suitable intervention studies, whether in relation to programs for pre-service or in-service teachers. The study by Bagaka (2011) is criticised by Philippou and Pantziara for its "alienation" from theory on the grounds that analytical categories are misaligned. However, the essential point is that lacking a coherent theory of affect and learning, the identification of some analytical categories with particular constructs will always either appear conceptually arbitrary or be wholly dependent on empirical association, lacking suitable theoretical connection. Similarly, the partitioning of efficacy into sub-categories such as "internet self-efficacy" suggests that we are unsure of the most appropriate grain-size at which to document, investigate and promote teacher self-efficacy.

In the remainder of the chapter by Philippou and Pantziara, teacher efficacy beliefs are discussed in relation to: reform efforts, coping with stress, and the school context, in addition to those above. The diversity of these considerations reflects the lack of structure that follows when a construct lacks a clear, well-grounded, theoretical foundation. In their conclusion, Philippou and Pantziara highlight the lack of conceptual clarity regarding the construct of teacher efficacy beliefs and state that "any attempt to incorporate all these meanings would result in an omnibus definition making the operational use of the construct too difficult to manage." Our capacity to measure teacher efficacy beliefs must reflect the theoretical framework through which we recognise both its integrity and its structure. The particular example of teacher efficacy beliefs reveals the apparent distinctiveness of the construct, the fuzziness of its sub-structure and the tangled nature of its connections with other aspects of affect. In these imprecisions, uncertainties and inconsistencies, it strongly resembles other constructs, such as beliefs, emotions, attitudes and motivations. This level of conceptual uncertainty undermines efforts to investigate the formation and development of teacher efficacy beliefs. Nonetheless, the case for utilizing teacher efficacy beliefs as an entry point for intervention or reform of practice is a plausible one, but the warrant for such a suggestion relies on arguments of perceived social or professional relevance rather than demonstrable connection to a theoretical framework in which such constructs are coherently located.

## Concluding Remarks

The five chapters in this section illustrate in their combination the challenge facing the education community (and mathematics education, in particular) to find coherence and structure within the affective domain. Despite the very different theoretical perspectives employed in each chapter and the differences in the focal constructs around which discussion has centred, there are some central themes that are all the more compelling because of their emergence from such diverse origins. It is useful to review these central themes.

## Integrative Perspectives

Radford asserted that emotions and thinking are not separate entities. Di Martino and Zan suggested that a central element in the tripartite model in which attitude had a cognitive, an affective, and a behavioural component was that the assumption of a link between attitude and behaviour became part of the construct definition itself. Skott's development of an action-oriented conception of affect identified both beliefs and practice as simultaneously and reflexively emergent and negotiated, rather than either being fixed and causally connected to the other. Philippou and Pantziara are critical of the internal coherence of the literature on teacher efficacy beliefs and seek improved connectivity in this specific area rather than the connection of teacher efficacy beliefs to other constructs in the affective domain. The advocacy of Rolka and Roesken-Winter might appear to be less integrative, since they advocate and illustrate the selective, purposeful and pragmatic utilisation of specific theories and their associated constructs, in ways that honour the separate areas of applicability of each theory. In this context, the metaphor of networking can be interpreted as integrative at a meta-theoretical level, without contesting the separate integrity of individual theories. Given this interpretation, the call for coherent, structured connection of constructs in relation to affect was a central theme of all chapters.

## Affect Categories

The combination of chapters makes reference to a variety of related constructs: Affect, feeling, emotions, attitudes, beliefs, motives and motivation. Their connection as constituent of the affective domain is assumed, the nature of their connection is unclear, and the lack of explicit structure by which they might be situated underlines the status of affect as an ill-defined or ill-structured domain. "An ill-structured domain is a conceptual arena in which the instances of knowledge application are both individually complex and in irregular relationship to each other" (Spiro et al. 2007, p. 93). Rolka and Roesken-Winter demonstrate the capacity of multi-theoretical analyses to interrogate and inform both settings and theories (see Clarke 2011). The aspiration here is the optimal correspondence of theory and methodology to context and purpose, without any aspiration to synthesising a grand theory. In such an approach, individual constructs are accorded their own individual status and relevance, dependent on the theory in which they are to be employed and the context to which they are to be applied. Each chapter calls for more coherent theoretical structuring of the affective domain, but adopts a different point of entry and provides a different rationale.

## Context and Affect

Claims as to the situatedness of beliefs (for example) ascribe a constitutive role to context (rather than just a qualifying or moderating role) and posit the possibility that contradictory beliefs might be enacted in differing contexts. This recognition of the situatedness of affective response is echoed in Radford's insistence on the historical and cultural antecedents of any social situation as constitutive of emotional expression in any given setting. Drawing on Di Martino and Zan's use of the work of Ruffel, Mason and Allen, our problem in theorising affect may derive from our confusion between categories of social performance and the mistaken attribution of those performances to individual traits or inclinations rather than to the aggregate of cultural and historical precedents that shape behaviour in specific social situations. Taken seriously, this proposal would free the researcher from the obligation to persistently treat every social action as reflective of some aspect of the individual's psyche.

## Faith, Hope and Charity

To return to my opening remarks: the combination of chapters provide a remarkable overview of the issues related to theorising affect. Despite their differences in orientation, the consistency of the prioritisation of clarity of structure and construct encourages one's faith that the scholarly community in mathematics education has the theoretical tools and meta-theoretical considerations necessary to address the perceived need for structure. Each chapter makes specific proposals regarding the way forward and these proposals can be read as complementary rather than competing. In this, each chapter offers hope for further advancement of our utilisation of affect and its constituent constructs in the interests of promoting the teaching and learning of mathematics. Underlying the various approaches outlined, and despite the evident frustrations, there is an encouraging commitment to theoretical tolerance, an avoiding of normative prescription, and a recognition that a domain as complex as that of affect will not be well-served by an insistence on any single theory. The charitable acknowledgement of the separate integrity of each theory and the need for its strategic use to meet the needs of locally determined context and purpose demonstrates a democratization of theory that may yet meet the challenges of the affective domain.

## References

Appleton, K. (1997). Analysis and description of students' learning during science classes using a constructivist-based model. *Journal of Research in Science Teaching, 34*(3), 303–318.

Artigue, M., Bartolini-Bussi, M., Dreyfus, T., Gray, E., & Prediger, S. (2006). Different theoretical perspectives and approaches in research in mathematics education. In M. Bosch et al. (Eds.), *Proceedings of the 4th CERME* (pp. 1239–1244). Sant Feliu de Guixols: Fundemi IQS – Universitat 2006.

Bagaka's, J. G. (2011). The role of teacher characteristics and practices on upper secondary school students' mathematics self-efficacy in Nyanza province of Kenya: A multilevel analysis. *International Journal of Science and Mathematics Education, 9*, 817–842.

Bandura, A. (1997). *Self-efficacy: The exercise of control*. New York: W. H. Freeman.

Betoret, F. D. (2009). Self-efficacy, school resources, job stressors and burnout among Spanish primary and secondary school teachers: A structural equation approach. *Educational Psychology, 29*(1), 45–68.

Clarke, D. J. (2001). Chapter 3: Untangling uncertainty, negotiation and intersubjectivity. In D. J. Clarke (Ed.), *Perspectives on practice and meaning in mathematics and science classrooms* (pp. 33–52). Dordrecht: Kluwer Academic Press.

Clarke, D. J. (2011, July 3–7). A less partial vision: Theoretical inclusivity and critical synthesis in mathematics classroom research. In J. Clark, B. Kissane, J. Mousley, T. Spencer, & S. Thornton (Eds.), *Mathematics: Traditions and [new] practices. Proceedings of the AAMT–MERGA conference held in Alice Springs* (pp. 192–200). Adelaide: AAMT/MERGA.

Clarke, D. J., & Helme, S. (1998). Context as construction. In O. Bjorkqvist (Ed.), *Mathematics teaching from a constructivist point of view* (pp. 129–147). Vasa: Faculty of Education, Abo Akademi University. ISBN 952-12-0151-7.

Clarke, D. J., Xu, L. H., Arnold, J., Seah, L. H., Hart, C., Tytler, R., & Prain, V. (2012). Multi-theoretic approaches to understanding the science classroom. In C. Bruguière, A. Tiberghien, P. Clément (Eds.), *Proceedings of the 2011 biennial conference of the European Science Education Research Association* (Part 3, 26–40), Lyon, France, September 7–11, 2011. (e-Book published March 23, 2012).

Cobb, P., & Yackel, E. (1996). Constructivist, emergent, and sociocultural perspectives in the context of developmental research. *Educational Psychologist, 31*(3/4), 175–190.

Daskalogianni, K., & Simpson, A. (2000). Towards a definition of attitude: The relationship between the affective and the cognitive in pre-university students. In T. Nakahara & M. Koyama (Eds.), *Proceedings of the 24th conference of the IGPME* (Vol. 2, pp. 217–224), Hiroshima, Japan.

Di Martino, P., & Zan, R. (2011). Attitude towards mathematics: A bridge between beliefs and emotions. *ZDM – The International Journal on Mathematics Education, 43*(4), 471–482.

Goldin, G. A., Epstein, Y. M., Schorr, R. Y., & Warner, L. B. (2011). Beliefs and engagement structures: Behind the affective dimension of mathematical learning. *ZDM – The International Journal on Mathematics Education, 43*(4), 547–560.

Haladyna, T., Shaughnessy, J., & Shaughnessy, M. (1983). A causal analysis of attitude toward mathematics. *Journal for Research in Mathematics Education, 14*, 19–29.

Hutchins, E. (1991). The social organization of distributed cognition. In L. B. Resnick, J. M. Levine, & S. D. Teasley (Eds.), *Perspectives on socially shared cognition* (pp. 283–307). Washington, DC: American Psychological Association Press.

Kulm, G. (1980). Research on mathematics attitude. In R. J. Shumway (Ed.), *Research in mathematics education*. Reston: NCTM.

Labone, E. (2004). Teacher efficacy: Maturing the construct through research in alternative paradigms. *Teaching and Teacher Education, 20*, 341–359.

Lave, J., & Wenger, E. (1991). *Situated learning: Legitimate peripheral participation*. Cambridge: Cambridge University Press.

Lerman, S. (1996). Intersubjectivity in mathematics learning: A challenge to the radical constructivist paradigm? *Journal for Research in Mathematics Education, 27*(2), 133–150.

Menon, U. (2000). Analyzing emotions as culturally constructed scripts. *Culture & Psychology, 6*(1), 40–50.

Pajares, M. F. (1992). Teachers' beliefs and educational research: Cleaning up a messy construct. *Review of Educational Research, 62*(3), 307–332.
Ruffel, M., Mason, J., & Allen, B. (1998). Studying attitude to mathematics. *Educational Studies in Mathematics, 35*, 1–18.
Schoenfeld, A. H. (1998). Toward a theory of teaching-in-context. *Issues in Education, 4*(1), 1–94.
Schoenfeld, A. H. (2010). *How we think. A theory of goal-oriented decision making and its educational applications*. New York: Routledge.
Sfard, A. (2008). *Thinking as communicating. Human development, the growth of discourses, and mathematizing*. Cambridge: Cambridge University Press.
Shulman, L. S. (1986). Those who understand: Knowledge growth in teaching. *Educational Researcher, 15*(2), 4–14.
Shulman, L. S. (2005). Signature pedagogies in the professions. *Daedalus, 134*(Summer), 52–59.
Spiro, R. J., Collins, B. P., & Ramchandran, A. R. (2007). Reflections on a post- Gutenberg epistemology for video use in ill-structured domains: Fostering complex learning and cognitive flexibility. In R. Goldman, R. D. Pea, B. Barron, & S. Derry (Eds.), *Video research in the learning sciences* (pp. 93–100). Mahwah: Lawrence Erlbaum Associates.
Thompson, A. G. (1992). Teachers' beliefs and conceptions: A synthesis of the research. In D. A. Grouws (Ed.), *Handbook of research on mathematics teaching and learning* (pp. 127–146). New York: Macmillan.
Vosniadou, S., & Verschaffel, L. (2004). Extending the conceptual change approach to mathematics learning and teaching. *Learning and Instruction, 14*(5), 445–451.

# Part II
# Relevance in the Field – Affective Systems of Individuals and 'Collectives'

The second part of the book focusses on empirical studies in the field of beliefs and affect in mathematics education. It raises the question about the various types of beliefs, whether held by students, teachers, or indeed groups of learners. It also asks whether different mathematical processes and different topic areas are connected to different beliefs and affective systems. International studies provide a window into the diverse contexts and a potentially rich field for examining the different relationships.

# Students' Non-realistic Mathematical Modeling as a Drawback of Teachers' Beliefs About and Approaches to Word Problem Solving

**Fien Depaepe, Erik De Corte, and Lieven Verschaffel**

**Abstract** Over the past decades numerous scholars have become aware of many compelling observations of students in mathematics classes abandoning their sense-making capabilities when doing word problems, and, in particular, carrying out arithmetic calculations that do not make sense in relation to the situations described. This led us, together with several other scholars, to embark upon an extended investigation of the phenomenon, the results of which are reported, among others, in two books (Verschaffel L, Greer B, De Corte E, Making sense of word problems. Swets & Zeitlinger, Lisse, 2000; Verschaffel L, Greer B, Van Dooren W, Mukhopadhyay S, Words and worlds: modelling verbal descriptions of situations. Sense Publishers, Rotterdam, 2009). The goal of the present chapter is to bring together and critically review the theoretical analyses and empirical studies that have focused on major aspects of teachers' instructional practices that affect – directly or indirectly – students' non-realistic approaches to and beliefs about word problem solving. Special attention will be given to the problems that appear in students' mathematical textbooks as well as to teachers' beliefs about word problems and what are appropriate ways to solve them, and to their instructional behavior, and how these factors affect students' beliefs about and approaches to word problems. While the focus is on research that has been done in our own center, we also integrate relevant studies by others.

**Keywords** Word problem solving • Mathematical modeling • Students' beliefs • Instructional approaches • Teachers' beliefs

---

F. Depaepe (✉) • E. De Corte • L. Verschaffel
Center for Instructional Psychology and Technology, University of Leuven,
Dekenstraat 2, P.O.B. 3773, 3000 Leuven, Belgium
e-mail: Fien.Depaepe@kuleuven-kulak.be; Erik.DeCorte@ppw.kuleuven.be;
Lieven.Verschaffel@ppw.kuleuven.be

 
B. Pepin, B. Roesken-Winter (eds.), *From beliefs to dynamic affect systems in mathematics education*, Advances in Mathematics Education,
DOI 10.1007/978-3-319-06808-4_7

## Word Problems as Exercises in Mathematical Modeling

Word problems have been assigned a central role in the mathematics curriculum in the elementary school (see e.g., National Council of Teachers of Mathematics 2010), not only because of their potential for motivating students and for the meaningful development of new mathematics concepts and skills, but also – and from a historical perspective even principally – to develop in students the skills of knowing *when* and *how* to apply their mathematics effectively in situations encountered in everyday life and at work (Boaler 1993; Hiebert et al. 1996; Verschaffel et al. 2000). Word problems are typically defined as essentially verbal descriptions of problem situations in which one or more questions are raised for which the answer(s) can be obtained by the application of one or more mathematical operations to the numerical data available in the problem statement (Verschaffel et al. 2000, p. ix). As they are composed of a mathematics structure embedded in a more or less realistic context, word problems can ideally serve as tools for mathematical modeling, which may be viewed according to Greer (1997) "as the link between the 'two faces' of mathematics, namely its grounding in aspects of reality, and the development of abstract formal structures" (p. 300).

Applying mathematics to solve problem situations in the real world can be usefully thought of as a complex process involving a number of phases. There are many different descriptions of this modeling process (e.g., Blum and Niss 1991; Burkhardt 1994; Mason 2001; Verschaffel et al. 2000), but, in essence, they all involve the following components (which do not necessarily follow a strictly linear order): (1) understanding and defining the problem situation leading to a situation model; (2) constructing a mathematical model of the relevant elements, relations, and conditions involved in the situation model; (3) working through the mathematical model using disciplinary methods to derive some mathematical results; (4) interpreting the outcome of the computational work in relation to the original problem situation; (5) evaluating the modeling process by checking if the interpreted mathematical outcome is appropriate and reasonable for its purpose; and (6) communicating the obtained solution of the original real-world problems.

For a long time, many teachers, textbook writers, and researchers in mathematics education assumed an unproblematic relationship between the situation and the mathematical model: Solving a word problem was considered as a direct translation process from the word problem text to mathematical symbols. However, more and more scholars have pointed to the difficulty of assuming a one-to-one relationship between mathematical models and real-world phenomena (Gerofsky 1997; Nesher 1980). This bridging problem became even clearer as empirical studies revealed that, after several years of schooling, many students have developed an approach to problem solving, whereby they ignore essential aspects of reality and whereby the mathematical actions they perform are based on a superficial analysis of the numbers and keywords provided in the problem text (Schoenfeld 1991). In this respect, we refer to the famous example of the captain problem: "There are 26 sheep and 10 goats on a ship. How old is the captain?". Confronted with this problem, many students were prepared to offer an answer to this absurd problem by combining the numbers given in the problem (e.g., 26+10) to produce an answer (i.e., 36) without showing any awareness of the meaninglessness of the problem and their solution (Baruk 1985). Inspired by this and some other striking examples of this phenomenon of "suspension of sense-making" (Schoenfeld 1991) when doing school word problems, Greer (1993) and Verschaffel et al. (1994) carried out two parallel studies in Northern Ireland and Belgium (Flanders). Paper-and-pencil tests were administered to upper elementary and lower secondary school students involving problems such as "Steve bought 4 planks of 2.5 m each. How many planks of 1 m can he saw out of these planks?" (=planks item), "John's best time to run 100 m is 17 s. How long will it take him to run 1 km" (=runner item). These authors termed each of these items "problematic" in the sense that they require the application of judgment based on real-world knowledge and assumptions rather than the straightforward application of one or more simple arithmetical operations. In both studies, students demonstrated a very strong tendency to exclude realistic considerations when confronted with these problematic items. For a more detailed overview of the design and the results of these two studies we refer to Greer et al. (2002).

The studies of Greer (1993) and Verschaffel et al. (1994) were replicated in several other countries, using a similar methodology and, to a considerable extent, the same items. The findings were strikingly consistent across many countries: Almost none of the problematic items was answered in a realistic fashion by more than a small percentage of students. The mean percentage of realistic answers on the problematic items across the studies varied from 12 % in Hidalgo's (1997) study to 30 % in Caldwell's (1995) study. Realistic reactions were typically higher on the division with remainder problems. The obtained results strongly surprised some of these other researcher(s) who had anticipated that the "disastrous" picture of the Irish and Flemish pupils would not apply to their students (for an extensive overview of these replication studies, see Verschaffel et al. 2000).

## Beliefs and Word Problem Solving

In search for an explanation of the students' non-realistic responses to word problems, Schoenfeld (1991) suggested that it is not a cognitive deficit as such that causes students' general and strong abstention from sense-making when solving mathematical word problems in a typical school setting. Rather, students seemed to be engaged in sense-making of a different kind: "such behavior is sense making of the deepest kind. In the context of schooling, such behavior represents the construction of a set of beliefs and behaviors that result in praise for good performance, minimal conflict, fitting in socially, etc. What could be more sensible than that? The problem, then, is that the same behavior that is sensible in one context (schooling as an institution) may violate the protocols of sense-making in another (the culture of mathematics and mathematicians)" (Schoenfeld 1991, p. 340). In other words, students' tendency to neglect real-world knowledge and realistic considerations when confronted with problematic word problems is assumed to be due to their beliefs about word problems and how to solve them, which they have gradually, implicitly, and tacitly developed in accordance with the "word problem game" (Verschaffel et al. 2000), or, as others would call it, the "didactical contract" (Brousseau 1998), or the "sociomathematical norms and practices" (Yackel and Cobb 1996) within "the culture of the mathematics classroom" (Seeger et al. 1998). Apart from some anecdotal indications collected in individual interviews, direct empirical evidence for the existence of these assumptions and beliefs is scarce. One exception is a study by Reusser and Stebler (1997) that provided some evidence for their existence based on interviews with students who gave explanations for their non-realistic behavior on problematic word problems. Reusser and Stebler (1997, pp. 324–325) identified the following assumptions that students typically develop through being immersed in the culture and practices of school mathematics:

- Assume that every problem presented by a teacher or in a textbook makes sense.
- Do not question the correctness or completeness of problems.
- Assume that there is only one "correct" answer to every problem.
- Give an answer to every problem presented to you.
- Use all numbers that are part of the problem in order to calculate the solution.
- If a problem is perceived to be indeterminate, equivocal, or unsolvable, go for an obvious interpretation given the information in the problem text and your knowledge of mathematical operations.
- If you do not understand a problem, look at key words, or at previously solved problems, in order to determine a mathematical operation.

In addition, indirect evidence for the existence of the previously described assumptions was obtained in a series of studies by Jiménez and colleagues. Jiménez and Ramos (2011) investigated the impact of four of these specific beliefs about word problems that develop in students as a result of traditional schooling: (1) every word problem is solvable, (2) there is only one numerical and precise correct answer to every word problem, (3) it is necessary to do calculations to solve a word problem, and (4) all numbers that are part of the word problem should be used in order to calculate the solution. Specifically, 22 second and 22 third graders were asked in the context of an individual interview to solve four word problems that each violate one of these four beliefs, i.e., (1) an unsolvable word problem, (2) a word problem with multiple solutions, (3) a word problem containing the solution in the problem statement, and (4) a word problem including irrelevant data. For instance, the word problem including irrelevant data was "Laura buys a box with 12 crayons for the Plastic arts class. Her friend Silvia gives her another box containing 3 pens and 9 crayons. How many crayons does Laura have now?". Results revealed, first, that only one third of all students responded correctly to the four problem types. Second, the percentage correct answers was higher for solutions in the statement and irrelevant data problems (resp. 45.5 % and 43.2 %) than for unsolvable and multiple solution problems (resp. 20.5 % and 23.9 %). Third, no differences were found between second and third graders. Fourth, the vast majority of the errors originated from doing one or more arithmetic operations on all given numbers in the problem.

Finally, many verbal explanations of erroneous responses contained spontaneous expressions of the above-mentioned beliefs about word problems. In a cross-sectional study, Jiménez and Verschaffel (2014) investigated the development of these beliefs from first to sixth grade. Using individual interviews they administered to students the same four problems as in the previous study, that respectively violate the belief that (1) every word problem has a solution, (2) there is only one numerical and precise answer to a word problem, (3) it is necessary to do calculations to solve a word problem, and (4) that all numbers mentioned in a word problem are relevant to its solution. The amount of correct responses on the distinct problem types was respectively 18 %, 30 %, 46 %, and 57 %. These results indicate, first, that accuracy scores were relatively low for all problem types. Second, the percentages correct answers suggest that some beliefs about arithmetic word problems were more established in students' thinking (e.g., every word problem has a solution) than others (e.g., all numbers mentioned in the word problem are needed for its solution). Third, this difference in performance across the distinct problem types was observed in all grades. Fourth, there was an increase in correct responses from grade 1 (15.5 %) to grade 6 (56 %), however, this increase was small in the upper grades (49.5 % in grade 4 and 55.5 % in grade 5). In general, the results of Jiménez and Verschaffel (2014) paralleled the findings of Jiménez and Ramos (2011). Overall students were weak at solving word problems that violate less appropriate beliefs – at least from a modeling perspective – about word problem solving. Moreover, the same pattern of differences in accuracy to solve the distinct problem types was observed in both studies, suggesting that students' belief that all numbers in a word problem are relevant to its solution is more prevalent than the belief that every word problem has a solution.

## Aspects of Teachers' Instructional Practices That Influence Students' Non-realistic Behavior

In an attempt to explain how these beliefs about and tactics for the solution of school word problems develop in students, it is assumed that mainly three aspects of the instructional practice and culture of traditional school mathematics are responsible, namely (1) the stereotyped and unrealistic nature of the problems used in classrooms, (2) the way in which teachers conceive word problems, and (3) the way in which teachers treat word problems in their daily practice (Mason and Scrivani 2004; Verschaffel et al. 1999). Even though it is generally accepted that the culture and practice in regular mathematics classrooms is responsible for the beliefs that students develop about word problems and for their non-realistic word problem-solving tactics, only rarely has attention been paid to whether, when, and how students are exposed to realistic modeling experiences in their daily mathematics classroom (Verschaffel et al. 2010). In what follows, we will give an overview of the studies that yield empirical evidence on these aspects of the instructional environment that may affect – directly or indirectly – students' non-realistic approaches to and beliefs about word problem solving. First, we will focus on the nature of the word problems that appear in mathematical textbooks and that teachers use in their instructional practice. Second, we will report studies on teachers' beliefs about problematic word problems and how they evaluate students' non-realistic approaches. Third, an overview will be given of the way in which teachers deal with word problems in their regular classroom practice.

## The Nature of Word Problems in Traditional School Mathematics

The research literature suggests two related criticisms regarding the nature of the problems to which students are exposed in regular mathematics classrooms. First, most problems can be solved by a simple and straightforward application of one or a combination of the four basic arithmetic operations (Davis-Dorsey et al. 1991; Gravemeijer 1997). Second, but related to the previous issue, problems that are closely related to students' experiential worlds, are rare (Gerofsky 1997; Palm 2002). In an attempt to give empirical grounding to this second criticism, Depaepe et al. (2009) investigated the nature of word problems in the most frequently used sixth-grade mathematics textbook in Flanders as well as the nature of the word problems actually selected and used by two typical teachers who used this textbook. We relied on Palm's (2002) conceptual framework for analyzing the realistic nature of word problems. The founding idea of his framework lies in the notion *simulation*: A word problem is considered to be realistic if its important aspects are taken under conditions representative for an out-of-school situation. The operationalization of the framework included 11 aspects that play an important role in the extent to which students may engage in similar mathematical activities in a school task as in an out-of-school situation: event, question, purpose in the figurative context, existence of data, realism of data, specificity of data, language use, availability of solution strategies, external tools, guidance, and solution requirements. Two classification levels were distinguished for all but one aspect. The two levels relate to whether a task was judged as simulating the aspects of a corresponding out-of-school situation to a reasonable degree (1) or not (0). For the aspect specificity of data three levels were distinguished. The operationalization of the different aspects of the framework is presented in Table 1.

The way in which we classified mathematical problems according to the aspects mentioned in Table 1 is illustrated in Fig. 1.

Overall, we found that the tasks from the textbook and those that were created by the teachers themselves were similar. The word problems seemed to simulate relatively well some aspects that are assumed to be important in designing realistic tasks according to Palm's coding scheme (e.g., event, language use), but failed to include others (e.g., specificity of data, purpose in the figurative context). Another important finding (that was however not revealed by Palm's coding scheme) was that almost all word problems could be solved straightforwardly by applying one or more arithmetic operation(s) with all numbers mentioned in the task.

**Table 1** Framework for analyzing the realism of word problems

| Aspect | Description |
| --- | --- |
| Event | 1 = The event in the school task could be encountered in real life outside school. |
| | 0 = The school task is about an imaginary event; the event includes objects from the real world, but is still a fictious event; or the school task is a pure mathematical task which is not embedded in a context. |
| Question | 1 = The question in the school task has been asked, or might be asked, in the stimulated event. The answer to the question is of practical value or of interest for others than just the people very interested in mathematics. |
| | 0 = The question in the school task is judged not to have been asked, and neither would be asked, in the event described in the task. |
| Purpose in the figurative context | 1 = The purpose of solving the task is explicitly mentioned in the school task and in concordance with the purpose of solving the task in the stimulated situation. |
| | 0 = The purpose of solving the task in the stimulated situation is unclear. The school context could be generally described, not pointing to a specific situation, resulting in many possible situations and purposes of the task solving. In other tasks the situation described in the task is more specific but still open for more than one purpose. |
| Existence of data | 1 = The relevant data that are important for the solution in the simulated situation coincide with the accessible data in the school task. |
| | 0 = The data that are important for the solution in the simulated situation are not the same as the accessible data in the school situation and/or this information is accessible only by applying other competencies that are different from those required in the simulated situation. |
| Realism of data | 1 = Numbers and values given are identical to or very close to the corresponding numbers and values in the simulated situation. |
| | 0 = Numbers and values given are not realistic. |
| Specificity of data | 2 = The text of the task describes a specific situation in which the subjects, objects, and places in the school context are specific. If graphs are used, the source is mentioned. |
| | 1 = The situation in the school task is not specific, but at a minimum the objects that are the foci of mathematical treatment are specific. |
| | 0 = The situation in the school context is a general situation in which the subjects and objects are not specified. |
| Language use | 1 = The task is linguistically similar to the corresponding simulated situation. Specific mathematical concepts which are not used in daily language are avoided. |
| | 0 = The terminology, sentence structure or amount of text in the school task is judged to affect more than an insignificant proportion of students in such a way that the possibility to use the same mathematics in the school task and in the simulated situation is greatly impaired. |
| Availability of solution strategies | 1 = The students' available solution strategies allow them to solve the task in the same way as the taken character in the simulated situation would have done. The textbook is not directing the student in a specific direction to solve the problem. |
| | 0 = The students' available solution strategies to solve the task are different than in the simulated situation. The textbook is directing the students into a specific solution strategy, which the problem solver would not necessary have used while solving a similar problem in real life. |

(continued)

**Table 1** (continued)

| Aspect | Description |
|---|---|
| External tools | 1 = The availability of external tools (i.e., concrete tools outside the mind: calculator, map, ruler…), important for the solution of a task, in the school task is similar to the simulated real situation. |
| | 0 = There is a discrepancy between the tools in the two corresponding situations. |
| Guidance | 1 = The same guidance is provided in the school task and in the corresponding out-of-school situation. |
| | 0 = The task does not match in the guidance given between the school task and the corresponding out-of-school situation. |
| Solution requirements | 1 = The explicit or implicit requirements on the solution to a task are considered to be similar to the corresponding situation in real life. |
| | 0 = The explicit or implicit requirements on the solution to a school task are not considered to be similar to the corresponding real-life situation. |

*Problem*

A jeweler makes a golden ring of 11 gram pure gold (12 euro each gram) and 4 gram silver (2.50 euro each gram).

Q [a] How much does the alloy cost?

C S

Gold g →

Silver g → OK

*Scoring*

| Event | Quest. | Purp. | Exist. | Real.[b] | Specif. | Lang. | Sol. strat. | Ext. tools | Guid. | Sol. req. |
|---|---|---|---|---|---|---|---|---|---|---|
| 1 | 1 | 0 | 1 | 1 | 1 | 1 | 0 | 0 | 0 | 1 |

[a] *The textbook writers explicitly refer to (parts of) the problem-solving process by means of letters which symbolize certain heuristics (S = scheme) and phases (Q = question; C = computation; OK = checking and interpreting the outcome). Since such hints do normally not occur in real life situations, a "0" was given for the aspect guidance.*

[b] *At the time of the analysis the value of the unit-price of gold and silver was realistic, therefore a code "1" was given for realism of data. However, it should be admitted that prices of objects involved in the task often automatically expires, which is a problem that is inherently connected to textbook tasks which are still used in a classroom years after their design.*

**Fig. 1** Illustration of a textbook word problem and its classification according to the framework for analyzing the realism of word problems

Very recently, Gkoris et al. (2013) analyzed the nature of the word problems of the compulsory national mathematics textbook for fifth grade in Greece, both before and after a major educational reform in 2003 that aimed among others at promoting critical reasoning in problem solving (Pedagogic Institute 2003). They also relied on Palm's framework for analyzing the authentic nature of word problems (Palm

2002), but additionally they investigated the problematic nature of the word problems in the old and the new textbook. The clearest difference between the old and the new textbook was observed for the aspect event. Whereas most of the new textbook problems related to students' personal interests and experiences, the majority of the problems in the old textbook provided only minimal contextual information. But also with respect to other aspects of Palm's framework the new textbook problems were more authentic than those from the old textbook: The purpose of solving the school problem was more in alignment with the purpose of solving the task in the simulated real-life situation, the use of external tools similar to situations in real life – such as the calculator – was allowed, and the information provided in the new textbook problems was more specific than in the old ones. However, like in Depaepe et al.'s analysis, both textbooks scored low on the aspect problematicity, meaning that Greek students were and are hardly confronted with problems that stimulate the use of real-life reasoning skills. Moreover, the few tasks in which the relation between the situation model (the problem context) and the mathematical model (the required mathematical operations) was neither straightforward nor simple were division with remainder problems, in which real-life considerations should be taken into account when interpreting the obtained results. However, for these type of problems, previous studies reported better realistic modeling results when compared to the other types of problematic tasks from Verschaffel et al.'s (1994) study (Hidalgo 1997; Reusser and Stebler 1997; Yoshida et al. 1997).

Similarly, Vicente et al. (2011) analyzed all word problems of two Spanish elementary textbooks (grade 1–6). The analysis focused on (1) the level of authenticity of the word problems (strongly relying on Palm's framework) and (2) the proportion of challenging problems (e.g., problems with irrelevant information or missing information that problem solvers must infer from their prior knowledge, problem posing activities). The results indicated that the Spanish word problems simulated well most aspects of Palm's framework such as language use, external tools, solution requirements, realism of data, question, event. Only the aspect purpose in the figurative context was only in 6 % of the word problems well simulated. However, their analysis of the problematic nature of word problems revealed that 95 % of the word problems were stereotyped, easy, and non-challenging.

In conclusion, these recent studies that analyzed the nature of word problems (Depaepe et al. 2009; Gkoris et al. 2013; Vicente et al. 2011) reveal, on the one hand, that the negative image of the unrealistic nature of the set of tasks students are confronted with in the mathematics classroom – as expressed in previous publications (e.g., Verschaffel et al. 2000) – does currently not anymore count to the same degree for regular classroom practices. This positive development may partly be a result of the past 15 years of research on students' suspension of sense-making. But, most probably, it has also been impacted by the global reform movement towards more realistic mathematics education in which policy makers, textbook writers and teachers generally believe that students should be confronted with realistic word problems. On the other hand, these studies also demonstrate that most word problems are characterized by only a restricted problematic nature. If one really wants students to develop appropriate beliefs towards solving word problems and to become competent problem solvers in real life, one should integrate more problematic problems into the mathematics curriculum, since most real word problems which one encounters in life beyond school are modeling problems in which the translation of the situation model into a mathematical model is neither straightforward nor simple.

## *Teachers' Knowledge and Beliefs About Mathematical Modeling Problems*

At least as important as the nature of the word problems is the way in which these problems are conceived and approached by the teacher. Hiebert et al. (1996, p. 16) argue: "given a different culture, even large-scale real-life situations can be drained of their problematic possibilities. Tasks are inherently neither problematic nor routine. Whether they become problematic depends on how teachers and students treat them". Accordingly, the teacher may play an important role in stimulating or discouraging students to take into account realistic considerations. In this section we will focus on teachers' knowledge (how do they solve problematic items themselves?) and beliefs (how do they value students' realistic considerations when solving mathematical word problems?) regarding realistic mathematical modeling.

Verschaffel et al. (1997) administered a paper-and-pencil test consisting of seven standard and seven problematic word problems to 332 Flemish prospective elementary teachers. The teachers were, first, asked to solve the word problems themselves. Afterwards, the test was given a second time to all prospective teachers and they were asked to score different answers from students to all word problems (including a typical non-realistic and a realistic answer). The results indicated that, similar to elementary and secondary students, prospective teachers demonstrated a strong overall tendency to exclude real-world knowledge and realistic considerations when confronted with the problematic word problems. Moreover, they valued students' non-realistic responses to these problematic items considerably more than realistic answers.

Similar results were obtained by replication studies in different countries. A study of Bonotto and Wilczewski (2007) with Italian prospective teachers revealed that their overall evaluations of the non-realistic answers were also considerably more positive than for the realistic ones, suggesting that these future teachers also seemed to believe that the activation of realistic context-based considerations should not be stimulated, rather, discouraged in elementary school mathematics. Xu (2005) asked 117 prospective and 72 in-service Chinese teachers to solve the seven problematic items from Verschaffel et al.'s (1997) study and to value students' realistic and non-realistic answers to these problematic items. The study indicated that Chinese (prospective) teachers, first, showed more realistic problem-solving behavior, and, second, evaluated students' realistic answers more positively than their Flemish and Italian peers. Another Chinese study (Chen et al. 2011) with 208 prospective teachers confirmed Chinese prospective teachers' more realistic disposition towards word problem solving, both in terms of their own problem-solving behavior and their valuation of students' responses to problematic items. This discrepancy between the performances of Chinese and Western teachers might be not so surprising, since studies revealed that Chinese teachers acquire more content and pedagogical content knowledge (e.g., Ma 1999; Zhou et al. 2006). Another factor that might have impacted the more realistic behavior of Chinese teachers compared to their Western peers is the increased emphasis on realistic mathematics education in the Chinese curriculum (Chen et al. 2011).

Inspired by the study of Verschaffel et al. (1997) Duan et al. (2011) asked 20 Chinese teachers of upper elementary school to value the educational suitability of six standard and six problematic word problems. Moreover, they were asked to justify their choices and to make suggestions to improve the word problems. First, the Chinese teachers favored – in alignment with the previously mentioned studies (Chen et al. 2011; Xu 2005) – realistic mathematical modeling approaches to the problematic items, even though their performance on these problematic items was considerably lower than on the standard items. Second, Chinese teachers evaluated the educational suitability of the problematic word problems clearly lower than of the standard word problems. Although some teachers acknowledged the possible additional value of the problematic items in word problem solving, most teachers expressed criticism on the ambiguous character of these problematic word problems which might mislead and confuse students (e.g., a teacher commented on the runner item mentioned in section "Word problems as exercises in mathematical modeling" "This is not realistic, because running 100 m is quite different from running 1,000 m"). In other words, these teachers assumed that the problematic word problems would have been better formulated in a clear, unambiguous way. This was also evident in their response to the question whether and how these problematic items could be improved: their typical reaction was to transform them into a standard format by eliminating all "problematic" aspects (e.g., for the runner item teachers suggested that it should be explicitly stated that the speed is fixed). Consequently, although the findings of Duan et al.'s (2011) study were in alignment with the previous Chinese studies (Chen et al. 2011; Xu 2005) indicating that some teachers performed very well on these problematic items and acknowledged that these items could help students to deal with complex and ambiguous problem situations, Duan et al.'s study additionally provided evidence that the same teachers seemed, in general, to attach little value to the opportunities that these word problems offer for students' realistic mathematical modeling.

Lee (2012) investigated how prospective elementary teachers perceive real-life connections in mathematical word problem solving. In this study 71 US prospective teachers were first asked to formulate at least three criteria for exemplary story problems, to collect two story problems, and to pose two word problems that, in their opinion, best represent real-life connections. Based on these prospective teachers' responses Lee selected ten word problems. In a second assignment prospective teachers were asked to value these ten word problems on a 5-point scale in terms of the quality of real-life connections, and to comment on the strengths and weaknesses of each word problem. The results of the first assignment revealed that 42 % of teachers' criteria for realistic word problems were not directly related to real-life connectedness but, rather, to mathematics problems in general (e.g., age or grade level appropriate, involve high-order and critical thinking, utilize multiple modes of representation). Moreover, the word problems posed by the prospective teachers were typical standard problems that can be solved by a straightforward application of one or more arithmetic operations with the given numbers. Based on the second assignment, Lee concluded that, in general, prospective teachers demonstrated positive beliefs about real-life connections in word problem solving, albeit with insufficient specifics. The majority of the participants held an utilitarian view on realistic problem solving: They stated that reality in word problems is important for enhancing students' interest and motivation, for making mathematics more meaningful, and for enhancing the application of mathematics concepts in real life. Discrepancies were observed between their positive beliefs on reality and word problem solving and the way they valued word problems. For instance, word problems that contained many details in order to make the word problem more real were typically negatively valued by arguing that these details could and should be ignored in terms of mathematical problem solving.

In general, the previous studies on teachers' beliefs about word problems and what are appropriate ways to solve them provide evidence for their lack of disposition towards realistic mathematical modeling. These studies suggest that also prospective teachers themselves seem to share the belief that realistic considerations about the problem context should be neglected when solving problematic word problems. Although this observation counts for a lesser extent for Chinese (prospective) teachers, they also seem – similar to their Western peers – to depreciate word problems in which there is a complex, ambiguous relationship between the mathematical model and real-world phenomena. This lack of disposition towards realistic mathematical modeling, most probably, also impacts teachers' classroom practice.

## *Teachers' Approaches to Word Problem Solving*

Arguably, also the way in which teachers actually treat word problems in their instructional practice can promote or inhibit students' realistic disposition towards word problem solving. As already mentioned in section "Word problems as exercises in mathematical modeling", we strongly recommend to conceive and use word problems as exercises in mathematical modeling, a non-trivial and non-linear complex process. This process aims at finding a proper balance between taking seriously into account the elements of the real world evoked by the problem statement, on the one hand, and finding an underlying mathematical structure that allows the use of the power of mathematics to efficiently understand and solve the problem, on the other hand. In modeling, not all aspects of reality can, nor should be modeled (Verschaffel 2002). In this regard, Ikeda and Stephens (2001) point to the pivotal task for the mathematical modeler of balancing appropriately between over-complication and over-simplification, taking into account the goals of the modeling task and one's personal and contextual constraints.

In this respect, Chapman (2006) made an interesting distinction – borrowed from Bruner (1985) – between two complementary modes of conceiving and treating word problems, namely a paradigmatic-oriented and a narrative-oriented mode. The paradigmatic approach towards a word problem is based on a focus on mathematical models and structures that are universal and context-free (e.g., fragmenting and translating the context into mathematical representations). The narrative mode, in contrast, deals with situational aspects of the word problem and, thus, focuses on context-sensitive explications (e.g., allowing students to resonate in the social context of the word problem to discuss specific aspects of it they were curious about or to critique it). Using audiotapes and field notes of two lessons related to word problems Chapman distilled different paradigmatic and narrative modes that emerged from the teaching of word problems of 14 experienced elementary, junior high, and senior high school teachers. Chapman's results revealed that the paradigmatic mode was more dominant, but also that it was combined with the narrative mode in different ways among the teachers. Based on her results, she made a plea for balancing between the paradigmatic and the narrative mode in order to realize realistic mathematical modeling.

Based on Chapman's (2006) distinction between a paradigmatic and narrative approach towards word problem solving we investigated whether two sixth-grade teachers used word problems as a vehicle for realistic mathematical modeling (Depaepe et al. 2010). Contrary to Chapman (2006) who studied the occurrence of the paradigmatic and narrative mode in the lessons of 14 teachers in general, we systematically analyzed to what degree the paradigmatic and narrative interventions were reflected in the teaching of each word problem in two classrooms over a period of 7 months. Inspired by Chapman's distinction in one of her narrative modes between the entry into the problem and the exit out of the problem, we distinguish in our analysis between the initial phases and the final phases of the problem-solving process. Our operationalization of teachers' paradigmatic and narrative interventions with regard to the entry and exit phase of the modeling process has resulted in a refinement of Chapman's scheme (see Table 2).

The results revealed, first, that both teachers adhered more to a paradigmatic than to a narrative approach towards word problem solving. Second, we observed that a strong focus on a paradigmatic approach does not exclude a strong narrative approach and vice versa. Indeed, notwithstanding the fact that the paradigmatic mode dominated in both classrooms, one teacher's approach towards problem solving reflected a substantially stronger combined paradigmatic *and* narrative focus than the other teacher. This finding reveals that a simultaneous emphasis on universal and context-free mathematical structures and models (a paradigmatic approach) and on contextual elements of the realistic situation to which the word problem refers (a narrative approach) is not only desirable (Chapman 2006), but also feasible. Third, it was observed that the relation between the mathematical model and the situation model (N4 and N7) was almost never addressed in both classrooms. Nevertheless, interventions that stress this relationship belong to the core of mathematical modeling and seem to be consistent with current perspectives on mathematics learning and teaching (e.g., Verschaffel et al. 2000, 2009).

**Table 2** Paradigmatic and narrative interventions towards word problem solving distinguished in Depaepe et al.'s (2010) study

| Approach | Phase | Intervention | Description |
|---|---|---|---|
| Paradigmatic | Entry | **P1:** Distinguishing relevant from irrelevant information | Differentiating between what does and does not "matter" for the problem solution and/or translating the "given" into mathematical terms. |
| | | **P2:** Applying a prototypical scheme | Transforming the information given in the problem context into a representational or solution scheme which enables the modeler to solve the problem. |
| | | **P3:** Addressing the underlying mathematical structure | Emphasizing the structural similarities of the problem with an analogous problem and/or labeling the problem in terms of a particular problem class. |
| | Exit | **P4:** Seeking confirmatory evidence for the solution being obtained | Checking whether no errors were made and/or whether all questions were answered. |
| | | **P5:** Addressing the underlying structure | Emphasizing the structural similarities of the problem with an analogous problem, and/or labeling the problem in terms of a particular problem class, and/or reviewing how a type of problems can or should be solved in general. |
| Narrative | Entry | **N1:** Rewording the problem | Rewording the problem into your own words based on the information given in the problem. |
| | | **N2:** Defining notions involved in the problem | Clarifying the meaning of objects, persons, occupations, and/or situations mentioned in the problem. |
| | | **N3:** Building on students' real-life experiences and prior knowledge | Linking the problem to a personal experience, and/or referring to a related event that happened in the real world, and/or elaborating on students' experiences with regard to objects mentioned in the problem text, and/or building on students' prior knowledge. |
| | | **N4:** Taking explicitly into account the realities of the problem context | Identifying the conditions and assumptions of the real-world context to which the modeler will attend as he or she mathematizes the situation. This may results in criticizing and/or reformulating the word problem as initially posed. |
| | Exit | **N5:** Interpreting the outcome | Interpreting the outcome with regard to the real-life situation and/or seeking for real-life explanations for the obtained solution. |

(continued)

**Table 2** (continued)

| Approach | Phase | Intervention | Description |
|---|---|---|---|
| | | **N6:** Thinking of corresponding real-life situations | Referring to corresponding real-world applications and/or indicating (practical) relevance for learning to solve a particular problem class. |
| | | **N7:** Taking explicitly into account the realities of the problem context | Identifying the conditions and assumptions of the real-world context to which the modeler will attend as he or she mathematizes the situation. This may results in criticizing and/or reformulating the word problem as initially posed. |

Also relying on the analytic distinction between a paradigmatic and narrative approach, Rosales et al. (2012) investigated the way in which 11 elementary school teachers (grade 3–5) approached two non-standard word problems that require additional mathematical and situational knowledge to solve them. An example of such a problem is the following: "A shepherd was taking care of a flock of sheep. The shepherd had a flock of 57 sheep. He wanted to increase the size of the flock because this year there was a good fodder. In order to do so the shepherd went to a market, where he decided to buy some more sheep. One evening the shepherd saw a pack of wolves in the area. The wolves were hungry and then they devoured 11 sheep and now there are 96 sheep left. How many sheep did the shepherd buy in the market?". Rosales et al. distinguished three different paradigmatic approaches: (1) data selection (contents devoted to selecting the data from the problem), (2) mathematical reasoning (contents related to a deep mathematical understanding of the problem, in terms of mathematical relations among the data involved), and (3) mathematical resolution (contents related to the selection and execution of mathematical algorithms). In addition, two narrative approaches were distinguished: (1) relevant situational knowledge (contents related to the intentions and goals of the characters, and the actions performed to reach the goals to link the situation to the mathematical model of the problem) and (2) irrelevant situational knowledge (contents not related to the causal chain generated by the character's goals, such as descriptions of characters, places, objects). The results can be summarized as follows. First, even with non-standard word problems in which additional context-based information should be taken into account, teachers only rarely relied on the narrative approaches. Second, teachers' paradigmatic approach to word problems was rather superficial involving selecting data and certain key words in the problem statement (data selection), followed by automatic triggering of the mathematical model and the execution of the calculations (mathematical resolution). Mathematical reasoning, typical for genuine processing of word problems, only rarely occurred in the observed lessons. Third, the results also suggest that teachers were willing to accept a mathematically correct, though situationally incorrect, problem-solving procedure (e.g., accepting the computation 96 – (57 – 11) to solve the above mentioned shepherd problem).

The previous studies indicate that, in alignment with their beliefs about the relationship between mathematics and reality (see section "Teachers' knowledge and beliefs about mathematical modeling problems"), teachers only rarely include and use situational information when engaging in problem-solving or modeling activities with their whole class or with groups or individual students, even when approaching non-routine problems that require the application of judgment based on real-world knowledge and assumptions. Interventions that are typical and necessary for genuine mathematical modeling – i.e., aiming at a deep understanding of the situation and the mathematical model, and of the mathematical relations among the involved data – remain scarce. Although more research is needed, these studies strongly suggest that (elementary) students are only rarely confronted with genuine mathematical modeling in today's teaching of mathematics.

## Conclusions and Discussion

Over the last decades numerous studies have revealed that, after several years of schooling, many students demonstrate a very strong tendency to exclude real-world knowledge and realistic considerations when confronted with problems that require – at least from the author's point of view – the application of judgment based on real-world knowledge and assumptions rather than the routine application of superficial solution strategies. It is assumed that this non-realistic problem-solving behavior is not "senseless" or "irrational". Indeed it is rather a result of students' beliefs about word problem solving that develop through being immersed in the culture and practices of traditional schooling. Especially three aspects of students' educational environment are assumed to directly or indirectly impact their beliefs about and approaches to word problem solving, namely (1) the nature of the word problems used in classrooms, (2) the nature of teachers' beliefs about word problems, and (3) the way in which teachers treat word problems in the classroom. This chapter has reviewed empirical research conducted at our own research center and by others, and focused on these three supposed causes of students' suspension of sense-making when solving mathematical word problems. Based on those studies we can, first, conclude that the caricature of unrealistic word problems in mathematics textbooks does not do justice to the reality of today's mathematics classrooms. But although many word problems are currently more closely connected to students' experiential worlds, most problems that students typically encounter in current mathematics lessons are still stereotyped in the sense that they require the routine application of simple arithmetical operations. Second, what concerns (prospective) teachers' beliefs towards realistic word problem solving, it was observed that they expressed positive beliefs regarding realistic connections in word problem solving. However, when they are confronted with students' answers to word problems they seem to value more non-realistic than realistic answers. Moreover, if they were asked to value word problems they are inclined to depreciate elements that make a word problem more realistic, such as a complicated relationship between real-world phenomena described in the problem and a mathematical model, and the addition of extraneous information from the story that is not necessarily needed to solve the problem. Third, the scarce studies that explicitly focused on the way in which teachers deal with word problems in regular classrooms revealed that students are offered only limited opportunities to exercise genuine mathematical modeling. Even when approaching non-routine word problems teachers rarely address situational information to provoke a deep understanding of the situation and the mathematical model. In conclusion, modifications on all three aspects mentioned above are needed to improve students' beliefs about and approaches to realistic mathematical modeling.

We acknowledge that the empirical evidence described in this chapter does not allow hard causal statements about the influence of the educational environment on students' non-realistic behavior, mainly due to a lack of an experimental component in the reviewed studies. To make such kind of causal statements further intervention research is needed which directly supports that modifications in the nature of the tasks, the beliefs of the teachers, and/or their instructional approach result in a change in students' beliefs and in more realistic problem-solving behavior. There are some intervention studies that indeed revealed positive effects of teachers' realistic modeling approaches in terms of students' performance, underlying processes, and motivational and affective aspects of learning (e.g., Mason and Scrivani 2004; Verschaffel et al. 1999). However, it should be admitted that these studies fall short in some aspects of either internal or external validity. After reviewing the available research evidence, Niss (2001, p. 8) concludes that "application and modeling capability can be learnt, and according to the above mentioned-findings has to be learnt, but at a cost, in terms of effort, complexity of task, time consumption, and reduction of syllabus in the traditional sense". Consequently, implementing these positive modifications in regular classroom practices is not an easy endeavor.

Pre-service and in-service teacher training can play an important role in preparing and equipping teachers to implement a realistic modeling approach towards word problem solving. Taking into account the findings of the reviewed studies, it is obvious that it is important to stress thereby, among others, the incorporation of modeling tasks in mathematics lessons. Moreover, it is necessary that pre-service as well as in-service teacher training address teachers' beliefs about the place and value of making realistic connections while solving word problems. Since these often hidden beliefs of the teacher are a major obstacle for change in school mathematics, only by explicitly addressing changing them, training will empower teachers to implement a genuine modeling approach to word problem solving (Ernest 1998). Meanwhile, changes are also needed at the meso- and macrolevel of the educational system that support a realistic modeling approach of school word problems. Textbooks should be revised in order to incorporate besides traditional word problems – that help students to master powerful schemes for identifying, understanding, and solving certain categories of problems (e.g., direct proportionality) – also genuine modeling problems – that may be used primarily as exercises in relating real-world situations to mathematical models and in reflecting upon that complex relationship between reality and mathematics. Finally, policy makers and school leaders may be supportive by creating working conditions that are helpful in teachers' implementation of a modeling approach.

## References

Baruk, S. (1985). *L'âge du capitaine. De l'erreur en mathématiques*. Paris: Seuil.
Blum, W., & Niss, M. (1991). Applied mathematical problem solving, modeling, applications, and links to other subjects – State, trends, and issues in mathematics education. *Educational Studies in Mathematics, 22*, 37–68.
Boaler, J. (1993). The role of contexts in the mathematics classroom: Do they make mathematics more real? *For the Learning of Mathematics, 13*(2), 12–17.
Bonotto, C., & Wilczewski, E. (2007). I problem di matematica nella scuola primaria: sull'attivazione o meno di conoscenze di tipo realistico. In C. Bonotto (Ed.), *Quotidianizzare la matematica* (pp. 101–134). Lecce: La Bibilioteca Pensa Multimedia.
Brousseau, G. (1998). *Théorie des situations didactiques*. Grenoble: La Pensée Sauvage.
Bruner, J. (1985). Narrative and paradigmatic modes of thought. In E. W. Eisner (Ed.), *Learning and teaching the ways of knowing* (84th yearbook (Part 2) of the National Society for the Study of Education, pp. 97–115). Chicago: University of Chicago Press.
Burkhardt, H. (1994). Mathematical applications in school curriculum. In T. Husén & T. N. Postlethwaite (Eds.), *The international encyclopedia of education* (2nd ed., pp. 3621–3624). Oxford: Pergamon Press.
Caldwell, L. (1995). *Contextual considerations in the solution of children's multiplication and division word problems*. Unpublished undergraduate thesis, Queen's University, Belfast, Northern Ireland.
Chapman, O. (2006). Classroom practices for context of mathematics word problems. *Educational Studies in Mathematics, 62*, 211–230.
Chen, L., Van Dooren, W., & Verschaffel, L. (2011). An investigation on Chinese teachers' realistic problem solving abilities and beliefs. *International Journal of Science and Mathematics Education, 4*, 80–96.
Davis-Dorsey, J., Ross, S. M., & Morrison, G. R. (1991). The role of rewording and context personalization in the solving of mathematical word problems. *Journal of Educational Psychology, 83*, 61–68.
Depaepe, F., De Corte, E., & Verschaffel, L. (2009). Analysis of the realistic nature of word problems in current elementary mathematics education. In L. Verschaffel, B. Greer, W. Van Dooren, & S. Mukhopadhyay (Eds.), *Words and worlds: Modeling verbal descriptions of situations* (pp. 245–263). Rotterdam: Sense.
Depaepe, F., De Corte, E., & Verschaffel, L. (2010). Teachers' approaches towards heuristic and metacognitive skills and its relationship with students' beliefs and problem-solving skills. *ZDM – The International Journal on Mathematics Education, 42*, 205–218.
Duan, X., Depaepe, F., & Verschaffel, L. (2011). Chinese upper elementary grade mathematics teachers' attitudes towards the place and value of problematic word problems in mathematics education. *Frontiers of Education in China, 6*, 449–469.
Ernest, P. (1998). The culture of the mathematics classroom and the relations between personal and public knowledge: An epistemological perspective. In F. Seeger, J. Voigt, & U. Waschescio (Eds.), *The culture of the mathematics classroom* (pp. 245–268). Cambridge: Cambridge University Press.
Gerofsky, S. (1997). An exchange about word problems. *For the Learning of Mathematics, 17*(2), 22–23.
Gkoris, E., Depaepe, F., & Verschaffel, L. (2013). Investigating the gap between real world and school word problems. A comparative analysis of the authenticity of word problems in the old and the current mathematics textbooks for the 5th grade of elementary school in Greece. *The Mediterranean Journal for Research in Mathematics Education, 12*(1–2), 1–22.
Gravemeijer, K. (1997). Commentary. Solving word problems. A case of modeling? *Learning and Instruction, 7*, 389–397.
Greer, B. (1993). The modeling perspective on wor(l)d problems. *The Journal of Mathematical Behavior, 12*, 239–250.

Greer, B. (1997). Modelling reality in mathematics classrooms. *Learning and Instruction, 7*, 293–307.

Greer, B., Verschaffel, L., & De Corte, E. (2002). "The answer is really 4.5": Beliefs about word problems. In G. C. Leder, E. Pehkonen, & G. Törner (Eds.), *Beliefs: A hidden variable in mathematics education?* (pp. 271–292). Dordrecht: Kluwer.

Hidalgo, M. C. (1997). *L'activation des connaissances à propos du monde reel dans la resolution de problèmes verbaux en arithmétique*. Unpublished doctoral dissertation, Université Laval, Québec, Canada.

Hiebert, J., Carpenter, T. P., Fennema, E., Fuson, K., Human, P., Murray, H., Olivier, A., & Wearne, D. (1996). Problem solving as a basis for reform in curriculum and instruction: The case of mathematics. *Educational Researcher, 25*(4), 12–21.

Ikeda, T., & Stephens, M. (2001). The effects of students' discussion in mathematical modelling. In J. F. Matos, W. Blum, S. K. Houston, & S. P. Carreira (Eds.), *Modelling and mathematics education. ICTMA 9: Applications in science and technology* (pp. 381–390). Chichester: Horwood.

Jiménez, L., & Ramos, F. J. (2011). El impacto negativo del contracto didáctico en la resolución realista de problemas. Un estudio con alumnus de 2° y 3° de Educación Primaria [The negative impact of the didactic contract in realistic problems: A study with second- and third-grade students]. *Electronic Journal of Research in Educational Psychology, 9*, 1155–1182.

Jiménez, L., & Verschaffel, L. (2014). Development of children's strategies for and beliefs about the solution of arithmetic word problems. *Revista Psicodidáctica* [Journal of Psychodidactics], *19*(1), 93–123.

Lee, J.-E. (2012). Prospective elementary teachers' perceptions of real-life connections reflected in posing and evaluating story problems. *Journal of Mathematics Teacher Education, 15*, 429–452.

Ma, L. (1999). *Knowing and teaching elementary mathematics. Teachers' understanding of fundamental mathematics in China and the United States*. Mahwah: Lawrence Erlbaum Associates.

Mason, J. (2001). Modelling modeling: Where is the centre of gravity of-for-when teaching modeling? In J. F. Mators, W. Blum, S. K. Houston, & S. P. Carreira (Eds.), *Modelling and mathematics education. ICTMA 9: Applications in science and technology* (pp. 39–61). Chichester: Horwood.

Mason, L., & Scrivani, L. (2004). Enhancing students' mathematical beliefs: An intervention study. *Learning and Instruction, 14*, 153–176.

National Council of Teachers of Mathematics. (2010). *Why is teaching with problem solving important to student learning?* Reston: National Council of Teachers of Mathematics.

Nesher, P. (1980). The stereotyped nature of school word problems. *For the Learning of Mathematics, 1*(1), 41–48.

Niss, M. (2001). Issues and problems of research on the teaching and learning of applications and modelling. In J. F. Matos, W. Blum, S. K. Houston, & S. P. Carreira (Eds.), *Modelling and mathematics education. ICTMA 9: Applications in science and technology* (pp. 72–89). Chichester: Horwood.

Palm, T. (2002). *The realism of mathematical school tasks. Features and consequences*. Unpublished doctoral dissertation, University of Umea, Sweden.

Pedagogic Institute. (2003). *Diathematikon Programma: Cross-thematic curriculum framework for mathematics*. Retrieved April 17, 2011, from http://www.pi-schools.gr/programs/depps/

Reusser, K., & Stebler, R. (1997). Every word problem has a solution: The social rationality of mathematical modeling in schools. *Learning and Instruction, 7*, 309–327.

Rosales, J., Vicente, S., Chamoso, J. M., Muñez, D., & Orrantia, J. (2012). Teacher-student interaction in joint word problem solving. The role of situational and mathematical knowledge in mainstream classrooms. *Teaching and Teacher Education, 28*, 1185–1195.

Schoenfeld, A. H. (1991). On mathematics as sense-making: An informal attack on the unfortunate divorce of formal and informal mathematics. In J. F. Voss, D. N. Perkins, & J. W. Segal (Eds.), *Informal reasoning and education* (pp. 311–343). Hillsdale: Lawrence Erlbaum Associates.

Seeger, F., Voigt, J., & Waschescio, U. (Eds.). (1998). *The culture of the mathematics classroom*. Cambridge: Cambridge University Press.
Verschaffel, L. (2002). Taking the modeling perspective seriously at the elementary school level: Promises and pitfalls (plenary lecture). In A. Cockburn & E. Nardi (Eds.), *Proceedings of the 26th annual conference of the international group for the psychology of mathematics education* (Vol. 1, pp. 64–82). Norwich: School of Education and Professional Development, University of East Anglia.
Verschaffel, L., De Corte, E., & Lasure, S. (1994). Realistic considerations in mathematical modeling of school arithmetic word problems. *Learning and Instruction, 4*, 273–294.
Verschaffel, L., De Corte, E., & Borghart, I. (1997). Pre-service teachers' conceptions and beliefs about the role of real-world knowledge in mathematical modelling of school word problems. *Learning and Instruction, 7*, 339–359.
Verschaffel, L., De Corte, E., Lasure, S., Van Vaerenbergh, G., Bogaerts, H., & Ratinckx, E. (1999). Learning to solve mathematical application problems: A design experiment with fifth graders. *Mathematical Thinking and Learning, 1*, 195–229.
Verschaffel, L., Greer, B., & De Corte, E. (2000). *Making sense of word problems*. Lisse: Swets & Zeitlinger.
Verschaffel, L., Greer, B., Van Dooren, W., & Mukhopadhyay, S. (Eds.). (2009). *Words and worlds: Modelling verbal descriptions of situations*. Rotterdam: Sense Publishers.
Verschaffel, L., Van Dooren, W., Greer, B., & Mukhopadhyay, S. (2010). Reconceptualising word problems as exercises in mathematical modelling. *Journal für Mathematik-Didaktik, 31*, 9–29.
Vicente, S., Orrantia, J., & Manchado, E. (2011, September). *Authenticity level of mathematic word problems solved by Spanish primary Education students*. Poster session presented at the 14th biennial conference EARLI 2011, Exeter, UK.
Xu, S. (2005). A research on student-teachers' and in-service teachers' realistic considerations of arithmetic word problems. *Psychological Science, 28*, 977–980.
Yackel, E., & Cobb, P. (1996). Sociomathematical norms, argumentation, and autonomy in mathematics. *Journal for Research in Mathematics Education, 27*, 458–477.
Yoshida, H., Verschaffel, L., & De Corte, E. (1997). Realistic considerations in solving problematic word problems: Do Japanese and Belgian children have the same difficulties. *Learning and Instruction, 7*, 329–338.
Zhou, Z., Peverly, S. T., & Xin, T. (2006). Knowing and teaching fractions: A cross-cultural study of American and Chinese mathematics teachers. *Contemporary Educational Psychology, 31*, 438–457.

# Students' Attitudes Towards Mathematics Across Lower Secondary Schools in Shanghai

**Liping Ding, Birgit Pepin, and Keith Jones**

**Abstract** Whilst students in Shanghai were top-ranked internationally in the 2009 and 2012 PISA (Programme for International Student Assessment) studies, much less is known about Shanghai students' attitudes towards mathematics and how this varies across the 11–15 age range. This chapter reports on a study of attitudes towards mathematics of Shanghai students in lower secondary schools, in terms of their emotional disposition towards mathematics and their perceived competence in mathematics. Data were obtained from 4,236 students across grades 6 to 9 in eleven schools across four school districts in SH. Our study found that across the grades an increasing proportion of Shanghai students reported a positive disposition towards mathematics, with only slight drop in Grade 8. We also found that the combinations "*I like it although I can't do it*" and "*I can do it but I dislike it*" were not rare. Interestingly, in seven out of the eleven schools surveyed, more than half of the students who disliked mathematics conveyed that they could do it. Noticeably, the seven schools were all top or above-average achieving schools (according to school district and city-level examinations). Our study also found that more boys expressed a positive emotional disposition than girls. Nevertheless, there was no difference between boys' and girls' perceived competence in mathematics.

**Keywords** Lower secondary schools • Students' attitudes towards mathematics • Emotional disposition towards mathematics • Perceived competence in mathematics • Shanghai

L. Ding (✉) • B. Pepin
Faculty of Teacher and Interpreter Education, Sør-Trøndelag University College, Trondheim, Norway
e-mail: liping.ding@hist.no; birgit.pepin@hist.no

K. Jones
Southampton Education School, University of Southampton, Southampton, UK
e-mail: d.k.jones@soton.ac.uk

 
B. Pepin, B. Roesken-Winter (eds.), *From beliefs to dynamic affect systems in mathematics education*, Advances in Mathematics Education,
DOI 10.1007/978-3-319-06808-4_8

## Introduction

The results of the 2012 PISA (Programme for International Student Assessment) study revealed that, amongst 65 countries, Shanghai-China (SH) was again ranked top for mathematics, reading and science by a clear margin (OECD 2010a, 2013a). Whilst the outstanding performance of SH students may lead the world to admire such success, OECD officials praised the "resilience" of students to succeed despite "tough backgrounds", and pointed to the "high levels of equity" between rich and poor students (OECD 2010a, p. 3). According to PISA coordinator Andreas Schleicher, "in China, the idea is so deeply rooted that education is the key to mobility and success" (quoted by Coughlan 2012). Nevertheless, inside the country a number of educators are worried about maintaining a positive learning culture in schools. As reported by Xie (2011), there has been critical public commentary that PISA 2009 did not, for instance, take into account the large amount of time that SH students tend to spend on studying at school, on completing homework, and at private tutorial classes outside school. Indeed, in mainland China one of the current challenges in school education is how to ensure positive student attitudes towards learning within an intensive examination-driven culture in schools. This is reflected in the latest edition of the Mathematics Curriculum Standards (MCS) (Ministry of Education [MoE] 2011) where affect-related issues and student attitude towards learning mathematics is highlighted. The document states that "assessment should not only concentrate on the results of learning, but also emphasize the learning process; it should pay attention to both the levels of students' mathematics learning and students' disposition and attitudes shown in mathematical activities; and it should help students to appreciate their own disposition and to establish confidence with the subject" (p. 3).

The 2003 PISA study (OECD 2004) revealed that students in countries of high average achievement in mathematics, such as Finland, Japan and Korea, did not express strong interest in mathematics. Moreover, students in high-attaining jurisdictions such as Japan, Korea and Hong Kong-China had amongst the lowest self-reported perceived competence in mathematics. PISA 2009 was the first major international survey in education that included a large sample of students from mainland China. In total, 5,115 SH 15-year-old students participated in PISA 2009 (OECD 2010a, p. 173). In earlier PISA surveys, the Chinese students that were surveyed were not from mainland China, but from jurisdictions such as Hong Kong, Macao and Taiwan (see Fan and Zhu 2004). Noticeably, the main focus of PISA 2009 was on students' skills in reading, rather than those in mathematics or science. As such there is not the data available to analyse the disposition, beliefs and attitudes towards mathematics of SH students. In PISA 2012 (OECD 2013b) students' mathematics self-beliefs and participation in mathematics-related activities was examined. One of the findings is that at the country level, mathematics self-efficacy (student belief in their own capacity to solve specific mathematics tasks) is strongly associated with mathematics performance (OECD 2013b).

Almost two decades ago Watkins and Biggs (1996) identified what they called the 'paradoxical phenomenon' of Chinese learners (from mainland China, Hong Kong, Taiwan): students were apparently taught in classroom conditions that were, according to prevailing views, not conducive to good learning: "large classes, expository methods, relentless norm-referenced assessment, and harsh classroom climate, yet they out-perform Western students at least in science and mathematics, and have deeper, meaning-oriented, approaches to learning" (Watkins and Biggs 1996, p. 3). The performance of SH students in PISA 2009 and 2012 adds to the 'puzzle', and is likely to prompt researchers and educators to pursue deeper and more comprehensive understandings of mathematics teaching and learning in China, and more particularly in SH.

Over the last two decades, a number of studies have investigated and identified a range of features of teaching and learning of mathematics in mainland China. For instance, "teacher as virtuoso" (Paine 1990) and "practice makes perfect" (Li 1999) have been heralded as general principles for teaching and learning. Other features that have been identified, and those that add to deeper knowledge of teaching and learning mathematics in China, include teachers' 'Profound Understanding of Fundamental Mathematics' (PUFM – see Ma 1999); the role of variation problems in Chinese textbooks (Sun 2011); students' solution strategies and the impact of teacher beliefs on these (Cai 2004); and the quality of teacher professional development through school-based activities (Ding et al. 2013; Huang and Bao 2006; Yang 2009). Chen (2010) found that, whilst the underlying philosophy of the reform-oriented mathematics curriculum in mainland China promotes a problem-solving and learner-focused view of teaching, most of the Chinese teachers who took part in the study believed in a teacher-centred approach to teaching mathematics. In terms of the apparent Chinese learners' paradox (see earlier), Chinese researchers and educators contend that it is more likely the result of viewing the cultural context of the Chinese classroom through an uninformed application of what, in the Western view, might constitute effective teaching and learning (e.g. Fan et al. 2004; Leung 2001; Li 2004; Zheng 2006).

Looking across the literature, it appears that little is known about attitudes of mainland Chinese students towards mathematics. Given the importance of ensuring that students remain positive about their continuing learning of mathematics (including beyond compulsory schooling), and the somewhat puzzling findings of existing studies regarding the relationship between students' achievement in mathematics and their attitudes towards mathematics, this chapter aims to make a contribution to the rich international research agenda on students' attitudes towards mathematics. In doing so, we also wish to develop a deeper understanding of the Chinese learners' paradox from the perspective of students' attitudes.

The study reported in this chapter is part of an international comparative project on students' attitude towards mathematics (SATM) conducted in China (SH), France, Norway and the UK (led by one of the authors in Norway). Here we focus in particular on students' emotional disposition towards mathematics (SEDM) and students' perceived competence in mathematics (SPCM), including the relationship between SEDM and SPCM, and both with respect to gender, for SH students in lower secondary school (Grades 6–9, 11- to 15-year-olds). Our research questions are:

Question 1: What are the trends of SATM (in terms of SEDM and SPCM) across Grades 6 to 9?
Question 2: What is the relationship between SEDM and SPCM?
Question 3: Is there a gender difference in SEDM and SPCM across Grades 6 to 9?

In what follows we present a review of the literature in terms of student affect in mathematics education; the theoretical concepts that underpin our study of SATM; the specific dimensions of SATM which were our focus; the results of the study; our discussion of the findings; and finally considerations for further study.

## Facets of Student Affect in Mathematics Education

Informed by work of Zan and Di Martino (2007) and Di Martino and Zan (2010), Pepin (2011) developed a qualitative questionnaire that can provide data on three dimensions of students' attitudes towards mathematics: emotional disposition ("*I like/dislike mathematics, because* ..."); perceived competence ("*I can/cannot do mathematics, because* ..."); and vision of mathematics ("*mathematics is* ..."). In a study of 278 Norwegian pupils (in Grades 7 to 9) and 194 English pupils (in Grades 6, 8 and 10), she found that the percentages of English pupils who said that they liked mathematics ranged from 42 % in Grade 6, to 39 % in Grade 8, and to 41 % in Grade 10. The percentage of Norwegian pupils who said that they liked mathematics varied from 44 % in Grade 7, to 32 % in Grade 8, and 46 % in Grade 9. As such, there was a rather similar pattern in the two countries' data sets; that is, pupils' interests in mathematics decreased between ages 12 to 14 and then appeared to increase after age 14. This apparent drop in interest between the ages of 12 and 14 was also reported by Hodgen et al. (2009) in their study of 3,000 lower secondary school students in England. They found that 63 % of English 12-year-olds responded that they enjoyed their mathematics lessons, but this fell to 54 % for 14-year-olds.

In terms of linking cognitive, conative and affective factors in students' mathematics learning, Op't Eynde et al. (2006) examined the relationship between students' mathematics-related beliefs, their emotions and their problem-solving behaviour in the mathematics classroom. Based on case studies of 16 Belgium students' affect and behaviour during mathematical problem solving, they found that the nature and intensity of emotion experienced during problem solving differed between students. That is, when confronted with the same kind of difficulty in the early stage of problem solving, one of the students felt 'hopeless' and immediately gave up, whilst another student apparently experienced this as a challenge and tried to solve the problem. Moreover, they found that those students who showed a high degree of confidence acted self-confidently throughout the problem-solving process, whilst others who scored low on confidence in their mathematics capabilities started questioning their capabilities the first minute they encountered a problem.

Regarding Chinese students, Luan and Tao (2008) investigated the views of 907 middle school students (Grades 7 to 9) across three cities in the east of China (Yangzhou, SH and Hefei). This team identified four dimensions of student affect in mathematics: mathematics-subject beliefs; learning interests; emotional experiences; and aesthetic feelings of mathematics. On a scale of 1–5, they reported that students' mathematics-subject beliefs were the highest (average 3.792); students' emotional experiences were positive (average 3.634); while students' aesthetic feeling of mathematics (average 3.458) and learning interests (average 3.375) were lower. Moreover, there were statistically significant differences in mathematics-subject beliefs, learning interests and emotional experiences by grade. For instance, Grade 7 students had the highest score of emotional experiences in mathematics, whilst Grade 9 students had the lowest score of emotional experiences in mathematics.

Another research team, Chen et al. (2011), investigated the views of 425 middle school students from a city in the south of China (Dongguan). The theoretical frame of this study included six dimensions of affect: self-confidence in mathematics learning (ML); attitude towards success in ML; parents' attitude towards mathematics; recognition of the usefulness of mathematics; the degree of anxiety in ML; and intrinsic motivation in ML. The findings showed a high correlation among the dimensions of students' self-confidence in ML, attitude towards success in ML, the degree of anxiety in ML and intrinsic motivation in ML. That is, a student who was confident in terms of mathematics learning rarely considered learning mathematics as a heavy burden, had considerably high interest in mathematics as a subject, and liked to solve challenging mathematics problems.

Exploring the development of students' epistemological beliefs about mathematics and the differences in mathematical beliefs amongst students at different performance levels, Zhou et al. (2011) used two instruments to investigate students' beliefs in the city of Liaocheng in the east of China: (1) problems in hypothetical situations (Huang 2002); and (2) a questionnaire (based on the framework in Liu and Chen 2004). The latter included three dimensions of affect: students' beliefs about the nature of mathematics; about the mathematical learning process; and about their experiences in mathematics. They found that students' attitude towards mathematics was closely related to the mathematical course content, as well as to the implicit beliefs which students acquired over the course of their learning of mathematics, including their learning outcomes. Perhaps more importantly, they reported on a v-shaped pattern of students' mathematical beliefs from primary school to college; that is, that students scored highly at primary schooling and at college level, but decreased at junior and senior high school level.

The above studies show that researchers have been engaged in examining various facets of student affect in the field of mathematics education, such as emotional disposition, belief, behaviour, interest, feeling, attitude, anxiety, and motivation, to name but a few. To date, however, these concepts remain relatively vague in terms of the multiple meanings used in the different studies and the complex relationships between them. In the next section we explain the theoretical concepts which underpin our study of SATM and clarify the specific dimensions on which we focus in this chapter.

## Conceptual Framework

McLeod (1992) identified attitudes as one of the three key affective constructs in mathematics education, together with beliefs and emotions. DeBellis and Goldin (1999) subsequently proposed *values* as a fourth construct. According to Ruffell et al. (1998), previous research on 'attitude' (e.g. Allport 1935) viewed attitude as a single dimension (i.e. as either belief or feeling), something that coincides with the colloquial meaning of the word. Since then the construct has evolved and is now regarded as multi-dimensional, comprising of cognitive, affective and conative dimensions (for more see Ajzen 1988; Triandis 1971; Di Martino and Zan 2010).

Di Martino and Zan (2010, p. 29) summarize three forms of definition of 'attitude towards mathematics', as follows:

- A *simple* definition that describes attitude as the positive or negative degree of affect associated with mathematics.
- A *bi-dimensional* definition that involves emotions and beliefs but in which behaviours related to mathematics do not appear explicitly.
- A *tripartite* definition that recognizes three components in attitude: emotional response towards mathematics, belief regarding mathematics and behaviour related to mathematics.

To make a contribution in terms of developing further insights into the different facets of the construct, we take the view of Di Martino and Zan (2010) and Pepin (2011) that we should not assume an a priori definition of 'attitude toward mathematics'. Hence, we investigated SATM by letting students write about their experiences with mathematics (see questions below), that is, from students' own perspectives in their school contexts, and tried to develop an understanding of the relationships between variables such as gender, grade and school environment that might be related to students' descriptions of their attitudes. In our study, we used the same qualitative 'questionnaire' developed by Pepin (2011), which focused on the following three key 'questions':

1. "I like/dislike mathematics because …";
2. "I can/cannot do mathematics because …";
3. "Mathematics is …".

The questionnaire enables students to narrate their emotional disposition towards mathematics (1); their perceived competence in mathematics (2); and their vision of mathematics (3). As pointed out by Pepin (2011), exploring these constructs in a culturally different learning context may help us to develop a deeper understanding of what we mean by 'affect in mathematics education'.

Exploring the relationships between these two dimensions of emotional disposition and perceived competence, Zan and Di Martino (2007) show that the connection between them is so strong that some students used expressions such as "I like" and "I can do it" as synonyms in their narratives about mathematics; likewise "I dislike" and "I can't do it" appeared as synonyms. In addition, Di Martino and

Zan (2010) point out that in their analysis of the students' narratives, where a negative emotional disposition was explicitly stated, this was always associated with either an instrumental view of mathematics (such as 'rules without reasons') or a low perceived competence.

In the present stage of our analysis, and in order to get an overview of the data, we decided to focus on two dimensions when analysing the responses to the questionnaire:

- Students' emotional disposition towards mathematics (SEDM) in the form of the students' responses to "I like or dislike mathematics";
- Students' perceived competence in mathematics (SPCM) in the form of the students' responses to "I can or cannot do mathematics".

Noticeably, PISA 2012 (2013b) used two terms to distinguish students' beliefs in their own mathematics skills: namely, self-efficacy (students believe in their own ability to solve specific mathematics tasks); self-concept (students' belief in their own mathematics capabilities). We applied the term 'perceived competence', as Di Martino and Zan (2010, p. 39) did. We considered that this term corresponded most appropriately to the colloquial meaning of the wording 'can/cannot' in the questionnaire, and hence was considered by the teachers and students in the study to be less abstract.

## The Study

### *Background Information on Schools and Students*

Data for the study were collected in SH between March and April in 2012. At that time students were in the second semester (late stage) of their school year (there are two semesters in each school year in SH). With the aim of developing a deeper understanding of the trends of students' attitudes across grades, as noted by researchers in other countries (e.g. Brown et al. 2010; Hodgen et al. 2009; Pepin 2011), we collected data from Grade (G) 6 students (age 11/12) to G9 students (age 14/15), these being recognized as the key grades of lower secondary education in SH. G8 was seen as particularly important in the data collection as we considered the possibility of linking our findings to those of TIMSS and PISA.

In total, our data consisted of 4,313 valid student questionnaires from eleven schools in four school districts in SH (there are 18 school districts in the SH region). The four school districts are located in four regions of the city:

1. The south-west of the city centre: part of this district was the French colonial area of the old city (1914–1945). It is now part of the business and commerce centres of the modern city;
2. The north-east of the city centre: this is now being transformed from an industrial to an 'academic' area of the city;

3. The west suburb of the city: this is the agricultural area of the city; and
4. The south-west suburb of the city: this is the new technology-development area of the city.

In terms of details of the 11 schools, we have chosen a diversity of types of school in SH: state schools; local private schools; and international private schools (for more official information of the types of school in SH see Chapter 4 in OECD 2010b). We considered that it was important to collect student data from different school contexts in the city, in order to gain a fuller picture of SH students' attitude towards mathematics. We collected data from four types of school within the 11 schools that were accessible to us in the study (the sizes of the 11 schools are given in Table 1):

1. State lower secondary schools (G6-9): schools coded as 11, 13, 15, 17, 19, 20, 21 in Table 1 (2,373 students, 61.5 % of the total students);
2. State secondary schools (G6-12): school coded as 14 in Table 1 (699 students, 18.1 % of the total students);
3. State schools (G1-9): schools coded as 16, 18 in Table 1 (508 students, 13.2 % of the total students);
4. Private lower secondary schools (G6-9): school coded as 12 in Table 1 (276 students, 7.2 % of the total students).

Thus, the majority of students in our study were from the state lower secondary schools. According to the school district and city-level standard examinations in SH, schools 12, 16, 18, 19 and 21 were among the top achieving schools; schools 11, 15 and 17 were above-average achieving schools; and schools 13, 14 and 20 were below-average achieving schools in their school district.

**Table 1** The codes of the school districts (SD) and the schools in SH

| SD locations | Codes | Number of students[a] | Percentage of students in total[a] | School codes | School size | Number of students[a] | Percentage of students in total[a] |
|---|---|---|---|---|---|---|---|
| South-west of the city centre | 1 | 1,569 | 40.7 % | 11 | 504 | 378 | 9.8 % |
| | | | | 12 | 1,011 | 276 | 7.2 % |
| | | | | 13 | 637 | 594 | 15.4 % |
| | | | | 14 | 407 | 321 | 8.3 % |
| West suburb of the city | 2 | 804 | 20.9 % | 15 | 516 | 474 | 12.3 % |
| | | | | 16 | 1,090 | 330 | 8.6 % |
| South-west suburb of the city | 3 | 639 | 16.6 % | 17 | 1,433 | 461 | 12.0 % |
| | | | | 18 | 210 | 178 | 4.6 % |
| North-east of the city centre | 4 | 844 | 21.9 % | 19 | 920 | 511 | 13.3 % |
| | | | | 20 | 449 | 244 | 6.3 % |
| | | | | 21 | [b] | 89 | 2.3 % |

[a]The number and percentage of students who participated in this study
[b]Not available at this time

In total, there were 1,058 G6 students (27.4 % of the total), 1,032 G7 students (26.8 % of the total), 1,073 G8 students (27.8 % of the total) and 693 G9 students (18.0 % of the total). The number of students in G9 was considerably smaller because G9 is the final year of lower secondary school and schools focus on the preparation for high stakes (and high pressure) city standard examinations (which are the key examinations to decide students' further study, e.g. in high schools, or vocational schools). As such, we were not able to collect data on G9 from two of the eleven schools. In terms of gender, 2,153 male students (55.8 % of the total) and 1,703 female students (44.2 % of the total) took part in the study.

It should be noted that our data cannot be viewed as a statistically representative sample of each school district; rather, as we wanted a diversity of schools (in terms of the characteristics described above), we selected schools on the basis of these characteristics and at the same time those that were accessible to us at the time of the study. Similarly, we collected data from the classes that were accessible to us in these schools at the time. Thus, rather than comparing SATM in different areas of the city, we concentrate on the trends and gender differences of SATM across each selected grade in these schools.

The students were asked to complete the questionnaires in school. Nevertheless, in some schools they were allowed to complete the questionnaires at home and outside school time (due to the intense pressure on study time in school).

## *Data Analysis and Procedures to Counter Threats to Validity and Reliability*

As discussed in the previous section, we collected data by using the same qualitative questionnaire developed by Pepin (2011) and we collected data from similar-aged students. Our reasoning was that this would enable us to further develop ideas and methods to examine the socio-culturally situated learning contexts, which, we believe, may be closely connected to the relationship between student attainment in mathematics and their attitudes towards mathematics.

We conducted the data analysis in two phases. In the first phase, we worked with five mathematics teachers from Shanghai Soong Ching Ling Schools in SH to take a first look at the whole questionnaire data. We (with the five mathematics teachers) took notes of the main key words or themes students expressed in their questionnaires. The first author first focused on the analysis of one school, and she then divided the whole data into parts, so the five teachers could respectively focus on parts of the data. Next, we had several meetings to discuss what we thought were the most significant themes in order to further examine our first reading of the questionnaires. Over the meetings and discussions, we faced several challenges in terms of inter-researcher reliability for category development. For example, we had to tackle the multiple perceptions that we had about the words students expressed about their emotional disposition ('like/dislike') and their perceived competence ('can/cannot do it'). For instance, it was not straightforward for us to identify the main categories that were connected to students' emotional disposition ('like') in mathematics according to the large number of questionnaires. Even in one single school, the first author identified 13 different categories of students' emotional disposition towards mathematics: (1) interesting; (2) successful; (3) logical thinking; (4) no need to commit to rote learning; (5) easy; (6) useful in real life; (7) do like the teacher; (8) specific sub-areas in mathematics (e.g. geometry); (9) textbooks; (10) key subject in examinations; (11) important subject for future career; (12) rigour; (13) improve one's thoughtfulness. During the discussions, some teachers considered some of the categories identified by the first author as non-significant in the data they studied, whilst other teachers suggested some new categories according to their data study. As a result it took a considerable amount of time for us to refine the categories.

Next, we also needed to be consistent with the words students used to convey their emotional disposition towards mathematics. For instance, students could use different words to express what they meant by 'interesting', such as 'playful', 'needs concentration', 'enjoyable' or 'fascinating'. With the 4,313 valid questionnaires, there was some uncertainty among us about which words meant exactly what, as the meanings of the word could be different for different people. Moreover, two aspects of liking that the students expressed could be detected: on the one hand, students liked mathematics because of their true interests or feeling of success in learning; on the other hand, some of them liked mathematics because it was a key subject in examinations or an important subject for future career.

In short, a decision had to be made with at least two options in mind in the initial analysis. One option was to conduct a qualitative analysis with a considerably smaller number of students in one single school. This would then enable us to refine the categories for analysing the complex constructs of students' attitudes. The other option was to conduct a quantitative analysis of all data. This would enable us to build up relationships between students' responses to the first two questions ('like/dislike' and 'can/cannot do it'); and to gain insights into variables such as grade, school and gender which were likely to relate to the change of students' attitude. Given the 4,313 valid qualitative questionnaires we collected, and in order to get an overview of the data, we considered it preferable to conduct a quantitative analysis of the first two questions across the 11 schools, rather than only focusing on part of the data in one or two schools.

For the second phase of our data analysis, we thus undertook a statistical analysis of the students' responses about whether they liked/disliked mathematics, and whether they perceived that they could, or could not, do mathematics. We applied the statistical codes to the analysis of our qualitative questionnaires. In so doing, we aimed to be less subjective on the meanings of the words students conveyed, and more objective in understanding the relationship between students' responses to the two questions and the connections with the variables of grade, school and gender. In the first place, we coded the four school districts and the 11 schools, as shown in Table 1.

Next, we had extensive discussions with a statistics researcher from Shanghai Soong Ching Ling Child Development Research Centre and the five teachers, to code students' gender and students' responses to the first two questions in the questionnaires. In our discussions, we were aware of the various degrees of like, or dislike, and the degree of certainty of perceived competence students expressed in their writing. For instance, in our initial analysis, we noted that students responded to the word 'like' with certain degrees such as 'like very much', 'considerably like', 'like a bit', and 'all right'. In our coding, we decided that these responses went to the category of 'like', for they showed positive responses to 'like'. Students' expressions to the word 'dislike' were generally clear. But there were some students whose responses were considerably mixed. For instance, some students responded 'I like it but sometimes I dislike it', or 'I do not like it but I do not hate it either'. Some students did not give a clear answer about whether they liked or disliked mathematics, but they did write several words to express their feeling about mathematics. In all of these mixed or vague cases, we coded them to 3 – others. We decided the following codes in our discussions:

- Gender: 1 – boy, 2 – girl, and 3 – if not provided information.
- Students' responses to the first question (T1 – like/dislike mathematics): T1-1 – like, T1-2 – dislike, and T1-3 – others.
- Students' responses to the second question (T2 – can/cannot do mathematics): T2-1 – can, T2-2 – cannot, and T2-3 – others.

Consequently, we selected 3,856 students (out of 4,313 students) from the 11 schools, as these students clearly responded to the first two questions in the questionnaire. The final results of the analysis were produced using SPSS17.

## Results

### *Attitudes, Grades and Schools*

Table 2 shows students' emotional disposition towards mathematics (their responses to 'like/dislike') by school grade level. In general, the proportion of students who liked mathematics increased across the grades from G6 (78.2 %) to G7 (84.1 %) and then to G9 (85.3 %), only slightly dropping down at G8 (82.2 %). In our study, we used a Chi-square test to examine both grade and SEDM ('like/dislike'). There was a statistically significant difference of SEDM across the grades ($p=0.000<0.01$).

**Table 2** Cross-analysis by grade for 'I like/dislike mathematics'

| Grade | Like # (%) | Dislike # (%) | In total |
|---|---|---|---|
| 6 | 827 (78.2 %) | 231 (21.8 %) | 1,058 (100 %) |
| 7 | 868 (84.1 %) | 164 (15.9 %) | 1,032 (100 %) |
| 8 | 882 (82.2 %) | 191 (17.8 %) | 1,073 (100 %) |
| 9 | 591 (85.3 %) | 102 (14.7 %) | 693 (100 %) |
| In total | 3,168 (82.2 %) | 688 (17.8 %) | 3,856 (100 %) |

**Table 3** Cross-analysis by grade in four schools for 'I like/dislike mathematics'

| Grade | School | Like # (%) | Dislike # (%) | In total # (%) |
|---|---|---|---|---|
| 6 | 12 | 51 (77.3 %) | 15 (22.7 %) | 66 (100 %) |
| | 15 | 93 (67.4 %) | 45 (32.6 %) | 138 (100 %) |
| | 18 | 51 (100 %) | 0 (.0 %) | 51 (100 %) |
| | 19 | 137 (74.9 %) | 46 (25.1 %) | 183 (100 %) |
| 7 | 12 | 58 (69.9 %) | 25 (30.1 %) | 83 (100 %) |
| | 15 | 104 (78.8 %) | 28 (21.2 %) | 132 (100 %) |
| | 18 | 56 (98.2 %) | 1 (1.8 %) | 57 (100 %) |
| | 19 | 120 (82.8 %) | 25 (17.2 %) | 145 (100 %) |
| 8 | 12 | 67 (85.9 %) | 11 (14.1 %) | 78 (100 %) |
| | 15 | 112 (100 %) | 0 (.0 %) | 112 (100 %) |
| | 18 | 30 (100 %) | 0 (.0 %) | 30 (100 %) |
| | 19 | 148 (80.9 %) | 35 (19.1 %) | 183 (100 %) |
| 9 | 12 | 39 (79.6 %) | 10 (20.4 %) | 49 (100 %) |
| | 15 | 70 (76.1 %) | 22 (23.9 %) | 92 (100 %) |
| | 18 | 35 (87.5 %) | 5 (12.5 %) | 40 (100 %) |

When further analysing the change of students' emotional disposition in mathematics in each of the 11 schools, we found that the change pattern in some of the schools was considerably different from the general trend shown in Table 2. We consider that this finding might indicate a significant difference amongst schools in the city. To illustrate such a variation, in Table 3 we separately show four schools correspondingly from the four school districts in the city (schools 12, 15, 18 and 19, which were top or above-average achieving schools). In School 12, there were 'double drops' of students' emotional disposition towards mathematics ('like') across the grades. The first drop was from G6 (77.3 %) to G7 (69.9 %), and the second drop was from G8 (85.9 %) to G9 (79.6 %). In School 15, the change of 'like' rose from G6 (67.4 %) to G7 (78.8 %) and to G8 (100 %), but then dropped in G9 (76.1 %). In School 18, students appeared to have very high emotional disposition towards mathematics, with little change across the three grades (100 % in G6, 98.2 % in G7, 100 % in G8). It only dropped down to 87.5 % in G9. In School 19, the change is correspondingly similar to the general trend across grades shown in Table 2, with 74.9 % in G6, 82.8 % in G7 and 80.9 % in G8. We were unable to collect the data in G9 in School 19 due to the high examination pressure upon teachers and students at the time.

Table 4 shows that students' perceived competence in mathematics (their responses to 'can/cannot') also changed by grade. We found similar changing 'waves' in students' perceived competence ('can') to those we found in their emotional disposition ('like') from G6 (76.7 %) to G7 (79.7 %) and from G8 (76.2 %) to G9 (79.4 %). In each grade (from G6 to G9), there was a considerably higher percentage of students who perceived competence in mathematics (ranging from 79.7 % in G7 to 76.2 % in G8) than those who did not perceive competence in

**Table 4** Cross-analysis by grade for 'I can/cannot do mathematics'

| Grade | Can # (%) | Cannot # (%) | In total # (%) |
|---|---|---|---|
| 6 | 811 (76.7 %) | 247 (23.3 %) | 1,058 (100 %) |
| 7 | 823 (79.7 %) | 209 (20.3 %) | 1,032 (100 %) |
| 8 | 818 (76.2 %) | 255 (23.8 %) | 1,073 (100 %) |
| 9 | 550 (79.4 %) | 143 (20.6 %) | 693 (100 %) |
| In total | 3,002 (77.9 %) | 854 (22.1 %) | 3,856 (100 %) |

**Table 5** Cross-analysis by grade in four schools for 'I can/cannot do mathematics'

| Grade | School | Can # (%) | Cannot # (%) | In total # (%) |
|---|---|---|---|---|
| 6 | 12 | 57 (86.4 %) | 9 (13.6 %) | 66 (100 %) |
| | 15 | 116 (84.1 %) | 22 (15.9 %) | 138 (100 %) |
| | 18 | 48 (94.1 %) | 3 (5.9 %) | 51 (100 %) |
| | 19 | 83 (45.4 %) | 100 (54.6 %) | 183 (100 %) |
| 7 | 12 | 69 (83.1 %) | 14 (16.9 %) | 83 (100 %) |
| | 15 | 116 (87.9 %) | 16 (12.1 %) | 132 (100 %) |
| | 18 | 54 (94.7 %) | 3 (5.3 %) | 57 (100 %) |
| | 19 | 72 (49.7 %) | 73 (50.3 %) | 145 (100 %) |
| 8 | 12 | 61 (78.2 %) | 17 (21.8 %) | 78 (100 %) |
| | 15 | 107 (95.5 %) | 5 (4.5 %) | 112 (100 %) |
| | 18 | 28 (93.3 %) | 2 (6.7 %) | 30 (100 %) |
| | 19 | 105 (57.4 %) | 78 (42.6 %) | 183 (100 %) |
| 9 | 12 | 42 (85.7 %) | 7 (14.3 %) | 49 (100 %) |
| | 15 | 78 (84.8 %) | 14 (15.2 %) | 92 (100 %) |
| | 18 | 33 (82.5 %) | 7 (17.5 %) | 40 (100 %) |

mathematics (ranging from 20.3 % in G7 to 23.8 % in G8). Using a Chi-square test we found no statistically significant difference of SPCM across grades ($p = 0.134 > 0.01$).

We also note some puzzling results when we purposefully traced the complex relationship of students' emotional disposition with their perceived competence in the top and above-average achieving schools. As shown in Table 3, on the one hand, one of the above-average schools appeared to enable those students who did not like mathematics to develop certain perceived competence in mathematics. For instance, Table 5 shows that in School 15 there were more students in three of the four grades (84.1 % in G6, 87.9 % in G7, 95.5 % in G8 and 84.8 % in G9) who considered that they could do mathematics compared with those who said that they liked mathematics (67.4 % in G6, 78.8 % in G7, 100 % in G8, and 76.1 % in G9). Similar results were found in another top achieving school (School 12 in Table 5). More detailed data of classroom observation and students' and teachers' interviews in these two schools are needed to help us better understand this phenomenon.

On the other hand, however, not every top achieving school had similar effects on students' perceived competence in mathematics. That is, a considerable number of students who liked mathematics did not think that they could do mathematics. In School 19, for instance, there was a considerably large number of students' who liked mathematics (see Table 3), yet these students' perceived competence in mathematics was polarized (see Table 5). Such a 'polarization' in students' responses to 'can/cannot' could also be identified across grades. For instance, in G6, 45.4 % students considered that they could do mathematics, yet 54.6 % students did not think so. In G7, 49.7 % students responded that they could do mathematics, yet 50.3 % responded 'cannot do it'. In G8, it seems that more students began to perceive more competence in mathematics (57.4 % responded that they 'can do mathematics', yet 42.6 % responded that they cannot do it).

## *The Connection Between SEDM and SPCM*

Table 6 shows that within the 3,002 students who considered that they could do mathematics, 88.4 % of them (2,655 students) liked mathematics, while 11.6 % of them (347 students) disliked it. Using a Chi-square test we found a statistically significant difference between SPCM ('can') and SEDM ('like/dislike') across the grades ($p=0.001<0.01$).

Moreover, we found that there were a considerable number of students who liked mathematics, yet they did not perceive that they had competence to do it. As Table 6 shows, there seems to be a trend that the older these students were, the less their perceived competence was in mathematics; see the relationship of 'cannot' and 'like' across G6 (57.1 %), G7 (62.7 %), G8 (59.6 %) and G9 (62.2 %). Using a Chi-square test we found that there was no statistically significant difference between SPCM ('cannot') and SEDM ('like/dislike') across the grades ($p=0.612>0.05$). This may be due to the sample of our study, and thus may not represent the trend of the whole population of students in SH.

**Table 6** Cross-analysis by grade for 'I like/dislike mathematics' and 'I can/cannot do mathematics'

| Grade | | Like # (%) | Dislike # (%) | Total # (%) |
|---|---|---|---|---|
| 6 | Can | 686 (84.6 %) | 125 (15.4 %) | 811 (100.0 %) |
| 7 | | 737 (89.6 %) | 86 (10.4 %) | 823 (100.0 %) |
| 8 | | 730 (89.2 %) | 88 (10.8 %) | 818 (100.0 %) |
| 9 | | 502 (91.3 %) | 48 (8.7 %) | 550 (100.0 %) |
| In total | | 2,655 (88.4 %) | 347 (11.6 %) | 3,002 (100.0 %) |
| 6 | Cannot | 141 (57.1 %) | 106 (42.9 %) | 247 (100.0 %) |
| 7 | | 131 (62.7 %) | 78 (37.3 %) | 209 (100.0 %) |
| 8 | | 152 (59.6 %) | 103 (40.4 %) | 255 (100.0 %) |
| 9 | | 89 (62.2 %) | 54 (37.8 %) | 143 (100.0 %) |
| In total | | 513 (60.1 %) | 341 (39.9 %) | 854 (100.0 %) |

**Table 7** Cross-analysis by school for 'I like/dislike mathematics' and 'I can/cannot do mathematics'

| School | | Can # (%) | Cannot # (%) | In total # (%) |
|---|---|---|---|---|
| 11 | Like | 277 (87.7 %) | 39 (12.3 %) | 316 (100 %) |
| 12 | | 190 (88.4 %) | 25 (11.6 %) | 215 (100 %) |
| 13 | | 419 (85.2 %) | 73 (14.8 %) | 492 (100 %) |
| 14 | | 216 (84.0 %) | 41 (16.0 %) | 257 (100 %) |
| 15 | | 360 (95.0 %) | 19 (5.0 %) | 379 (100 %) |
| 16 | | 256 (91.4 %) | 24 (8.6 %) | 280 (100 %) |
| 17 | | 338 (86.4 %) | 53 (13.6 %) | 391 (100 %) |
| 18 | | 159 (92.4 %) | 13 (7.6 %) | 172 (100 %) |
| 19 | | 219 (54.1 %) | 186 (45.9 %) | 405 (100 %) |
| 20 | | 159 (84.1 %) | 30 (15.9 %) | 189 (100 %) |
| 21 | | 62 (86.1 %) | 10 (13.9 %) | 72 (100 %) |
| In total | | 2,655 (83.8 %) | 513 (16.2 %) | 3,168 (100 %) |
| 11 | Dislike | 35 (56.5 %) | 27 (43.5 %) | 62 (100 %) |
| 12 | | 39 (63.9 %) | 22 (36.1 %) | 61 (100 %) |
| 13 | | 40 (39.2 %) | 62 (60.8 %) | 102 (100 %) |
| 14 | | 31 (48.4 %) | 33 (51.6 %) | 64 (100 %) |
| 15 | | 57 (60.0 %) | 38 (40.0 %) | 95 (100 %) |
| 16 | | 27 (54.0 %) | 23 (46.0 %) | 50 (100 %) |
| 17 | | 38 (54.3 %) | 32 (45.7 %) | 70 (100 %) |
| 18 | | 4 (66.7 %) | 2 (33.3 %) | 6 (100 %) |
| 19 | | 41 (38.7 %) | 65 (61.3 %) | 106 (100 %) |
| 20 | | 25 (45.5 %) | 30 (54.5 %) | 55 (100 %) |
| 21 | | 10 (58.8 %) | 7 (41.2 %) | 17 (100 %) |
| In total | | 347 (50.4 %) | 341 (49.6 %) | 688 (100 %) |

Next, we found that in seven out of the 11 schools (schools 11, 12, 15, 16, 17, 18 and 21), more than half of those students who disliked mathematics conveyed that they could do it (see Table 7). Noticeably, the seven schools were all from the top achieving schools (numbers 12, 16, 18 and 21) or above average schools (numbers 11, 15, 17) in their school district. As discussed in the foregoing section, school 19 was one of the top achieving schools in the school district, yet a considerable number of students across Grades 6 to 8 who liked mathematics did not think that they could do it. These findings further lead us to recognize the significant difference amongst schools in SH.

## *Attitudes, Gender and Grades*

Table 8 shows the difference between boys and girls in their responses to T1: 85.8 % of 2,153 boys in the study expressed that they liked mathematics, while 77.5 % of 1,703 girls did so. Using a Chi-square test we found a statistically significant difference of SEDM ('like/dislike') between genders across the grades ($p=0.000<0.01$).

**Table 8** Cross-analysis by gender for 'I like/dislike mathematics'

| Gender | Like # (%) | Dislike # (%) | In total # (%) |
|---|---|---|---|
| Boy | 1,848 (85.8 %) | 305 (14.2 %) | 2,153 (100 %) |
| Girl | 1,320 (77.5 %) | 383 (22.5 %) | 1,703 (100 %) |
| In total | 3,168 (82.2 %) | 688 (17.8 %) | 3,856 (100 %) |

**Table 9** Cross-analysis by grade and gender for 'I like/dislike mathematics'

| Grade | Gender | Like # (%) | Dislike # (%) | In total # (%) |
|---|---|---|---|---|
| 6 | Boy | 526 (84.3 %) | 98 (15.7 %) | 624 (100 %) |
| | Girl | 301 (69.4 %) | 133 (30.6 %) | 434 (100 %) |
| | In total | 827 (78.2 %) | 231 (21.8 %) | 1,058 (100 %) |
| 7 | Boy | 498 (88.0 %) | 68 (12.0 %) | 566 (100 %) |
| | Girl | 370 (79.4 %) | 96 (20.6 %) | 466 (100 %) |
| | In total | 868 (84.1 %) | 164 (15.9 %) | 1,032 (100 %) |
| 8 | Boy | 508 (84.7 %) | 92 (15.3 %) | 600 (100 %) |
| | Girl | 374 (79.1 %) | 99 (20.9 %) | 473 (100 %) |
| | In total | 882 (82.2 %) | 191 (17.8 %) | 1,073 (100 %) |
| 9 | Boy | 316 (87.1 %) | 47 (12.9 %) | 363 (100 %) |
| | Girl | 275 (83.3 %) | 55 (16.7 %) | 330 (100 %) |
| | In total | 591 (85.3 %) | 102 (14.7 %) | 693 (100 %) |

**Table 10** Cross-analysis by gender for 'I can/cannot do mathematics'

| Gender | Can # (%) | Cannot # (%) | In total # (%) |
|---|---|---|---|
| Boy | 1,691 (78.5 %) | 462 (21.5 %) | 2,153 (100 %) |
| Girl | 1,311 (77.0 %) | 392 (23.0 %) | 1,703 (100 %) |
| In total | 3,002 (77.9 %) | 854 (22.1 %) | 3,856 (100 %) |

When further analysing the trends of gender in responding to 'like/dislike' (T1) across the grades, we found that boys' emotional disposition in mathematics appeared to remain high and stable by grade (84.3 % in G6, 88.0 % in G7, 84.7 % in G8 and 87.1 % in G9) while girls seemed to develop their disposition towards mathematics from G6 (69.4 %) to G7 (79.4 %), remained stable at G8 (79.1 %), and then increased to 83.3 % at G9 (see Table 9).

Interestingly, there was no difference in these boys' and girls' perceived competence in mathematics; 78.5 % of the 2,153 boys and 77.0 % of the 1,703 girls believed that they could do mathematics (see Table 10). Using a Chi-square test we found no statistically significant difference of SPCM ('can/cannot') between genders ($p=0.222>0.01$).

Table 11 shows that boys' and girls' perceived competence in mathematics changed at G8. Girls' perceived competence in mathematics increased faster than

**Table 11** Cross-analysis by grade and gender for 'I can/cannot do mathematics'

| Grade | Gender | Can # (%) | Cannot # (%) | In total # (%) |
|---|---|---|---|---|
| 6 | Boy | 484 (77.6 %) | 140 (22.4 %) | 624 (100 %) |
| | Girl | 327 (75.3 %) | 107 (24.7 %) | 434 (100 %) |
| | In total | 811 (76.7 %) | 247 (23.3 %) | 1,058 (100 %) |
| 7 | Boy | 451 (79.7 %) | 115 (20.3 %) | 566 (100 %) |
| | Girl | 372 (79.8 %) | 94 (20.2 %) | 466 (100 %) |
| | In total | 823 (79.7 %) | 209 (20.3 %) | 1,032 (100 %) |
| 8 | Boy | 459 (76.5 %) | 141 (23.5 %) | 600 (100 %) |
| | Girl | 359 (75.9 %) | 114 (24.1 %) | 473 (100 %) |
| | In total | 818 (76.2 %) | 255 (23.8 %) | 1,073 (100 %) |
| 9 | Boy | 297 (81.8 %) | 66 (18.2 %) | 363 (100 %) |
| | Girl | 253 (76.7 %) | 77 (23.3 %) | 330 (100 %) |
| | In total | 550 (79.4 %) | 143 (20.6 %) | 693 (100 %) |

that of boys from G6 (girls 75.3 %, boys 77.6 %) to G7 (girls 79.8 %, boys 79.7 %). At G7 and G8, there were more girls than boys who perceived their competence in mathematics positively. At G9, there were more boys (81.8 %) than girls (76.7 %) who perceived their competence in mathematics positively.

## Discussion of the Findings

In the foregoing sections, we addressed our three research questions regarding SATM across 11 lower secondary schools in SH. In this section, we discuss the results and relate these to the relevant literature.

As a first result, we assert that SEDM and SPCM changed with grade. As PISA 2003 (OECD 2004) found, students' intrinsic motivation ('interest in and enjoyment of mathematics', p. 116) tends to be lower at later stages of schooling, and they seem to lose interest in and the enjoyment of mathematics. Findings from Brown et al. (2010) also show that student attitude towards mathematics drops for older students. Interestingly, our findings from SH lower secondary schools show a different trend. In our study SEDM increased across the grades from G6 (78.2 %) to G7 (84.1 %) and then to G9 (85.3 %), only slightly dropping down at G8 (82.2 %). We are aware of two factors that are likely to explain the difference in findings. First, different questions were asked in Brown et al.'s (2010) study and in the OECD PISA studies. For instance, in Brown et al.'s (2010) study students were asked the question 'Do you enjoy maths lessons?', whilst in our study students were asked 'Do you like/dislike mathematics?' Second, there are likely to be different perceptions of 'studying hard' and 'pleasurable learning' between the East and the West (Leung 2001). In other words, students might perceive different meanings of the key words in the questionnaires, such as 'enjoy' and 'like' which may be differently interpreted in different cultures. Di Martino and Zan (2010) highlighted the dichotomy between positive and negative emotional disposition in the 'simple' definition of attitude. Op't Eynde et al. (2006) contend that emotions are social by nature and situated in a specific socio-historical context. There is a need for us to further clarify the constructs of the emotional disposition according to students' descriptions in the qualitative questionnaires.

As a second result, in our study we noted the drop of students' responses to T1-1 ('like') from G7 (84.1 %) to G8 (82.2 %), and then an increase from G8 (82.2 %) to G9 (85.3 %). Our study thus supports Pepin's (2011) observation that G8 (students aged 13–14 years) is a critical school year for developing students' emotional disposition towards mathematics.

Nevertheless, differently from Pepin's findings of English and Norwegian classes, in our study students' responses to T1-1 did not drop down from G6 to G7. In fact, in our study the percentage of T1-1 increased from G6 (78.2 %) to G7 (84.1 %). Moreover, unlike the 'polarization' in the English classes in Pepin's study (2011), there were more students who claimed that they liked mathematics (ranging from 78.2 % in G6 to 85.3 % in G9) compared with those who said that they disliked mathematics (ranging from 21.8 % in G6 to 14.7 % in G9). We also consider that the change pattern of SATM across the 11 schools may indicate a difference amongst schools in SH.

A third result is that our study also supports the results of Zan and Di Martino (2007) that there is a positive correlation between SEDM and SPCM. Notably, the combinations "*I like it although I can't do it*" and "*I can do it but I dislike it*" observed by Zan and Di Martino were not rare cases in our study. First, we found that there were a considerable number of students who liked mathematics (T1-1) yet did not consider that they could do it (T2-2). This phenomenon did not reduce when they got older, rather it increased. Unfortunately, and because of our SH samples, this result cannot be regarded as statistically significant across the whole student population of SH. Second, we found that in seven out of the 11 schools, more than half of the students who disliked mathematics (T1-2) conveyed that they could do it (T2-1). Notably, the seven schools were all top or above-average achieving schools in their school district (according to school district and city level standard examinations). We consider that we need to take account of the contextual particularities of the different schools in order to explain the complex relationships both between T1-1 and T2-2 and between T1-2 and T2-1.

As a final result, our study indicates that boys expressed significantly higher SEDM than girls. The results of our study are consistent with those of PISA 2003 (OECD 2004) in the sense that male students expressed significantly higher interest in, and enjoyment of, mathematics compared with female students. Nevertheless, there is no difference in SPCM of these SH boys and girls.

When further analysing the trends of gender in responding to T1 ('like/dislike') across the grades, we found that boys' SEDM appeared to remain high and stable by grade (84.3 % in G6, 88.0 % in G7, 84.7 % in G8 and 87.1 % in G9). These findings are consistent with those of Brown et al. (2010). However, our findings of girls' SEDM across grades are different from those of Brown et al. (2010). In our study, girls seemed to increase their SEDM from G6 (69.4 %) to G7 (79.4 %), remain stable at G8 (79.1 %), and then increase to 83.3 % at G9. In addition, boys' and girls' SPCM changed at G8. The percentage of girls' SPCM increased faster than that of boys from G6 to G7. At G8, however, there were fewer boys than girls who dropped down in terms of SPCM. At G9, there were more boys who considered that they could do mathematics than girls. These results of our study thus differed from those of PISA 2003 (OECD 2004) and TIMSS 2007 (Mullis et al. 2008). PISA 2003 (OECD 2004) reported that whilst on average across OECD countries 36 % of males agreed or strongly agreed that they were not good at mathematics, the average for females was 47 %. Mullis et al. (2008) highlighted that on average across TIMSS 2007 countries, 45 % of Grade 8 boys were at the high level of the self-confidence index, compared with 41 % of eighth grade girls, while 22 % of girls were at the low level, compared with 18 % of boys.

## Considerations for Further Study

Our analyses and findings lead us to consider the following three aspects in terms of the design of a further study of SATM across a large sample, such as the one in SH/China. Firstly, we should examine the key factors that affect SATM. For instance, in our data we have rich descriptions from students' own perspective about why they liked mathematics, such as (1) interesting; (2) successful; (3) logical thinking; (4) no need to commit to rote learning; (5) easy; (6) useful in real life; (7) do like their teacher; (8) specific sub-areas in mathematics (e.g. geometry); (9) textbooks; (10) key subject in examinations; (11) important subject for future career; (12) rigour; (13) improve one's thoughtfulness. These key words appear to link to the multidimensional definition of attitude (Hart 1989): emotional response (e.g. interesting); beliefs regarding the subject (e.g. logical thinking, useful in real life, important subject for future career, rigour, improve one's thoughtfulness); and behaviour related to the subject (e.g. successful, rote learning, easy). In a future study, we aim to further refine the categories we identified in this study to develop further insights into the multiple dimensions that make up students' attitude towards mathematics.

Next, in a larger-scale study, such as ours, school-level variables (of the area/city) should be taken into account. Our sample in the study consisted of 11 schools with different school characteristics – state lower secondary school, state secondary school, state school and private lower secondary school – and in each school district of the city there were, and are, various types of schools (OECD 2010a, b). Pepin (2011) highlights the influence of the socio-cultural context on the formation of beliefs, which in Lester's (1996) view goes "beyond the narrow confines of the classroom to include the total school environment, the educational system and society in general" (p. 353). Ma and Kishor (1997) contend that the school-level variables – for example, school size, school mean socio-economic status, and the variables describing school policies and practices – affect the attitude toward mathematics achievement in the mathematics relationship of students.

Finally, our study found a positive correlation between SEDM and SPCM; that is, if a student likes mathematics, it is likely that s/he considers that s/he can do mathematics. However, the present study was not able to relate this to student achievement in mathematics. This leads us to consider the possibility of linking the school district, or the city standard examinations, to SATM in a follow-up study. Another option would be to select the sample according to a ranked list of schools (according to the school district or the city standard examinations). We should also consider the sample of students with varied achievements from top, above average, average, below average, and low in mathematics in a future study. In doing so, we would aim to contribute to overcoming the obstacle of uniform application of partial views of effective teaching and learning to children in the specific cultural context of China and to make a contribution towards developing new insights into the 'Chinese Learners' Paradox' from the students' attitude perspective.

**Acknowledgements** As authors of this paper we would like to acknowledge the support of China Welfare Institute and Shanghai Soong Ching Ling School for funding this study in Shanghai. We also recognize the contributions made by MS Yihong Zhu in Shanghai Soong Ching Ling Child Development Research Centre and by teachers Wenjie Pu, Ruyi Jin, Yunzhu Liu, Huilin Hu and Zhouyi Yao in Shanghai Soong Ching Ling School in the data analysis.

## References

Ajzen, I. (1988). *Attitudes, personality and behaviour*. Milton Keynes: Open University Press.

Allport, G. W. (1935). Attitudes. In C. Murchison (Ed.), *Handbook of social psychology* (pp. 798–844). Worcester: Clark University Press.

Brown, M., Hodgen, J., Küchemann, D., Coe, R., & Pepper, D. (2010). Attitudes, gender and attainment: Evidence from a large-scale survey in England. In *Proceedings of the 34th conference of the international group for the psychology of mathematics education* (Vol. 1, pp. 217–224). Belo Horizonte: PME.

Cai, J. (2004). Why do U.S. and Chinese students think differently in mathematical problem solving? Impact of early algebra learning and teachers' beliefs. *Journal of Mathematical Behavior, 23*, 135–167.

Chen, Q. (2010). *Teachers' beliefs and mathematics curriculum reform: A comparative study of Hong Kong and Chongqing*. Doctoral thesis at the University of Hong Kong, Hong Kong.

Chen, X., Yu, Y., Wu, J., Yuan, Y., Shao, J., Yu, J., & Huang, J. (2011). An investigation of middle school students' learning affection in mathematics. *Journal of Mathematics Education, 20*(4) (Chinese Version). [陈熙仁, 庾艳昭, 吴佳荣, 袁月明, 邵继勇, 余娟, 黄俊华 (2011 年8 月). 初中生数学学习情感的调查研究. 数学教育学报, 第 20 卷第 4 期]

Coughlan, S. (2012, May 9). *China: The world's cleverest country?* BBC News online. http://www.bbc.co.uk/news/business-17585201. Accessed 29 Jan 2014.

DeBellis, V., & Goldin, G. A. (1999). Aspects of affect: Mathematical intimacy, mathematical integrity. In *Proceedings of the 25th conference of the international group for the psychology of mathematics education* (Vol. 2, pp. 249–256). Haifa: PME.

Di Martino, P., & Zan, R. (2010). 'Me and maths': Towards a definition of attitude grounded on students' narratives. *Journal of Mathematics Teacher Education, 13*, 27–48.

Ding, L., Jones, K., & Pepin, B. (2013). Task design in a school-based professional development programme. In A. Watson et al. (Eds.), *Task design in mathematics education. Proceedings of ICMI study 22* (pp. 447–455). Oxford: University of Oxford.

Fan, L., & Zhu, Y. (2004). How have Chinese students performed in mathematics? A perspective from large-scale international comparisons. In L. Fan, N. Wong, J. Cai, & S. Li (Eds.), *How Chinese learn mathematics: Perspectives from insiders*. River Edge: World Scientific.

Fan, L., Wong, N., Cai, J., & Li, S. (Eds.). (2004). *How Chinese learn mathematics: Perspectives from insiders*. River Edge: World Scientific.

Hart, L. (1989). Describing the affective domain: Saying what we mean. In D. B. McLeod & V. M. Adams (Eds.), *Affect and mathematical problem solving* (pp. 37–45). New York: Springer.

Hodgen, J., Küchemann, D., Brown, M., & Coe, R. (2009). Lower secondary school students' attitudes to mathematics: Evidence from a large-scale survey in England. *Proceedings of the British Society for Research into Learning Mathematics, 29*(3), 49–54.

Huang, Y. (2002). State-of-art in conception of mathematics research. *Journal of Mathematics Education* (Chinese Version), *1*, 1–8.

Huang, R., & Bao, J. (2006). Towards a model for teacher professional development in China: Introducing *Keli*. *Journal of Mathematics Teacher Education, 9*, 279–298.

Lester, F. K. (1996). Implications of research on students' beliefs for classroom practice. In G. C. Leder, E. Pehkohnen, & G. Törner (Eds.), *Beliefs: A hidden variable in mathematics education?* (pp. 345–353). Dordrecht: Kluwer.

Leung, F. K. S. (2001). In search of an East Asian identity in mathematics education. *Educational Studies in Mathematics, 47*, 35–51.

Li, S. (1999). Communications: Does practice make perfect? *For the Learning of Mathematics, 19*(3), 33–35.

Li, J. (2004). A Chinese cultural model of learning. In L. Fan, N. Wong, J. Cai, & S. Li (Eds.), *How Chinese learn mathematics: Perspectives from insiders*. River Edge: World Scientific.

Liu, R., & Chen, H. (2004). Conception of mathematics in primary and secondary school students. *Journal of Beijing Normal University* (Social Science), *5*, 39–43.

Luan, Q., & Tao, Q. (2008, March). An investigation and analysis of middle school students' affect in mathematics. *Education Academic Monthly*. [栾庆芳, 陶琼 (2008). 初中生数学 情感现状的调查与分析. 教育学术月刊, 2008年, 3月]. [In Chinese]

Ma, L. (1999). *Knowing and teaching elementary mathematics: Teachers' understanding of fundamental mathematics in China and the United States*. Mahwah: Lawrence Erlbaum.

Ma, X., & Kishor, N. (1997). Assessing the relationship between attitude toward mathematics and achievement in mathematics: A meta-analysis. *Journal for Research in Mathematics Education, 28*(1), 26–47.

McLeod, D. (1992). Research on affect in mathematics education: A reconceptualization. In D. Grows (Ed.), *Handbook of research on mathematics teaching and learning* (pp. 575–596). New York: McMillan.

Ministry of Education of People's Republic of China. (2011). *Mathematics curriculum standards of the compulsory education*. Beijing: Beijing Normal University Press. [中华人民 共和国教育部制订,义务教育数学课程标准(2011 版). 北京:北京师范大学出版社]

Mullis, I. V. S., Martin, M. O., & Foy, P. (with Olson, J. F., Preuschoff, C., Erberber, E., Arora, A., & Galia, J.). (2008). *TIMSS 2007 international mathematics report: Findings from IEA's trends in international mathematics and science study at the fourth and eighth grades*. Chestnut Hill: TIMSS & PIRLS International Study Center, Boston College.

OECD. (2004). *Learning for tomorrow's world: First results from PISA 2003*. http://www.oecd.org/education/school/programmeforinternationalstudentassessmentpisa/34002216.pdf. Accessed 29 Jan 2014.

OECD. (2010a). *PISA 2009 results: What students know and can do – Student performance in reading, mathematics and science* (Vol. I). http://dx.doi.org/10.1787/9789264091450-en. Accessed 29 Jan 2014.

OECD. (2010b). *Strong performers and successful reformers in education: Lessons from PISA for the United States*. http://www.oecd.org/pisa/46623978.pdf. Accessed 29 Jan 2014.

OECD. (2013a). *PISA 2012 results: What 15-year-olds know and what they can do with what they know* (Vol. I). Paris: OECD. http://www.oecd.org/pisa/keyfindings/pisa-2012-results-overview.pdf. Accessed 17 Feb 2014.

OECD. (2013b). *PISA 2012 results: Ready to learn: Students' engagement, drive and self-beliefs* (Vol. III). Paris: OECD. http://www.oecd.org/pisa/keyfindings/pisa-2012-results-volume-iii.htm

Op't Eynde, P., De Corte, E., & Verschaffel, L. (2006). "Accepting emotional complexity": A socio-constructivist perspective on the role of emotions in the mathematics classroom. *Educational Studies in Mathematics, 63*, 193–207.

Paine, L. W. (1990). The teacher as virtuoso: A Chinese model for teaching. *Teachers College Record, 92*(1), 49–81.

Pepin, B. (2011). Pupils' attitude towards mathematics: A comparative study of Norwegian and English secondary students. In B. Bösken, G. Törner, & B. Pepin (Eds.), Beliefs and beyond: Affecting the teaching and learning of mathematics. *ZDM – The International Journal on Mathematics Education, 43*(4), 535–546.

Ruffell, M., Mason, J., & Allen, B. (1998). Studying attitude to mathematics. *Educational Studies in Mathematics, 35*(1), 1–18.

Sun, X. (2011). "Variation problems" and their roles in the topic of fraction division in Chinese mathematics textbook examples. *Education Study in Mathematics, 76*, 65–85.

Triandis, H. C. (1971). *Attitude and attitude change*. New York: Wiley.

Watkins, D. A., & Biggs, J. B. (Eds.). (1996). *The Chinese learner*. Hong Kong: Comparative Education Research Centre.

Xie, X. (2011, March 4). *The golden reward Shanghai students attained in PISA: What does it tell?* [PISA 测试上海夺冠回答了什么?] Chinese Youth News [中国青年报]. (In Chinese). http://zqb.cyol.com/html/2011-03/04/nw.D110000zgqnb_20110304_5-08.htm. Accessed 29 Jan 2014.

Yang, Y. (2009). How a Chinese teacher improved classroom teaching in Teaching Research Group: A case study on Pythagoras theorem teaching in Shanghai. *ZDM – The International Journal on Mathematics Education, 41*, 279–296.

Zan, R., & Di Martino, P. (2007). Attitude toward mathematics: Overcoming the positive/negative dichotomy. *The Montana Mathematics Enthusiast*, Monograph 3, 157–168.

Zheng, Y. (2006). Mathematics education in China: From a cultural perspective. In F. K. S. Leung, K. Graf, & F. J. Lopez-Real (Eds.), *Mathematics education in different cultural traditions* (pp. 381–390). New York: Springer.

Zhou, Y., Tan, D., & Zhu, Y. (2011). A study on the development of students' epistemological beliefs about mathematics. *Journal of Mathematics Education, 4*(2), 115–129.

# Domain-Specific Belief Systems of Secondary Mathematics Teachers

**Andreas Eichler and Ralf Erens**

**Abstract** This chapter focuses on belief systems of secondary mathematics teachers as part of teachers' mathematics-related affect. Our particular interest concerns teachers' belief systems that represent the teachers' instructional planning. Further we focus briefly on the impact of the teachers' belief systems on their classroom practice and their professional development. In this paper we discuss our theoretical approach in relation to the international discussion on mathematics-related affect. After a brief outline of methodological considerations, the structure of calculus teachers' belief systems is analyzed with regard to the issue of central and peripheral beliefs and the relationships of belief clusters. Secondly we comment on patterns found in the belief systems of teachers thinking about different mathematical domains. An identification of distinctive features of beliefs regarding different mathematical domains is followed by an analysis of the impact of teachers' beliefs on their classroom practice and their professional development.

**Keywords** Teachers' beliefs • Teachers' goals • Belief systems • Central and peripheral beliefs

## Setting the Field

> How teachers make sense of their professional world [...], and how teachers' understanding of teaching, learning, children, and the subject matter informs their everyday practice are important questions that necessitate an investigation of the cognitive and affective aspects of teachers' professional lives. (Calderhead 1996, p. 709)

---

A. Eichler (✉)
Institute of Mathematics, University Kassel, Kassel, Germany
e-mail: eichler@mathematik.uni-kassel.de

R. Erens
Institute of Mathematics Education, University of Education Freiburg, Freiburg, Germany
e-mail: erens@ph-freiburg.de

 
B. Pepin, B. Roesken-Winter (eds.), *From beliefs to dynamic affect systems in mathematics education*, Advances in Mathematics Education,
DOI 10.1007/978-3-319-06808-4_9

The vast and still increasing amount of research into teachers' mathematics-related affect, that sometimes seems to fulfill the demand to investigate the cognitive and affective aspects of teachers' professional lives, makes it reasonable to clarify the potential benefit of a further contribution to this field of research. For this reason, we will integrate our research approach that we have pursued for 10 years into the body of research resulting in substantial findings in recent decades. However, we do not want to add a further review of the entire body of research into teachers' mathematics related affect (e.g. Thompson 1992; McLeod 1992; Philipp 2007), but rather highlight three issues to clarify the aim of our contribution, i.e. *models*, *external influences* and *impacts*.

## *Models of Teachers' Mathematics-Related Affect*

In recent years several models have been proposed for positioning the parts of teachers' sense-making. These models enable us to relate a specific research approach to already existing approaches and to describe possible relations among different parts of teachers' sense-making (Schoenfeld 1998). For example Hannula et al. (2007, p. 204) proposed a model in which mathematics-related affect consists of three overlapping constructs, i.e. motivation, cognition and affect, which in turn consist of further constructs. For example the construct of motivation is used to integrate the constructs of goals, needs and, in the intersection of the three overarching constructs, beliefs and belief systems. This model could be understood as further development of McLeod's model (1992), in which emotions, attitudes and beliefs are positioned on a continuum from least stable and cognitive to most stable and cognitive. Further, in the model of Hannula et al. (2007) knowledge and belief are distinct parts of the construct of cognition, which is in line with other models regarding teachers' decision-making (e.g. Ball 1990; Borko and Putnam 1996).

Using the model of Hannula et al. (2007), our own research on upper secondary teachers' mathematics related affect considers the motivational and cognitive aspects, since we investigate primarily teachers' instructional goals and beliefs, which also fits in some sense the approach of Schoenfeld (1998), who proposes a distinction between knowledge, goals and beliefs. However, in contrast to Schoenfeld, we put less emphasis on the teachers' knowledge.

## *External Influences on Mathematics-Related Affect*

A lot of research into teachers' mathematics related affect – particularly when teachers and their thinking began to be an important issue for educational research – does not consider external influences. However, this research yielded important results, e.g. the seminal case study of Thompson (1984), who reported the beliefs of three teachers described as having instrumental, formal and conceptual understanding of mathematics that are immanent to the three teachers beliefs and observable into their classroom practice. Also other researchers provide empirical or theoretical driven categorizations of beliefs about mathematics or mathematics teaching and learning (e.g. Dionne 1984; Ernest 1989; Grigutsch et al. 1998) that still impact on research in teachers' mathematical affect.

However, particularly in the recent decade, researchers increasingly consider external influences on teachers' mathematics-related affect. The broadest scope in this line of research is constituted by the cultural dimension. For instance, the large scale study TALIS (OECD 2009) reported striking differences of different countries referring to teachers' beliefs about their teaching orientation representing a direct transmission and constructivist understanding of teaching mathematics. This finding partly agrees with the results of TEDS-M (Felbrich et al. 2012) that makes a distinction between countries with a culture of individualism and countries with a culture of collectivism.

A further external influence is represented by a social context. For example, the case study of "Larry at Mellemvang" Skott (2009, p. 31) defines the social context as a school specific setting "construed by individuals as they participate in praxis that evolve in interaction". Based on this framework, he explains the influence of social norms in a traditional private school on the beliefs and the classroom practice of Larry. Also Sztajn (2003) explains the differences of the classroom practices of two teachers holding similar beliefs by the social setting of two different schools.

Finally, Schoenfeld (1998) defines the context in a narrow sense regarding teachers' knowledge, beliefs and goals and their teaching. Thus, Schoenfeld analyzed in depth four teachers' "moment-to-moment decision making and acting" (ibid., p. 1) in the context of a specific mathematical "instructional segments" (ibid., p. 78) in a specific class.

Our research approach refers partly to the latter two aspects of influence, while we did not consider cultural differences since we restrict our sample to German teachers. We further acknowledge the influence of the teachers' social context primarily concerning the formation of our sample that we describe in a later section. Secondarily we regard the social context, when we analyze the teachers' professional development to which we refer briefly in the sixth section. Our main focus is, however, on the context that Schoenfeld (1998) describes. In contrast to Schoenfeld, we define context in a broader sense. Thus, we opt for a broad scope regarding the context, i.e. the teachers' beliefs about the entire mathematics curriculum of upper secondary schools lasting from grade 5 to grade 12 or 13. In this broad scope we further define the context by a specific mathematical domain like data and chance (e.g. Eichler 2011).

## *Impacts of Teachers' Goals and Beliefs*

As in the quoted works of Schoenfeld (1998) and Skott (2009), the impact of teachers' beliefs on their classroom practice is an important research question (e.g. Philipp 2007). However, the relation of espoused and enacted beliefs still seems to be far from a conclusion. Thus, it is firstly not clear if the classroom practice impacts on the teachers' beliefs or if the relation is inverse (Franke et al. 1997). Further different researchers reported inconsistencies while others report a consistency between beliefs and classroom practice (e.g.Philipp 2007). From the suggested assertions for observed inconsistencies, i.e. the inexperience of the observed teachers (Artzt and Armour-Thomas 1999), the specific social context of a classroom (Skott 2009) or the grade of intensity with which a teacher holds a belief (Putnam and Borko 2000), we will make a contribution to the latter aspect. For this reason, a main focus in this paper is the identification of central and peripheral beliefs of a teacher about a specific mathematical discipline.

Since there is on the one side a potential impact of teachers' beliefs on their classroom practice, a further function of beliefs is to be a filter that impacts on a person's perception (e.g. Franke et al. 1997; Philipp 2007). For this reason, teachers' beliefs potentially have an impact on their professional development (Chapman 1999). We also refer briefly to this aspect referring to the development from pre-service to in-service teachers.

### *Concluding Remarks*

Based on the brief overview of important issues of the research in teachers' mathematics-related affect the aim of our research approach is to make a contribution to the following research questions:

1. What is the structure of a teachers' system of beliefs and goals referring to a mathematical discipline and the teaching and learning of this discipline?
2. How do these systems of beliefs and goals differ with regard to different mathematical disciplines?
3. How do the teachers' beliefs and goals impact on their classroom practice and their professional development?

We refer to these questions after outlining the central constructs and the method of our research approach. We discuss the results primarily referring to the first two research questions and with less detail on the third.

## Theoretical Framework

Stein et al. (2007) define a model to distinguish the possible phases of a curriculum that impact on teachers' beliefs.

The term *written curriculum* involves both instructional content, and teaching objectives, or, more recently, standards, often prescribed by national governments. The teachers' interpretation of the written curriculum – that is, the individual teacher's transformation of the written curriculum – is called the *intended curriculum*. The interactions of a teacher, his students, and the instructional content "bring the curriculum to life and, in the process, create something different than what could exist [...] in the teacher's mind" (Stein et al. 2007, p. 321). This transformation of the intended curriculum is called the *enacted curriculum*.

Finally, the students transform the content addressed in the enacted curriculum into their own personal subjective knowledge and develop their own beliefs about the content. This is the *students' learning*.

A teacher's own experiences with his classroom practice (enacted curriculum) as well as his awareness of the beliefs and knowledge attained by the students (students' learning) in turn have an impact on the teacher's intended curriculum (Hofer 1986) so that it actually develops over time. In this chapter we focus on different parts of the curriculum model. For this reason, a possible aspect of the consistency of teachers' espoused and enacted beliefs could potentially be explained with the teachers' grade of experience (Artzt and Armour-Thomas 1999).

Further, we understand the term *beliefs* as an individual's personal conviction concerning a specific subject, which shapes an individual's ways of both receiving information about a subject and acting in a specific situation (Pajares 1992; Thompson 1992; Furinghetti and Pehkonen 2002). Knowledge and beliefs could be seen as "inextricably intertwined" (Pajares 1992, p. 325). For this reason we distinguish knowledge and beliefs theoretically by understanding beliefs as more individual convictions and by understanding knowledge as more inter-individual (or objective) convictions (Pajares 1992; Borko and Putnam 1996).

An individual's organization of beliefs we call *belief system* following Green (1971) or Thompson (1992). The individual's organization of beliefs involves the distinction of central beliefs, i.e. strongly held beliefs, and peripheral beliefs referring to an individual's belief system of lesser importance. Further, belief systems consist of belief clusters that are quasi-logically interconnected and, thus, different beliefs in an individual's belief system may be contradictory. We discuss the two aspects of centrality and quasi-logicalness later when analyzing mathematics teachers. However, we avoid the theoretical distinction of primary and derivative beliefs, which is the third aspect of the structure of belief systems (Green 1971), since we have no empirical evidence concerning this aspect in our research (c.f. for this aspect also Liljedahl 2010).

As stated in the introductory section, we regard both teachers' beliefs and teachers' goals that are understood as different constructs (e.g. Schoenfeld 1998; Hannula 2012). Schoenfeld describes beliefs as a mental orientation that shape the way of establishing a specific goal. Accordingly, in a further development of his model, Schoenfeld (2010, p. viii) distinguishes *goals* and *orientations* that include beliefs in addition to dispositions or values. This is in line with the consideration of Hannula (2012) about the psychological dimension to state and trait of the motivational aspect of teacher mathematics-related affect. Referring to this distinction, he suggests goals to represent the state and (motivational) beliefs to represent the trait of this motivational aspect. Thus, in both theoretical frameworks goals are necessarily connected with an observable behavior (Cobb 1986). However, our research approach is based on a model of teachers' action that is described in the so-called rubicon-model (Heckhausen and Gollwitzer 1987). In this model, a person defines goals before an observable behavior (pre-behavioral phase; motivation), decides when and how she or he wants to establish the goals (pre-behavioral phase; volition), establish goal-oriented behavior (behavioral phase), and finally evaluate for example if the goals were achieved (post-behavioral phase). Based on this theory, a teacher's intended curriculum consists of goals that are closely connected to his beliefs. For example, a teacher believes that both frequentist and an axiomatic approach to probabilities are important mathematical concepts (belief). However, the teacher plans to achieve his students' understanding of probabilities by choosing the introduction to probabilities according to the frequentist approach (goal).

Teaching goals could represent overarching beliefs representing "world views" (Grigutsch et al. 1998) or epistemological beliefs (Hofer and Pintrich 1997) about mathematics (or different mathematical disciplines), about school mathematics or about teaching and learning mathematics (Grossman 1990). However, teaching goals could also concern, for instance, specific content or issues of a mathematical discipline, representations of mathematical objects or students' difficulties with specific content. Thus teaching goals exist with different grain sizes (Schoenfeld 1998, p. 21) or rather ranges of influence. We discuss teaching goals of a lower range of influence later. For analyzing overarching teaching goals, we use the construct of mathematical world views proposed by Grigutsch et al. (1998):

- a formalist (world) view that stresses that mathematics is characterized by a strongly logical and formal approach. Accuracy and precision are most important and a major focus is put on the deductive nature of mathematics.
- a process-oriented view that is represented by statements about mathematics being experienced as a heuristic and creative activity that allows solving problems using different and individual ways.
- an instrumentalist view that places emphasis on the "tool box"-aspect which means that mathematics is seen as a collection of calculation rules and procedures to be memorized and applied according to the given situation.
- an application oriented view that accentuates the utility of mathematics for the real world and the attempts to include real-world problems into mathematics classrooms.

Concluding our theoretical framework (c.f. Eichler 2011), we understand a teacher's *intended curriculum* as an individual's belief system including

- an individual's world view consisting of beliefs about the nature of mathematics or a mathematical discipline represented by overarching teaching goals,
- beliefs represented by teaching goals of different ranges of influence that a teacher takes into account when planning (in his view) appropriate classroom practices. These goals (beliefs) might concern content, the best way to teach mathematics or a specific mathematical discipline, or the way students learn mathematics.

Further, teachers' *enacted curricula* involve the observable part of the teachers' intended curricula transformed by the interaction of teachers, their students, and the content within the classroom practice. Finally, *students' learning* is represented by students' knowledge and beliefs concerning mathematics.

## Method

For different parts of our research program, we used different methods. We briefly discuss these methods structured by the curriculum model (Fig. 1).

In this report, we refer to a sample of 51 secondary teachers. 30 teachers' were interviewed in respect to calculus, 13 teachers were interviewed in respect to stochastics (statistics and probability), and 8 teachers were interviewed in respect to geometry. Regarding the selection of teachers, different degrees of teaching experience were considered as well as a balanced proportion concerning gender (Hannula 2012). Teachers who were interviewed about geometry or stochastics are all in-service teachers. The "calculus sample" consists of 30 teachers divided into three subsamples: pre-service teachers, teacher trainees and experienced teachers. The first subsample includes 10 experienced teachers who have been teaching calculus for at least 5 years. Data concerning the intended curricula of experienced teachers that are assumed to be relatively stable (McLeod 1992) were collected once. The other subsamples consist of each 10 prospective teachers. The data for these subsamples were collected twice within one and a half years in a quasi-longitudinal design.

In order to capture both the need of contextualizing beliefs and the notion of belief enactment in a locally social approach (c.f. Skott 2009, p. 29), the teachers who participated in this study were recruited from different universities, teacher training colleges and secondary schools across Germany. Every (in-service) teacher in this study teaches all domains of mathematics from grade 5 to 12. The domains of stochastics and calculus are a central part of the curriculum at upper secondary level (grade 10–12). However, our sample is a theoretical, not a representative sample.

To investigate *teachers' intended curricula* referring to one discipline we use intensive semi-structured interviews (Witzel 1982) lasting about 2 h following a qualitative case study approach and questionnaires for a quantitative analysis.

The interviews consist of several clusters of questions that mostly concern intended curricula referring to a specific mathematical discipline (e.g. calculus) but also to mathematics in general, e.g. instructional content, teaching objectives, reflections on the nature of mathematics (as a discipline generally) and of school mathematics, the students' views, or textbook(s) used by the teachers. Further, we use prompts to

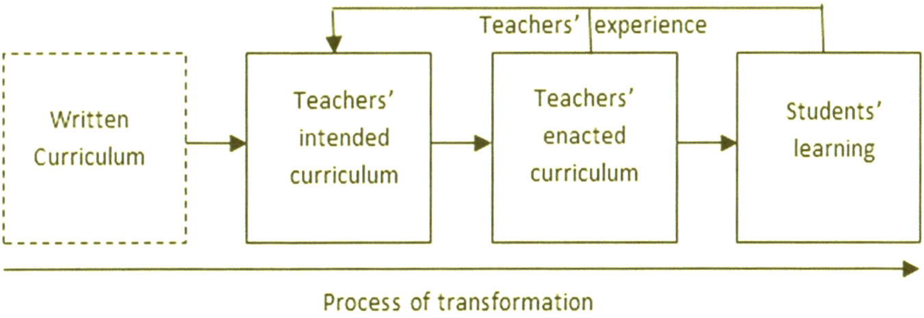


**Fig. 1** Model of a curriculum

(a) I like calculus, because there is a connection to real life problems.
(b) I like calculus, because hard nuts must be cracked and difficult problems can be solved.
(c) I like calculus, because many exercises can be solved by similar procedures and patterns.
(d) I like calculus, because the logic is clear and it follows strict mathematical rules.

**Fig. 2** Fictive statements of students concerning calculus

provoke teachers' beliefs, e.g. by making teachers comment on different parts of a textbook. We show one of these prompts in the results sections. In other parts of the interviews, the teachers were asked to comment on fictive or real statements of teachers or students. One of these prompts is shown in Fig. 2.

Each of the various prompts represents a specific view concerning mathematics or a mathematical discipline, e.g. a formalist view (see above). Further we employed two different questionnaires including adapted scales concerning mathematical beliefs (Grigutsch et al. 1998) and teaching orientation (Staub and Stern 2002) of which we refer to the former in the results sections.

For analysis of qualitative and quantitative data we used mixed methods including coding methods (Strauss and Corbin 1998; Mayring 2003), and statistical methods. A qualitative coding method was used for analysis of the interview data that is close to grounded theory (Glaser and Strauss 1967). The codes gained by interpretation of each episode of the verbatim transcribed interviews indicate goals of calculus teaching. We used deductive codes to the above mentioned mathematical views as well as the teachers' teaching orientation. The latter is not the focus of this report and will therefore not be discussed in the results section. Inductive codes for those goals we did not deduce from existing research such as the integration of technology into classrooms or the impact of authorities were developed from the interview data. The codes were conducted by at least two persons. The interrater reliability shows an accordance above 80 %.

To investigate *teachers' enacted curricula*, we used videography and protocols to document the teachers' classroom practices. In this chapter, we refer to a subsample of stochastics teachers that were observed in their classroom practice for half a year. These teachers were selected due to the differences that these teachers show referring to their intended curricula.

## Structure of Teachers' Intended Curricula: Calculus Teachers

In this paragraph, we analyze calculus teachers' belief systems representing their intended curricula. For this, we firstly discuss the issue of central and peripheral beliefs (Green 1971), and afterwards the issue of (quasi-)logical relationships of clusters of beliefs.

## *Central and Peripheral Beliefs*

In order to categorize and illustrate teachers' beliefs concerning the planning and teaching of calculus by means of qualitative analysis, the deductive aspects of four different views (see above) were chosen. This involves the subjective teachers' definition of a specific view that represents the teachers' overarching teaching objectives.

First, we illustrate a coherent view, in this case a formalist view concerning the subjective definition of Mr. $C_{Calc}$.

> Mr. $C_{Calc}$: In general, exactness is crucial for me. That means to fit a necessary formalism as I know from my university studies. This also means that it must be possible to recognise a logical rigor. Sometimes I do more in that sense than the textbook actually demands.

Taking this teacher as a paradigmatic example, he did not mention aspects such as to apply mathematics in real world problems or to learn problem solving, which means to emphasize the process of developing mathematical concepts. By contrast, for Mr. $C_{Calc}$, the main goal of calculus teaching seems to be emphasizing the stringent and logical construction of a mathematical domain.

The identification of specific teachers' views is always established in various parts of a single interview with either questions regarding the teaching of calculus or teaching orientation in general or prompts to provoke teachers' beliefs (see section "Method") and we report only teachers' views that are in some sense coherent throughout the whole interview. We illustrate this concerning this exemplified teacher. When Mr. $C_{Calc}$ was asked to regard the expectations and needs of his students, he agrees consistently with a formalist view.

Mr. $C_{Calc}$ further explains his goals concerning his students' beliefs towards calculus:

> Interviewer: How should your students characterize calculus?
> Mr. $C_{Calc}$: Precise mathematics. Thus, on the one side that it is possible to understand how one develops mathematical ideas and how it is possible to build up a theory on the foundation of few basic ideas.

The coherence of the beliefs of Mr. $C_{Calc}$ is also apparent in his responses to several prompts used in the interview regarding decisions on instructional content and the above described views concerning teaching calculus. For example, when Mr. $C_{Calc}$ was asked to evaluate four tasks that represent the four different views, he valued the task representing the formalist view (Fig. 3) higher than the other tasks.

Summarizing the beliefs of Mr. $C_{Calc}$ concerning the teaching and learning of calculus, there exist a lot of other unambiguous examples of evidence for Mr. $C_{Calc}$'s formalist view. The high degree of coherence in different parts of the interview leads to the hypothesis that this formalist view is dominant and thus *central* in the belief system on calculus. This hypothesis is supported by reported examples and tasks of Mr C's classroom practice. Furthermore the hypothesis of centrality is supported by the evaluation of the questionnaires which consistently confirm the qualitative codings (Erens and Eichler 2013a).

Calculate the area between the graph of the function $f(x)=\frac{1}{5}\cdot x^3$ and the x-axis between $0 \le x \le 3$.

Solution:
As the function f is continuous, we obtain the same result for the limit of upper sum and lower sum of the area of rectangles. It is thus sufficient to regard the upper sum of rectangles.

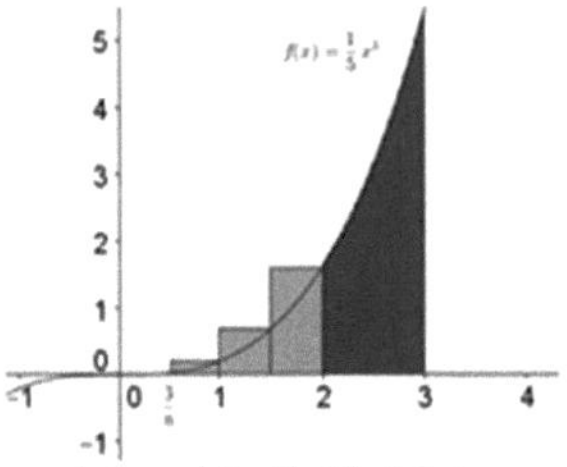

The given interval is divided into n parts of length $\frac{3}{n}$ and the corresponding area of the sum of upper rectangles is

$$O_n=\frac{3}{n}\left[\frac{1}{5}\cdot\left(\frac{3}{n}\right)^3+\frac{1}{5}\cdot\left(2\cdot\frac{3}{n}\right)^3+\ldots+\frac{1}{5}\left(n\cdot\frac{3}{n}\right)^3\right]$$
$$=\frac{3^4}{n^4}\cdot\frac{1}{5}\cdot\left[1^3+2^3+3^3+\ldots+n^3\right]$$

Because

$$1^3+2^3+3^3+\ldots+z^3=\frac{1}{4}\cdot z^2\cdot(z+1)^2$$

one can simplify $O_n$ to

$$O_n=\frac{81}{n^4}\cdot\frac{1}{5}\cdot\frac{1}{4}\cdot n^2(n+1)^2=\frac{81}{20}\cdot\frac{(n+1)^2}{n^2}$$
$$=\frac{81}{20}\cdot\left(1+\frac{1}{n}\right)\cdot\left(1+\frac{1}{n}\right)$$

The limit of $O_n$ for $n\to\infty$ is $\lim_{x\to\infty} O_n=\frac{81}{20}$

which is the area we wanted to calculate.

**Fig. 3** Task representing the formalist view

In addition to central beliefs that teachers like Mr. $C_{Calc}$ show in different parts of the interview in a coherent way, most of the teachers also provide insights into some peripheral goals. For example, some teachers indicate a *peripheral* goal that calculus and the teaching of calculus is a collection of rules and procedures although their beliefs can neither qualitatively nor quantitatively be categorized globally as an instrumentalist view.

> Mr. $F_{Calc}$: The main goal of every student is to perform well in his final exams – therefore calculation rules and procedures have to be thoroughly practiced in class. Especially the calculus part of final exam tasks are alike in some respect, so practicing is a substantial guideline for my course.

Like Mr. $F_{Calc}$, several calculus teachers show a connection of an instrumentalist view and considerations referring to normative aspects such as final exam tasks that represent the teacher's acknowledgement to a social context (c.f. Skott 2009). However, it is apparent that for most of the teachers in our sample teaching goals representing an instrumentalist view are not central for mathematics or calculus instruction per se, but are important in terms of preparing students for the final exam, which supposedly is a particularity due to the situation of German teachers (cultural dimension; e.g. OECD 2009).

## *(Quasi-)Logical Relations into Belief Systems*

In contrast to Mr. $C_{Calc}$, most of the teachers show a mixture of different views that individually are also coherent. In particular, if teachers' hold beliefs that represent different views, we analyze relationships within and among the different views. The codes gained by interpretation of each episode of the interviews allow a more differentiated analysis of the different views and can be warranted with substantial reactions to content and teaching goals. We describe this analysis exemplarily by considering Mr. $A_{Calc}$ and Mr. $B_{Calc}$ starting at the subjective definitions of their possibly central beliefs. In contrast to a (central) formalist view, these two teachers delivered an insight into their views on applications:

> Mr $A_{Calc}$: I quite agree with the emphasis on applications in the given example. That is certainly a way to motivate them (students), but nevertheless one should not reduce genuine calculus or the teaching of calculus to that topic.
> Mr. $B_{Calc}$: Examples for applications are quite suitable here, and with applications I always associate modeling of real data, [...] increasingly introducing relevant applications into lessons may, for the students, succeed in a deeper insight into the concepts and ideas of calculus.

For Mr. $B_{Calc}$, beliefs representing an application oriented view seem to be central, since other teaching goals are peripheral or of no importance if school mathematics is regarded:

> Mr $B_{Calc}$: "...because I think that the formal derivation of integrals by limits is of no avail for secondary level students. It's just too complex for most of them."

By contrast Mr. $A_{Calc}$ supports the integration of applications as a principle of learning calculus at school for reasons of (student) motivation. Using applications in his teaching represents an additive goal to achieve a teaching goal of higher importance, i.e. students' motivation. Consistently Mr. $A_{Calc}$ mentions other possibly more central teaching goals that represent a formalist view illustrated in the following quotation:

> Mr. $A_{Calc}$: "Calculus is more than just dealing with application-oriented tasks. Then, for example, one would not regard the precision and exactness of calculus and use applications as a means to an end."

The difference between the instructional goals of motivation on the one side, and solving real problems on the other, is stated by Förster (2011) concerning teachers who teach modeling. Both views on applications can be found several times in our sample.

Our hypothesis on the basis of the present data is the following: If teachers hold a consistent formalist view on calculus like Mr. $C_{Calc}$, they do not mention any applications. The converse conclusion, however, is not possible. Teachers who favour applications in their calculus courses, e.g. Mr. $A_{Calc}$ and Mr. $B_{Calc}$ (see above), cannot necessarily be described as non-formalist. This example already demonstrates the abundance of calculus teachers' beliefs and the need to differentiate the views of teachers on calculus as well as relations between different views. It demonstrates further, that qualitative analysis of the data hence enables us to discern these relations in a sophisticated manner.

Referring to the teachers regarded so far, both the formalist view and the application-oriented view is identified to be central for some of the teachers. By contrast, for a majority of teachers in our sample, the process-oriented view is subordinated to other views, namely an application-oriented view. Thus, many teachers in our sample noted application-oriented tasks as illustrative approaches for relevant mathematical methods, often manifested by giving appropriate examples from their own lessons. In close connection with these evidential classroom episodes teachers often used key words such as "understanding", "comprehension", "problem-solving strategies" and "students finding out" or "discovering by themselves". The emphasis on problem-solving strategies and student activity in the classroom discourse suggests a rather close connection between an explicit preference of experiencing calculus methods as a heuristic and creative activity (process-oriented view) and attempting to accentuate the utility of calculus for the real world. This result agrees with the findings of Felbrich et al. (2012) referring to the teaching orientation of German mathematics teachers.

While we have described above relations between beliefs that are logically connected, we also found contradictory belief clusters that we call conflicts of instructional goals that represent the quasi-logicalness of a belief system. We illustrate contradictory clusters of beliefs with the paradigmatic example of Mrs. $E_{Calc}$. Throughout the whole interview she speaks about the central role of logic in calculus lessons offering her perspective that exactness and logical rigour are necessary ingredients of secondary level calculus courses. Again, the degree of coherence of favoring formalist elements could provide an indication for a central belief. Yet, as she describes representative classroom situations, her subjective experience surfaces a conflict between her belief system about calculus teaching and pedagogical processes in her calculus course.

> Mrs. $E_{Calc}$: In my view it is quite important that there are formal definitions of concepts because you need them for proofs later on and it's the tiny details that are particularly important.
>
> In my class I clearly notice that students come to their limits concerning the degree of abstraction. [...] Remembering my own calculus course at school I can't remember any bad experience with these formal aspects. So far I haven't seen such a mismatch between teacher and students in maths.

Mrs $E_{Calc}$ can be identified favoring a formalist view, but probably will not enact her formalist view on calculus teaching in the classroom in a predominant way because there is a conflict with the real situation she encounters in the classroom i.e. the students' ability to understand the formal way of developing calculus ideas. This teacher shows a high awareness and consciousness of how these conflicting forces are affecting her curricular and pedagogical decisions with respect to the differences between teaching and learning calculus. Therefore this situation can be characterized as a conflict of goals between her view on calculus and her teacher authority and responsibility. In particular when teachers are asked to reflect on representative examples of their actual teaching processes of specific elements of their calculus courses, the interview transcripts provide a deep and concrete insight into teachers' subjective notions of their intended curricula and sometimes yield conflicts of a teacher's system of instructional goals. We hypothesize that a conflict of teaching objectives gives evidence for a central belief since peripheral beliefs might be superimposed if they show a conflict with central beliefs.

## Differences Among teachers' Belief Systems of Different Mathematical Disciplines

We firstly illustrate patterns found in the belief systems of teachers thinking about specific mathematical disciplines structured by the mathematical views outlined above. However, the teachers sometimes compared the focused mathematical discipline like calculus with other disciplines like geometry. We emphasize these comparisons at the end of this section.

### *Teaching Mathematics with an Application Oriented View*

The eight teachers in our sample mostly tended to neglect application-oriented goals when they think about geometry. One paradigmatic example for this assertion is the case of Mr. $B_{Geo}$:

> Mr. $B_{Geo}$: Geometry as a tool to get access to the real world is not fundamentally important, and it is deservedly not in the first place. An application is useful to introduce a new subject, to legitimize it, and to test the competencies of this field by realistic tasks in the end. But in between, a lot has to be done without any reference to the real world, detached from these accessory parts which are not important to the mathematical model.

Like Mr. $B_{Geo}$, most of the investigated geometry teachers understood real applications at most as a strategy to motivate students, but not as an important aspect of this mathematical discipline. By contrast, for geometry teachers, geometry is rather seen as a language that can be used to describe reality, but doesn't have to. The predominant goal of teaching geometry is to learn this language, wherefore real situations are mainly used just as illustrations, and not as interesting occasions to gain insight into realistic problems and to learn model building. Hence, the real situations, their data and empirical challenges are of minor interest and in principle interchangeable: The context is suspended in favor of the theory (Girnat 2009).

In contrast to the geometry teachers, the *role of the context* (Shaughnessy 2007) is omnipresent for the teachers interviewed on their intended stochastics curricula. Whereas the geometry teachers doubted whether geometry is an adequate discipline to emphasize the applied aspect of mathematics, the participating German stochastics teachers did not question that applications play a significant role in stochastics teaching (Eichler 2011):

Mrs. $B_{Stoch}$: Which objectives I have? That students were enabled not to fail when they were confronted with challenges or allurements in their daily life, but to develop the possibility to evaluate things for themselves.

The consideration of Mrs. $B_{Stoch}$ represents a common idea concerning the application-oriented view of the stochastics teachers: Applications are used to gain the insight that mathematics can be useful for real world problems; and therefore, the real situations have to be treated more seriously than in geometry. However, the stochastics teachers differ in their way to highlight stochastics as an applied domain of mathematics (e.g. Eichler 2011).

Since both geometry teachers evaluate an application-oriented view to be peripheral and stochastics teachers evaluate this view to be central, the findings referring calculus teachers are ambiguous. Thus, we found some teachers like Mr. $C_{Calc}$, who neglect an application oriented view (see above), as well as some teachers, who stress real applications as tool for motivating students (e.g. Mr. $A_{Calc}$, see above) or stress calculus as discipline to emphasize modeling (e.g. Mr. $B_{Calc}$, see above).

## *Teaching Mathematics with a Process Oriented View*

Most of the teachers articulated a process-oriented view by thinking about problem solving according to the approach of Pólya (1949). Referring to this orientation, geometry teachers tended to emphasise problem solving as the main idea of a geometry curriculum in school. We illustrate this orientation by quoting the typical statement of Mrs. $G_{Geo}$:

Mrs. $G_{Geo}$: Besides proof abilities, problem solving is in fact the most important thing I want to convey in my lessons on geometry. To pose students problems.

For geometry teachers "problems" mean mathematical problems that need not have a connection to a real world situation and that are posed to enhance properties in reasoning, and not to gain empirical knowledge or to conceive mathematics as being useful.

By contrast, stochastics teachers also mentioned that to learn problem-solving strategies has to be a teaching objective. However, these teachers identified the problem of stochastics tasks to find an appropriate model for a realistic situation:

Mr. $E_{Stoch}$: To learn problem-solving in stochastics is to learn to argue mathematically on the basis of a specific realistic context.

In the same way, calculus teachers showed a process-oriented view in connection with other views, e.g. a formalist view or an application-oriented view emphasizing creativity in the students' individual ways of modeling real situations, working on main concepts of calculus, and, more peripheral, to solve mathematical problems.

Thus, since problem solving predominantly appears to be a central goal for geometry teachers, it seems to be a more peripheral or subordinated one for both stochastics teachers and calculus teachers. For stochastics teachers in our sample the process oriented view is subordinated to the main objective to translate real world problems into stochastics and to interpret stochastical results by referring to a real world situation. For calculus teachers the process-oriented view is subordinated to different main objectives that represent a formalist or an application-oriented view.

## *Teaching Mathematics with a Formalist View*

Some geometry teachers tended to emphasize the formalist view mentioning for example the integration of phenomena investigated in mathematics lessons into a formal and abstract mathematical structure following a deductive approach that Girnat (2009) calls classical Euclidean view on geometry. Mr. $C_{Geo}$ formulates this view mentioning a meaningful example:

> Mr. $C_{Geo}$: If someone asserted in case of the Pythagorean Theorem "Proved by measuring, the theorem holds", then something valuable would disappear, something which is genuinely mathematical. [...] If geometry just consisted of measuring, calculations, drawing, constructing, and land surveying, then I would regard it as poor.

Although some of the stochastics teachers hold beliefs representing a formalist view, they mostly seem to understand these goals as peripheral ones. The case of Mrs. $B_{Stoch}$ shows a paradigmatic example of more or less neglecting the formalist view:

> Mrs. $B_{Stoch}$: Formalism is out. Indeed, there are some colleagues, who say that it is not the right way to show, for instance, the theorem of Bayes by using an example. I think let them teach in this way. In my opinion, for students it is better to show them the theorem of Bayes using examples or using a probability tree.

At first, beliefs representing a formalist view seem to be central for geometry teachers. However, taken into account all interviewed geometry teachers the formalist view seems to be subordinated in comparison to the process-oriented view. By contrast, our stochastics teachers mostly neglected the formalist view in favor of the application-oriented view.

Although the calculus teachers differed concerning their beliefs representing a formalist view, it is striking that only calculus teachers like Mr. $C_{Calc}$ hold a coherent belief system that represents a nearly pure formalist view.

## *Teaching Mathematics with an Instrumentalist View*

None of the geometry teachers and the stochastics teachers emphasized an instrumentalist view, i.e. highlighting teaching formulas and rules to enable students to solve a category of specific tasks. Only the calculus teachers tended to value an instrumentalist view in respect to their students' final exams and, thus, refer to the social context (Skott 2009). So, do secondary teachers mostly neglect the instrumentalist view that Thompson (1992) described? We hypothesize that the absence of this view is the consequence of the mathematical domains we are focused on. If we regard the beliefs of teachers thinking about mathematics instruction in primary schools and the first grades of secondary schools (Bräunling and Eichler 2011), in which arithmetics is the core subject, these beliefs represent in their majority an instrumentalist view.

## *Differences of Teachers or Differences of Mathematical Disciplines?*

Since we investigated the teachers' beliefs only referring to one mathematical domain, the differences between the teachers have to be interpreted carefully. However, our purpose in this report is to illustrate the fundamentally distinct views towards the teaching and learning of mathematics in different mathematical domains. Further, almost all the teachers were asked to comment on the comparison of different mathematical disciplines to highlight characteristics of that discipline. We illustrate three of the mentioned comparisons:

> Mr. $A_{Geo}$: I think the better applications can be found in algebra or stochastics, per cent calculations, linear optimization. It is important to get a deeper insight into reality by modeling. In geometry, there are such things as dividing a pizza by a compass. I saw a trainee teacher do so. That's ridiculous.
>
> Mr. $J_{Stoch}$: One goal is to know that stochastics has a high relevance in real life [...]. I have to say, we have neglected this aspect of mathematics for a long time. We have emphasized geometry and transformation geometry and have put application to the side. However application oriented mathematics is very important and more important for stochastics than calculus.
>
> Mr. $T_{Calc}$: I think in geometry it is just a different, constructional kind of approach: vectors, lines, reflection with respect to a plane and so on. [...] Stochastics is rather based on our living environment, statistical investigations, polls, all of these topics that come from real life [...]. Of course that is more challenging for students as they can't apply the schematic tools from calculus.

These three quotations provide evidence that teachers have different views regarding different mathematical domains. Particularly, teachers seem to emphasize an application-oriented view when they consider stochastics. By contrast they seem to emphasize a process oriented view when they consider geometry. Since both stochastics teachers and geometry teachers showed a consistent predominance referring to one view, calculus teachers differed concerning their predominance in respect to an application view or a formalist view.

## Possible Impacts of Teachers' Beliefs on Their Classroom Practice and Professional Development

In the last two paragraphs, we discussed on the one side findings referring to the identification of central and peripheral beliefs. On the other side we provided evidence that the teachers' central beliefs vary when different mathematical disciplines are regarded. We took these findings into account when discussing possible impacts of the teachers' beliefs on their classroom practice and on their professional development. Due to the limited space in this report, we restrict discussion on results that we reported elsewhere (Erens and Eichler 2013b).

### *Impact of Teachers' Beliefs on Their Classroom Practice*

From the sample of stochastics teachers, of which we analyzed their intended curricula and observed these teachers in their classroom practice in a stochastics course lasting a half year, we examine only the case of Mr. $D_{Stoch}$ (for greater detail see Eichler 2008). For this teacher the application-oriented view is central. We illustrate this view only by the following episode of the interview:

> Mr. $D_{Stoch}$: That's what I am trying to illustrate, that you will of course get quite far with relative frequency, but that if you have similar situations afterwards, such as elections or opinion polls, you will need to develop the use of confidence intervals. This means showing them [students], as well, that mathematics really has applications that there are quite often problems which you can solve with maths. Students should be enabled to better categorize mathematical models which determine our economic condition.

Actually, Mr. $D_{Stoch}$ did not show his central goal (or belief) in every lesson or instructional segment (Schoenfeld 1998). However, he enacted his central belief over the period of half a year consistently. Thus, his students predominantly worked on realistic problems comprising real data sets. The students were asked to look at statistics-related broadcasts on TV, e.g. concerning polls. Afterwards, Mr. $D_{Stoch}$ discussed the main information in his lessons and often introduced new concepts from these discussions. It is further interesting that Mr. $D_{Stoch}$ also referred to a formalist view concerning his intended curriculum that is a central goal in calculus or analytical geometry for him. However, except for a brief oral presentation referring to Kolmogoroff's axioms, there is no evidence that Mr. $D_{Stoch}$ enacted his peripheral beliefs in his stochastics course. Thus, Mr. $D_{Stoch}$ enacts his central goals but not his peripheral ones if the entire course lasting half a year is regarded.

### *Impact of Teachers' Beliefs on Their Professional Development*

One of the main questions in teachers' professional development was the potential change of central and peripheral goals. We investigated this question concerning the 'calculus sample' referring to teachers from their final exams at university and their 2-year-period as teacher trainees until their start as qualified teachers (for greater details see Erens and Eichler 2013b). The teacher trainees were strongly schooled over a period of about 2 years and assessed after this period by their trainers. The grade in this final exam may determine the teachers' possibility of getting employment. Accordingly, all the teacher trainees like Mr. $G_{Calc}$ tried to meet the demands of their trainers:

> Mr. $G_{Calc}$: In conceptualizing new content I always use a task-oriented approach, which is a guideline given by our teacher trainers. In my opinion it's not bad, but I think it's too stringently guided like our trainers want it to be implemented. [...] From time to time I vary a little bit, but at the moment I must keep in mind my demonstrative exam lessons with my students.

However, the exemplary quotation of Mr. $G_{Calc}$ involves an illustration of a striking result: The teachers in our subsample tend to retain their central beliefs regardless of the influences of either teacher trainers or the first intense classroom experience. Of course, we will neither suggest that it is impossible to change teachers' central beliefs nor suggest that trainee teachers' beliefs show no changes at all. We find, for example, considerable changes in the teachers' rationales of their beliefs, e.g. a change from justifying their beliefs by considering their university studies to justifying their beliefs by the needs of their students. We further find that these teachers seem to adopt many aspects of teaching and learning referring to their peripheral beliefs. However, analyzing the intended curricula of these teacher trainees after their teacher training phase, we did not find any fundamental change in their previously held central beliefs (c.f. ibid.).

## Discussion and Conclusion

In this report, we focused on different parts of a research program aiming to investigate mathematics teachers' beliefs referring to different mathematical domains. The main aim of this report was not to give a deep insight into the teachers' beliefs concerning a specific discipline – we reported about this aspect elsewhere (e.g. Eichler 2011), but to emphasize several aspects that might be important for research in teachers' mathematics-related affect in general. We will highlight three aspects in this concluding section.

**A Qualitative Interview Design Enables an Identification of the Structure of Teachers' Belief Systems** The research-approach we reported in this chapter facilitates the identification of a teacher's belief system including beliefs representing overarching teaching objectives (world views). In the case of Mr. $C_{Calc}$, we identified his coherent formalist view. This qualitative result agrees with the result of the teachers' individual responses to questionnaires (Erens and Eichler 2013a). Thus, a predominant view could also be gained through a quantitative survey. However, in addition to overarching teaching goals, the qualitative approach could disclose teaching goals of a lower range of influence including even the selection of specific content or a specific task. For the calculus teachers (like Mr. $C_{Calc}$), the selection of tasks used in prompts were consistent with their predominant view. Further, this approach enables us to identify predominant views or central beliefs on the one side, but also to analyze relationships among different beliefs or belief clusters that sometimes match each other, but sometimes are contradictory (quasi-logicalness; Green 1971).

**Teachers' Beliefs Seem to Differ Referring to Different Mathematical Domains** In fact, the comparison of mathematics teachers' thinking about various mathematical disciplines gave evidence that teachers hold different beliefs about different mathematical domains. Although these teachers were mainly interviewed concerning one specific domain, i.e. calculus, geometry or stochastics, the differences in the teachers' belief systems are striking: It seems that the teachers- each of them teaches all the mentioned mathematical disciplines in upper secondary school- think considerably differently about mathematics when a specific discipline is concerned. Whereas an application oriented view seems to characterize teachers' beliefs concerning stochastics, it seems to be a process oriented view concerning geometry and, less specific, a formalist view concerning calculus. It is possible that this finding is a particular characteristic of German secondary teachers, who teach different mathematical disciplines. However, for these teachers, it is hard to claim for mathematical beliefs in general, but only for beliefs concerning a specific mathematical domain (c.f. Franke et al. 2007).

**Teachers' Central Beliefs Impact on Their Enacted Curricula and Influence Their Professional Development** We do not suggest clarifying completely the difficult relation between the teachers' espoused and enacted beliefs. However, we hypothesize that under specific conditions the teachers' espoused beliefs could explain the teachers' enacted beliefs. The first condition concerns the distinction between central and peripheral beliefs, since central beliefs seems to be more clearly enacted than peripheral beliefs (c.f. Putnam and Borko 2000). The second condition concerns a global perspective on a teacher's intended curriculum instead of a local perspective referring to one or few lessons. For instance, a teacher like Mr. $D_{Stoch}$ does not enact his central beliefs in every lesson. However, regarding a teaching period of a half year, this teacher showed predominately the enacting of those beliefs that we identified to be central in his intended curriculum. A third condition is to analyze conflicts of goals represented by contradictory beliefs about mathematics and mathematics teaching. Actually, the formalist view of calculus is central for Mrs. $E_{Calc}$. However, we expect she will not enact this view in her classroom practice, since it is in contradiction to further beliefs referring to the teaching and learning of calculus that might be more relevant for her actual teaching.

The third condition seems to us close to Skott (2009) since the teaching-related beliefs of Mrs. $E_{Calc}$ refer to the social context of her teaching practice.

Finally, the brief discussion of the professional development of teacher trainees highlights the robustness of deep-seated central beliefs. Whereas peripheral beliefs seemed to be modified in the period of a teacher traineeship, partly caused by teacher educators, partly caused by the first intense practical experience of these teachers, the central beliefs of these teachers seem to be stable. This result is partly in line with Franke et al. (1997) and is also in compliance with theoretical considerations about the stability of teachers' beliefs (e.g. McLeod 1992).

To conclude, the careful examination of mathematics teachers' beliefs is on the one hand a crucial challenge of educational research to understand "the cognitive and affective aspects of teachers' professional lives" (Calderhead 1996, p. 709), it is, on the other hand, a mandatory research field since "the nature of mathematics teachers' thinking becomes a key factor in any movement to reform the teaching of mathematics" (Chapman 1999, p. 185). For both teachers' professional lives, and a change of teachers' beliefs, a long-term and discipline-specific approach referring to teachers' intended curricula – involving the investigation of teachers' systems of beliefs including central and peripheral beliefs, coherent and contradictory belief clusters – could be a reasonable contribution to the research in teachers' mathematics-related affect.

## References

Artzt, A. F., & Armour-Thomas, E. (1999). A cognitive model for examining teachers' instructional practice in mathematics: A guide for facilitating teacher reflection. *Educational Studies in Mathematics, 40*, 211–235.

Ball, D. L. (1990). The mathematical understandings that prospective teachers bring to teacher education. *The Elementary School Journal, 90*(4), 449–466.

Borko, H., & Putnam, R. (1996). Learning to teach. In R. C. Calfee & D. C. Berliner (Eds.), *Handbook of educational psychology* (pp. 673–708). New York: Macmillan.

Bräunling, K., & Eichler, A. (2011). Subjektive Theorien von Lehrerinnen und Lehrern zum Lehren und Lernen von Arithmetik. In R. Haug & L. Holzäpfel (Eds.), *Beiträge zum Mathematikunterricht* (pp. 927–929). Münster: WTM.

Calderhead, J. (1996). Teachers: Beliefs and knowledge. In D. C. Berliner (Ed.), *Handbook of education* (pp. 709–725). New York: Macmillan.

Chapman, O. (1999). Researching mathematics teacher thinking. In O. Zaslavsky (Ed.), *Proceedings of the 23rd conference for the psychology of mathematics education* (Vol. 2, pp. 385–392). Haifa: PME.

Cobb, P. (1986). Context, goals, beliefs, and learning mathematics. *For the Learning of Mathematics, 6*(2), 2–9.

Dionne, J. J. (1984). The perception of mathematics among elementary school teachers. In J. M. Moser (Ed.), *Proceedings of the 6th conference of the North American chapter of the international group for the psychology of mathematics education* (pp. 223–228). Madison: University of Wisconsin-Madison.

Eichler, A. (2008). Teachers' classroom practice in statistics courses and students' learning. In C. Batanero, G. Burrill, C. Reading, & A. Rossman (Eds.), *Joint ICMI/IASE study: Teaching statistics in school mathematics. Challenges for teaching and teacher education. Proceedings of the ICMI Study 18 and 2008 IASE round table conference.* Monterrey: ICMI and IASE. Online: www.stat.auckland.ac.nz/~iase/publicatons

Eichler, A. (2011). Statistics teachers and classroom practices. In C. Batanero, G. Burril, & C. Reading (Eds.), *Teaching statistics in school mathematics-challenges for teaching and teacher education* (New ICMI study series, Vol. 15, pp. 175–186). Heidelberg/New York: Springer.

Erens, R., & Eichler, A. (2013a). Reconstructing teachers' beliefs on calculus. In *Proceedings of the 8th Conference on the European Society for Research in Mathematics Education* (CERME). http://www.mathematik.uni-dortmund.de/~erme/

Erens, R., & Eichler, A. (2013b). Belief systems' change – From preservice to trainee high school teachers on calculus. In A. M. Lindmeier & A. Heinze (Eds.), *Proceedings of the 37th conference of the international group for the psychology of mathematics education* (Vol. 2, pp. 281–288). Kiel: PME.

Ernest, P. (1989). The knowledge, beliefs and attitudes of the mathematics teacher, a model. *Journal of Education for Teaching, 15*(1), 13–33.

Felbrich, A., Kaiser, G., & Schmotz, C. (2012). The cultural dimension of beliefs: An investigation of future primary teachers' epistemological beliefs concerning the nature of mathematics in 15 countries. *ZDM, 44*(3), 355–366.

Förster, F. (2011). Secondary teachers' beliefs about teaching applications design and selected results of a qualitative case study. In G. Kaiser, W. Blum, R. Borromeo Ferri, & G. Stillman (Eds.), *Trends in teaching and learning of mathematical modelling* (pp. 65–74). Dordrecht: Springer.

Franke, M. L., Fennema, E., & Carpenter, T. (1997). Examining evolving beliefs and classroom practice. In E. Fennema (Ed.), *Teachers in transition* (pp. 252–282). Mahwah: LEA.

Franke, M. L., Kazemi, E., & Battey, D. (2007). Understanding teaching and classroom practice in mathematics. In F. K. Lester (Ed.), *Second handbook of research on mathematics teaching and learning* (pp. 225–256). Charlotte: Information Age Publishing.

Furinghetti, F., & Pehkonen, E. (2002). Rethinking characterizations of beliefs. In G. C. Leder, E. Pehkonen, & G. Törner (Eds.), *Beliefs: A hidden variable in mathematics education?* (pp. 39–58). Dordrecht: Kluwer.

Girnat, B. (2009). Ontological beliefs and their impact on teaching elementary geometry. In M. Tzekaki, M. Kaldrimidou, & H. Sakonidis (Eds.), *Proceedings of the 33rd conference of the international group for the psychology of mathematics education* (Vol. 3, pp. 89–96). Thessaloniki: PME.

Glaser, B., & Strauss, A. (1967). *The discovery of grounded theory.* Chicago: Aldine.

Green, T. (1971). *The activities of teaching.* New York: McGraw-Hill.

Grigutsch, S., Raatz, U., & Törner, G. (1998). Einstellungen gegenüber Mathematik bei Mathematiklehrern. *Journal für Mathematikdidaktik, 19*(1), 3–45.

Grossman, P. L. (1990). *The making of a teacher: Teacher knowledge and teacher education.* New York: Teacher College Press.

Hannula, M. S. (2012). Exploring new dimensions of mathematics-related affect: Embodied and social theories. *Research in Mathematics Education, 14*(2), 137–161.

Hannula, M. S., Op't Eynde, P., Schlöglmann, W., & Wedege, T. (2007). Affect and mathematical thinking. In D. Pitta-Pantazi & G. Philippou (Eds.), *Proceedings of the fifth congress of the European Society for Research in Mathematics Education* (pp. 202–208). Larnaca: Department of Education: University of Cyprus.

Heckhausen, H., & Gollwitzer, P. M. (1987). Thought contents and cognitive functioning in motivational versus volitional states of mind. *Motivation and Emotion, 11*(2), 101–120.

Hofer, M. (1986). *Sozialpsychologie des erzieherischen Handelns.* GŁottingen: Hogrefe.

Hofer, B. K., & Pintrich, P. R. (1997). The development of epistemological theories: Beliefs about knowledge and knowing in their relation to learning. *Review of Educational Research, 67*, 88–140.

Liljedahl, P. (2010). Changing beliefs as changing perspective. In V. Durand-Guerrier, S. Soury-Lavergne, & F. Arzarello (Eds.), *Proceedings of sixth conference of the European Group for Research in Mathematics Education* (pp. 44–53). Lyon: INRP.

Mayring, P. (2003). *Qualitative inhaltsanalyse. Grundlagen und Techniken.* Weinheim: Beltz.

McLeod, D. B. (1992). Research on affect in mathematics education: A reconceptualization. In D. A. Grouws (Ed.), *Handbook of research on mathematics learning and teaching* (pp. 575–596). New York: Macmillan.

OECD. (2009). *Creating effective teaching and learning environments: First results from TALIS*. Paris: OECD Publishing.

Pajares, F. M. (1992). Teachers' beliefs and educational research: Cleaning up a messy construct. *Review of Educational Research, 62*(3), 307–332.

Philipp, R. A. (2007). Mathematics teachers' beliefs and affect. In F. K. Lester (Ed.), *Second handbook of research on mathematics teaching and learning* (pp. 257–315). Charlotte: Information Age Publishing.

Pólya, G. (1949). *Schule des Denkens: vom Lösen mathematischer Probleme*. Bern: Francke.

Putnam, R. T., & Borko, H. (2000). What do new views of knowledge and thinking have to say about research on teacher learning? *Educational Researcher, 29*(1), 4–15.

Schoenfeld, A. H. (1998). Toward a theory of teaching-in-context. *Issues in Education, 4*(1), 1–94.

Schoenfeld, A. H. (2010). *How we think. A theory of goal-oriented decision making and its educational applications*. New York: Routledge.

Shaughnessy, M. (2007). Research on statistics learning and reasoning. In F. K. Lester (Ed.), *Second handbook of research on mathematics teaching and learning* (pp. 957–1010). Charlotte: Information Age Publishing.

Skott, J. (2009). Contextualising the notion of 'belief enactment'. *Journal of Mathematics Teacher Education, 12*, 27–46.

Staub, F., & Stern, E. (2002). The nature of teacher's pedagogical content beliefs matters for students' achievement gains: Quasi-experimental evidence from elementary mathematics. *Journal of Educational Psychology, 94*(2), 344–355.

Stein, M. K., Remillard, J., & Smith, M. S. (2007). How curriculum influences student learning. In F. K. Lester (Ed.), *Second handbook of research on mathematics teaching and learning* (pp. 319–369). Charlotte: Information Age Publishing.

Strauss, A., & Corbin, J. (1998). *Basics of qualitative research: Techniques and procedures for developing grounded theory*. Thousand Oaks/London/New Delhi: Sage Publications.

Sztajn, P. (2003). Adapting reform ideas in different mathematics classrooms: Beliefs beyond mathematics. *Journal of Mathematics Teacher Education, 6*(1), 53–75.

Thompson, A. G. (1984). The relationship of teachers' conceptions of mathematics teaching to instructional practice. *Educational Studies in Mathematics, 15*, 105–127.

Thompson, A. G. (1992). Teachers' beliefs and conceptions: A synthesis of the research. In D. A. Grouws (Ed.), *Handbook of research on mathematics teaching and learning* (pp. 127–146). New York: Macmillan.

Witzel, A. (1982). *Verfahren der qualitativen Sozialforschung*. Frankfurt am Main/New York: Campus.

# Meta-emotion and Mathematical Modeling Processes in Computerized Environments

**Inés M[a] Gómez-Chacón**

**Abstract** Integrating technology into teaching mathematics is a complex issue whose inter-related components must be addressed holistically. The research on the interaction between affect and cognition proposed in this chapter focuses on a number of understudied areas in problem-solving: visualization, affect, meta-emotion and the identification of students' affective pathways. The two studies described revealed the existence of several emotional phenomena associated with technology-assisted learning: (a) an initially positive attitude toward computer-aided mathematics learning and a preference for visual reasoning; (b) instrumental genesis associated with social and contextual dimensions of emotion and cognition; and (c) the effect of meta-emotion on task performance and the development of visual processes.

**Keywords** Visual thinking • Teacher training • Geometry • Technology • Beliefs • Meta-emotion • Mathematical modeling

## Introduction

In recent decades, building on McLeod's groundbreaking studies (1992) on affect in mathematical education, a number of authors have conducted in-depth research on the definition of affect with a view to developing more consistent theoretical frameworks (Zan et al. 2006; Leder et al. 2002; Goldin et al. 2009), while others have analyzed the interaction between cognition and affect (DeBellis and Goldin 2006; Goldin 2000; Gómez-Chacón 2000a, b; Malmivuori 2001, 2006; Hannula 2002; Schlöglmann 2002).

Scientific output on the latter issue has been less abundant, although the findings are promising and encourage further research. For example, studies on learning and affect tend to refer either to the affective reactions that may have a bearing on

I.M[a] Gómez-Chacón (✉)
Facultad de Ciencias Matemáticas e Instituto de Matemática Interdisciplinar, Universidad Complutense de Madrid, Madrid, Spain
e-mail: igomezchacon@mat.ucm.es

 
B. Pepin, B. Roesken-Winter (eds.), *From beliefs to dynamic affect systems in mathematics education*, Advances in Mathematics Education,
DOI 10.1007/978-3-319-06808-4_10

cognitive and conative processes (DeBellis and Goldin 1997; Goldin 2000, 2004; Gómez-Chacón 2000, 2011; McLeod 1994; Liljedahl 2005) or to the so-called directive processes (metacognitive and meta-affective processes) involved in the development of mathematical thinking (creativity and intuition, attribution, visualization, generalization processes and similar) (De Corte et al. 2011; Gómez-Chacón 2008). Others address the ways that emotions impact cognitive processing, such as the bias introduced in attention and memory and the encouragement of a tendency to act (Schlöglmann 2002). Emotions have also been seen to play a key role in human coping and adaptation (Evans 2000; DeBellis and Goldin 2006; Hannula 2002; Gómez-Chacón 2011).

These studies focus, on the one hand, on students' emotions during problem solving and on the other on the importance of cognition-affect interaction pathways in the construction of mathematical knowledge. Nonetheless the conceptual structure underlying this interaction has yet to be addressed in any depth. The present research sought just such depth, focusing on the cognition-affect inter-relationship and an understanding of the role played by emotion and meta-emotion in personal learning in which technology and visualization are involved. Affect is believed to play an important role in the conversion of artifact into mathematical instrument, inasmuch as a positive or negative attitude toward computers (for instance) may influence how cognitive and instrumental schemes develop. Whereas most studies adopting an instrumental approach analyze the technological use of such schemes in their cognitive and institutional dimensions (Artigue 2002; Monaghan 2004), the present research stresses the individual and affect in their generation. The study hinged on the observation of cognitive-affect processes in learning situations involving Dynamic Geometry Systems (DGS) that prioritize visualization.

Two studies, whose subjects were Spanish mathematics undergraduates planning to become secondary school math teachers are described in this chapter. The question posed was: what affective or belief systems inform mathematical visualization processes when using DGS in mathematical learning? The first study characterizes cognitive-emotional interactions in a context of technology-assisted learning, identifying the emotional typologies and phenomena experienced by subjects. The second focuses on meta-emotion and the cognitive-emotional processes that characterize interactive visualization in technology-assisted problem-solving situations.

The specific research questions posed in each study were as follows. In Study I: What are subjects' initial attitudes toward technology-mediated mathematics teaching? What cognitive-emotional processes lead to subjects' positive or negative appraisal of the use of GeoGebra to learn mathematics? And in Study II, focusing on visualization processes: What conceptual structures underlie visualization and affect during the use of DGS to learn mathematics? What information on meta-emotion and visualization can be gleaned from the productive affective pathways reported by students in locus problems? (See items 3.1. and 3.2 for further information).

The present research is primarily exploratory for two reasons: (1) meta-emotion has been scantly analyzed in mathematics classroom contexts; and (2) previous studies on visualization and technology have yielded divergent findings (e.g. Eisenberg 1994; McCulloch 2011; Presmeg and Bergsten 1995; Stylianou 2001).

The section below reviews the theoretical background and literature related to the subject addressed. This is followed by a description of the methodology, the results and discussion, and the conclusions.

## Theoretical Considerations

Given the complexity of the subject, a number of theoretical considerations were addressed to establish a consistent interpretative framework: integration of technology, visualization processes and the emotional dimension.

### *Technology and Instrumental Genesis*

Over the last two decades the impact of technologies on learning and teaching processes in mathematics has been widely studied, as attested to by the ICMI Study on the subject. According to some of these studies, attitudes toward mathematics and technology occupy different domains, while other reports contend that even when students hold very positive attitudes toward technology, they find that it tends to interfere with mathematical understanding (Galbraith and Haines 2000; Forgasz 2006; Pierce et al. 2007). New approaches suggest that much remains to be learned about the human side of technology-assisted mathematics learning, and especially about cognitive, metacognitive and affective interaction.

The theory of instrumentation (Rabardel 1995) has been acknowledged by several authors (Drijvers et al. 2010) to be an assistive framework for research. Set into a socio-cultural framework, the instrumental approach (Artigue 2002) combines the anthropological approach in didactics with the theory of instrumentation developed in cognitive ergonomics.

For Rabardel, the individual plays a key role in the process of conceiving, creating, transforming and using instruments. This process evolves in terms of both behavior and knowledge. Instruments are not spontaneously generated, but are rather the outcome of instrumental genesis. Each subject constructs artifact usage schemes for the task at hand at any given time. A scheme is defined as the invariable organization of an activity for a specific series of situations.

Artigue (2002) stressed the need for instrumental genesis that transforms artifacts into tools in two directions relevant to the present study. Upward transition, from artifact to the construction of geometric configuration, known as instrumentation, describes users' manipulation and mastery of drawing tools. The downward process, called instrumentalization, runs from configuration to the proper choice and use of an instrument and refers to geometric construction.

Research has shown that instrumental genesis is a complex, (Guin et al. 2004), primarily individual process in which cognitive, metacognitive and affective channels carry heavy specific weight. Nonetheless, it is also characterized by a social dimension, because students develop their mental schemes in the context of their class community.

This chapter analyzes instrumental genesis in subjects in a DGS environment and the impact of affect on the manner that such genesis integrates artifact and visualization in mathematical learning. The stress is on instrumental phenomena, analyzing the interaction and overlapping between the development of mathematical knowledge (visualization processes) and the understanding of artifact during teaching experiments conducted over long genesis periods.

## Visualization

Any analysis of the psychological (cognitive and affective) processes involved in working with (internal and external) representations or images in reasoning and problem solving requires a holistic definition of the term visualization. Hence, Arcavi's (2003: 217) proposal was adopted here: "the ability, the process and the product of creation, interpretation, use of and reflection upon pictures, images, diagrams, in our minds, on paper or with technological tools, with the purpose of depicting and communicating information, thinking about and developing previously unknown ideas and advancing understandings". In the present study the focus was on interactive visualization, a term used to mean an interactive approach to learning concepts such as DGS in which users receive feedback within a few seconds of entering their input. This term stresses the object's changed appearance and the dynamic view of functional dependencies, which are readily attained when the user creates figures using interactive tools in DGS.

Image typology and the use of visualization were analyzed as per Presmeg (2006) and de Guzmán (2002). Presmeg describes images as both functional distinctions between types of imagery and products (concrete pictorial, kinesthetic, dynamic and patterned imagery; memory images of formulae). In Guzmán they are categorized conceptually: the use of visualization as a reference and its role in mathematization, and the heuristic function of images in problem solving (isomorphic, homomorphic, analogical and diagrammatic visualization). This final category was adopted in the present study in connection with the use of tools in problem solving and research and the precise distinction between the iconic and heuristic functions of images (more closely related to non-iconic visualization) (Duval 2006) in analyzing students' performance (see Table 7). According to Duval, iconic visualization, the recognition of what forms represent, is based on resemblance to the (real) object represented, or by comparison with a type-model. Non-iconic visualization, in turn, is a heuristic series of operations through which geometric properties are recognized when certain configurations cannot be obtained or the configuration obtained cannot be varied.

## Emotion and Meta-emotion in Cognition-Affect Interaction

If learning is perceived essentially as interaction among cognition, conation and emotion, emotion must be broached multi-dimensionally (Op't Eynde and Turner 2006). The nature of these interactions varies, rendering their interpretation highly complex.

On the one hand, emotional experience consists of multiple inter-relations among affect, cognition (appraisal) and motivation. And on the other, in learning situations, emotional experience is closely linked to learning objectives and the control of behavior (i.e., volition), and most particularly to cognitive and metacognitive knowledge acquisition processes and heuristic problem-solving strategies.

The approach adopted here was to view affect through the lens of a representational system. The reference framework for studying affective processes has been described by a number of authors (Debellis and Goldin 2006; Goldin 2000; Gómez-Chacón 2000a, 2011), who propose regarding affect as one of several internal, mutually-interacting systems of human representation that encode meaning for the individual and can be externalized to communicate meaning to others. Affect includes changing states of emotional feeling during mathematical problem solving (*local affect*). It also includes more stable, longer-term constructs that establish contexts for and can be influenced by local affect. Known generically as *global affect*, such constructs include attitudes, beliefs and values.

The present hypothesis is that affect is fundamentally representational, rather than a system of frequently involuntary effects on cognition. Affective pathways are sequences of (local) emotional reactions that interact with cognitive configurations in problem solving. Such pathways provide solvers with useful information, favoring the learning process and suggesting heuristic problem-solving strategies. Prior research (Gómez-Chacón 2000b) identified interactions between affect and reasoning (geometrical visualization as an aspect of mathematical reasoning). The potential for affective pathways was shown to be at least partially inherent in the individual, although the effect of social and cultural conditions was also discussed. The present study focuses on the individual and any local or global affect appearing in classroom mathematical problem solving or observed by questionnaires. That same procedure was adapted for use in technological environments, where characterizing affective competencies is meaningful (affective competencies=capabilities that depend on appropriate affective encoding of strategically significant information and instrumental and cognitive skills).

Cognitive and affective self-control constitute another key factor in the cognition-affect interaction. Meta-emotion, along with meta-cognition, is regarded as "an organized and structured set of emotions and cognitions about the emotions, both one's own emotions and the emotions of others" (Gottman et al. 1997). In mathematics education, this term can be found in recent papers by the De Corte team (such as De Corte et al. 2011), and in earlier proposals under the term meta-affect (DeBellis and Goldin 1997, 2006; Goldin 2000; Gómez-Chacón 1997, 2000; Schlöglman 2005).

Meta-affect is a conceit introduced by DeBellis and Goldin (1997). Goldin describes it as: "a central notion … to refer to affect about affect, affect about and within cognition that may again be about affect, monitoring of affect both through cognition and affect" (Goldin 2002: 62).

Several studies have improved this initial intuitive definition by refining factors referring to meta-emotional understanding and meta-emotional skills. This has contributed conceptually to how meta-affect arises in the formation of an individual's cognitive and affective schemes (Gómez-Chacón 2000a, b, 2008; Malmivuori 2001; Schlöglmann 2005). Cognitive understanding of affect enables individuals to control their actions in affective situations. Successful handling of affective situations stabilizes affect schemata, and consequently beliefs, through simulation, as a cognitive window to emotions (Schlöglmann 2005). Prior research has shown that the stability of the individuals' beliefs is closely related to the interaction among belief structures. These include not only affect (feelings, emotions) but also and especially meta-affect (emotions about emotional states, emotions about cognitive states, thinking about emotions and cognitions, regulation of emotions) (Gómez-Chacón 2000b). Those findings reveal the personal and social dimensions of the affective constructs and self-control of emotions.

## Research Design and Methodology

The decision to discuss two studies here was informed by questions of approach and methodology.

1. *Comprehension of the subject of the research at different levels:* the first study, which explored cognitive processes and their interaction with affect through surveys (attitude scales and questionnaires), revealed a positive relationship among attitudes, emotions and visualisation processes in computer-aided mathematics learning. The second aimed to explore the Study I findings in greater depth, in particular the conceptual structure underlying visualisation processes and productive routes and the role of meta-emotion in the solution of problem types using DGS.
2. *The combination of qualitative and quantitative methods* to broach the subject of the study: while survey studies and questionnaires may be suitable methods for measuring trait-like variables, design experiments provide the means for addressing the complexity of educational settings. They afford a fuller understanding of learning ecology (a complex, interacting system involving many elements of different types and at different levels) by designing the elements involved and anticipating how they interact to support learning (Cobb et al. 2003: 10). Qualitative research was conducted by observing the subjects as they solved a problem during training. Students were asked to discuss their approach to solving the problem on protocols covering the following items: problem-solving process, step-by-step, description of the difficulties they might face and strategies deployed. They were also asked to record the emotions and difficulties experienced in writing. Their motivational and emotional processes were assessed via performance analysis and video-recorded, semi-structured interviews.

   The methodology deployed in each study is described below.

### *Methodology of Study I*

The research questions posed in the study were as follows. What are subjects' initial attitudes toward technology-mediated mathematics teaching? What cognitive-emotional processes lead to subjects' positive or negative appraisal of the use of GeoGebra to learn mathematics?

The subjects surveyed in this study were 98 (65 women and 33 men) Spanish undergraduates working toward a B.Sc. in mathematics with a view to becoming secondary school math teachers.[1] They comprised four classroom groups engaging in training in the analysis and instrumental aspects of teaching situations. This is what is known as a convenience sample, a category of selected samples in which the accessible population is representative of the theoretical population (Gliner et al. 2009). In this case, the population included all the students in the Faculty of Mathematics taking this subject, from which a smaller group was selected and asked by the researcher to participate in the study.

The survey used an adapted version of the instruments (Likert-type attitudes scales) developed by other researchers (Galbraith and Haines 2000) to evaluate attitudes toward mathematics and technology, along with a specific questionnaire to determine preferences for visual reasoning and feelings about computers. These instruments covered both feelings and opinions about the use of technology to learn and use mathematics. The questionnaire posed questions such as the following. Is visual reasoning central to mathematical problem solving? Justify your reply and provide examples. Describe your feelings about the use of problem representations or visual imagery. Describe your emotional reactions and specify whether you hit a mental block when doing the problem with pencil and paper or with a computer. Do computer graphics help you learn mathematics?

Different groups of items required different statistical methods. (1) Likert-type scale attitudes were analyzed with SPSS software, which computed the means, standard deviation and internal consistency (Cronbach's $\alpha$) for each of these sub-scales of the survey (based on a 5-point Likert scale, from 1 to 5); the correlation between attitude scales; the factor pattern matrix; and clusters. (2) The open-ended questions concerning the most and least preferred method of visual reasoning, computer-related emotions and cognitive learning difficulties in technology-assisted mathematics work were coded by qualitative data processing using content analysis to define the categories listed below. Frequency values were computed by two researchers. (3) Similarly, all categories were compiled and coded in a matrix for implicative analysis performed using CHIC (Bodin et al. 2000) software (see item 3.3).

---

[1] The three requirements to teach mathematics in Spanish secondary schools are: (a) a B.Sc. in mathematics or science, (b) an M.Sc. in education for secondary school, and (c) passing a series of public exams.

The subjects of this study are working toward a B.Sc. in mathematics. Spanish Faculties of Mathematics offer specialised training in secondary school mathematics education as part of the undergraduate mathematics curriculum. The subjects presently available include: "Mathematics Education in Secondary Schools", "Mathematics for Teaching" and Practicum (practice teaching in secondary schools).

Most future teachers participating in this program believe that they have sound mathematics training, after having taken advanced courses in several areas of geometry, algebra and analysis. Two-thirds of these students acquire teaching experience prior to the training plan (Practicum) as private tutors or in tutoring schemes for secondary school students.

The M.Sc. in Education (Secondary) is a post-graduate course that builds on prior learning and develops advanced professional knowledge, practice and relationship skills relevant to teaching.

The data was coded into categories:

*Emotions regarding computer (GeoGebra) use*: Positive (EmoP), Negative (EmoN), depend on the task and activity (Emodep).

*Preference for visual reasoning:* VisualA (like); VisualN (indifferent); VisualD (dislike)

*Attitudes toward mathematics and technology*: self-confidence in mathematics (mathconf), mathematical motivation (mathmot), mathematics engagement (matheng), computer motivation (compmot) and interaction between mathematics and computers (mathcompuint).

*Cognitive learning difficulties in technology-assisted mathematic work*: Diff_Proc (understanding and interpreting the problem in the initial phase of problem solving); Diff_Ins (instrumental genesis (software commands and mathematical meaning or the dependencies between objects in geometry dynamics); Diff_Block (blockage in overall control of the geneses involved in geometric work (blockage in the switch from discursive to instrumental and from discursive to visual)).

In Study I, survey-based assessments of attitudes and emotions were supplemented by observation during teaching experiments (Gómez-Chacón and Kuzniak 2011) designed to explore the interaction between cognition and affect in mathematics learning. The teaching experiment findings supported the survey results, revealed the complexity of interpreting the data (contextual nature of emotions) and contributed to defining better instruments. In these experiments, the causal relationships among emotions, cognitive processes expressed as cognitive difficulties and attitudes toward technology are highly context-dependent. The results also highlighted the importance of visualization in understanding and solving problems and showed that visualization may be associated with different emotions and beliefs. Based on these analyses and results, and with the classification of the data into the aforementioned categories, new conjectures and hypotheses were defined and tested in Study II (section "Methodology of Study II") and better instruments were developed.

## *Methodology of Study II*

The subjects for this study were 32 mathematics undergraduates (future secondary school math teachers) classed in groups whose attitudes and belief systems were representative of the profiles identified in the first study (see Table 5 for the individual characteristics).

The study was mainly based on Design Experiment methodology (Cobb et al. 2003). Qualitative research was conducted by observing subjects during training, performance analysis sessions, video-recorded, semi-structured interviews and two questionnaires.

Geometric locus training was conducted in three 2-h sessions. In the first two sessions, subjects were asked to individually solve six non-routine geometric locus problems using GeoGebra in accordance with a proposed problem-solving procedure. A description of the problems, their graphic solutions and design are given in Gómez-

Chacón and Escribano (2011). That paper discusses the results of the following problem (P4) *the top of a 5-m ladder rests against a vertical wall, and the bottom on the ground. Define the locus generated by midpoint M of the ladder when it slips and falls to the ground. Define the locus for any other point on the ladder.*

The wording provides no explicit instructions for constructing the *geometric locus*. The situation is realistic and easy to understand, but its translation to a GeoGebra construct is not straightforward. Once that initial difficulty is overcome, *visual reasoning* is deployed, and the ladder is drawn with the aid of an auxiliary object, GeoGebra accurately represents the locus. Finding the algebraic answer entails defining five points on the locus and then the conic passing through all five. The result is an algebraic equation. Using *instrumental reasoning*, the two keys to the problem are: (1) construction of the ladder with an auxiliary circle, and (2) the precise definition of the point (mid-point, ¼ point) to study the locus of the positions occupied by the ladder.

The subjects were also asked to describe and record their emotions, feelings and mental blocks when solving problems on protocols designed for that purpose. The third session was devoted to discussing joint approaches and the difficulties experienced in problem solving.

Data were collected from the subjects' problem-solving protocols mentioned above, as well as with two questionnaires, one on beliefs and emotions about visual reasoning completed at the outset and the other on the interaction between cognition and affect in a technological context filled in after each problem was solved.

A first questionnaire focused on identifying subjects' beliefs about visualization and computers to study their global affect and determine whether a given belief can elicit different emotions from different individuals. A second questionnaire was completed at the end of each problem. The main questions are listed in Table 1.

Two types of analyses were conducted in this study. The first was exploratory, descriptive and interpretational, involving mainly inductive data analysis, with categories and interpretation building on the information collected (section "Theoretical Considerations"). This analysis used a qualitative approach based on cross-checking the solutions by three researchers. The first step in data analysis was to classify the methods used by the students and identify classes of answers in accordance with a set of visualization variables and emotions. For instance, subjects' visualization, use of imaging and emotions as reported on their solution protocols were identified for each problem. This information was summarized in schemes such as outlined in Table 7. The second was based on implication analysis, for which the *following categories* were defined.

1. *Emotion associated with visual reasoning* in the ladder exercise: P4EviP (like), P4EviN (dislike), P4EviM (mixed emotions), and P4viInd (indifferent).
2. *Instrumental difficulties*: the focus in this category was on two types of difficulties arising around the six problems (Table 2 and Gómez-Chacón and Escribano 2011). *Typology 1: Static constructions (discrete)* (DT1P4). In this typology, subjects used GeoGebra as an advanced blackboard, but they did not use dynamic properties. They repeated the constructions for a number of points. To draw the

**Table 1** Student questionnaire on the interaction between cognition and affect

| Please answer the following questions after solving the problem: | |
|---|---|
| 1. | Was this problem easy or difficult? Why? |
| 2. | What did you find most difficult? |
| 3. | Do you usually use drawings when you solve problems? When? |
| 4. | Were you able to visualize the problem without a drawing? |
| 5. | Describe your emotional reactions and specify whether you hit a mental block when doing the problem with a pencil and paper or with a computer. |
| 6. | Which of the following routes best describes your emotional pathway when solving the problem? If you identify with neither, please describe your own pathway. |
| | Affective pathway 1 (enabling problem solving): curiosity → puzzlement → bewilderment → encouragement → pleasure → elation → satisfaction → global structures of affect (specific representation, general self |
| | Affective pathway 2 (constraining or hindering problem solving): curiosity → puzzlement → bewilderment → frustration → anxiety → fear/distress → global structures of affect (general self |
| 7. | Specify whether any of the aforementioned emotions was related to problem visualization or representation and the exact part of the problem concerned. |

**Table 2** Examples of subjects' difficulties with the "ladder" exercise

| | |
|---|---|
| 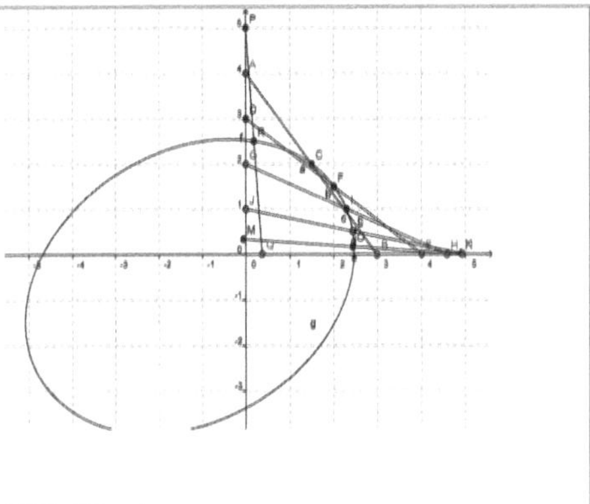 | 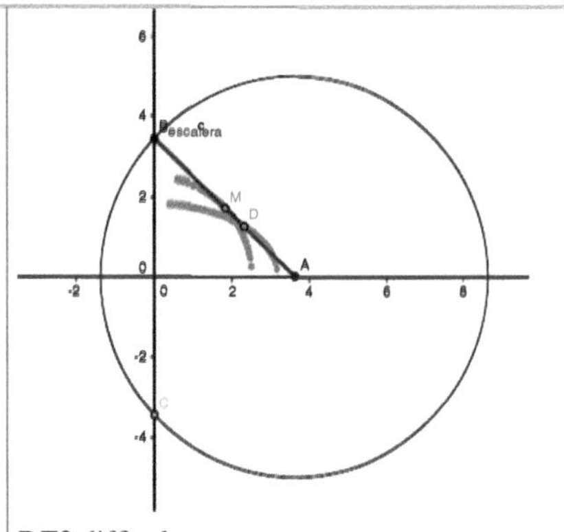 |
| DT1 difficulty | DT2 difficulty |

geometric locus, they used the "5-point conic" tool. *Typology 2: Incorrect definition of the construction* (DT2P4). The subjects solved the problem (albeit inaccurately), but the GeoGebra tools are unusable in such an approach. To use the "locus" tool, the points must be precisely defined (they may not be free points). The best subjects could do when broaching the exercise from this perspective was to obtain a partially valid construction, but, without the GeoGebra tools, the algebraic answer could not be found. In this exercise, the difficulty consisted of precisely defining a point other than the mid-point. Using an undetermined point on the ladder would preclude using the locus tool.

3. Initial problem visualization: VisiP4

4. Beliefs about visual reasoning: BeviP (positive), BeviN (negative)
5. Preferences and emotions around the use of visualization: EviP (like), EviN (dislike), EviInd (indifference)
6. Beliefs about computer-aided learning: BeGeoP (positive), BeGeoN (negative)
7. Emotions around computer use: EGeoP (like), EGeoN (dislike), EGeInd (indifference)
8. Affective-cognitive pathways R1 and R2 (explained in the questionnaire in Table 1) and R3 (subject-formulated, as described in Table 7).

Each researcher conducted a separate analysis, using two dimensions of the genesis of mathematical work as a guide. The use and role of instruments and techniques were used for instrumental genesis, while the basis for visual-figurative genesis was the use and role of figures and images. The researchers' findings were compared and discussed where disagreements were detected. The identification of possible links among affective-cognitive pathways, emotions and meta-emotion was the object of joint analysis.

## *Implicative Data Analysis*

The implicative data analysis method (Gras et al. 1997) used in both Study I and Study II to supplement qualitative analysis is described briefly below. This procedure begins with a group of individuals (the 98 subjects in the first study, for instance) described by a finite set of binary variables (computer attitudes and emotions, preference for visual reasoning, cognitive learning difficulties). The question posed is: to what extent is variable $b$ true when variable $a$ is true? In other words, do subjects known to be characterized by $a$ tend also to exhibit $b$? In real-life situations deductive theorems of the logical form $a \rightarrow b$ are often difficult to establish due to the existence of exceptions. The dataset must consequently be mined to extract rules reliable enough to conjecture causal relationships around which population is structured. At the descriptive level, they can be used to detect a certain degree of stability in the structuring, while for predictive purposes, they provide the grounds for assumptions.

This statistical analysis was then used to establish rules of association for data series in which variables and individuals were matched to define trends in sets of properties on the grounds of inferential, non-linear measurement. This non-symmetrical statistical approach draws from the notion of implication, borrowed from Boolean algebra and artificial intelligence. Knowledge is formed inductively after a number of successful attempts ensure a certain level of confidence in a given rule. As soon as this (subjective) level is reached, the rule is accepted and implemented.

According to Gras (Gras et al. 1997), learning begins with inter-related facts and rules that progressively form learning structures. That is precisely the aim of the present study, to find rules that reduce the number of categories (listed in items 3.21 and 3.1) while furnishing information on the factors involved in the cognition-affect structure. Gras defines three important rules that can be described in learning processes: (1) $a \rightarrow b$, where a and b may be categories or rules; (2) $a \rightarrow (b \rightarrow c)$; and (3) $(a \rightarrow b) \rightarrow (c \rightarrow d)$. These rules describe a hierarchical, oriented and non-symmetrical learning structure, that can be obtained with cohesive hierarchical implicative classification (CHIC) software (Bodin et al. 2000). The result is three types of diagrams that contain different types of information. (a) Similarity trees group variables on the grounds of their uniformity, allowing for interpretation of the groupings with which the variables are handled. Each level on the resulting graph contains groups arranged in descending order of similarity. (b) Hierarchy trees are used to interpret classes of variables defined in terms of significant levels along the lines of similarity, identifying association rules and levels of cohesion among variables or classes. (c) Implication graphs are constructed around both an intensity index and a validity index to show associations among implications that are significant at specific levels.

# Results

## *Study I: Identifying Cognitive-Affective Interaction Phenomena in a Technological Context*

The phenomena identified to characterize cognition-affect interaction in technology-mediated learning were: (a) initial medium-high acceptance of computer-aided mathematical learning; (b) the situational and contextual nature of emotions and cognitive processes.

### Initial Attitude Towards Learning Mathematics in a Technological Context

This group showed an initially satisfactory (medium-high) attitude, i.e., appropriate disposition from the outset, characterized by dimensions such as self-confidence in mathematics (mathconf), mathematical motivation (mathmot), mathematics engagement (matheng) and positive beliefs about mathematics and computer-enhanced mathematics learning. The mean values for these dimensions were similar in this group, although statistically significant differences were found among the standard deviations at a 95.0 % confidence level (Table 3).

Many subjects provided details on their initial attitude, noting that they felt joy when they were able to solve a problem with the computer and let go of their apprehension. The reasons given were as follows. (a) Working with a computer enhanced their pleasurable classroom experience and made mathematics more interesting, less abstract, by helping them to find a "meaning". (b) Computers favored learning and success in mathematics, strengthening visualization skills and the accuracy of calculations. (c) Computers helped them establish connections between the

**Table 3** Mean, standard deviation and internal consistency (Cronbach's α) for each sub-scale

| | Mathconf | Mathmot | Matheng | Compmot | Mathcompuint |
|---|---|---|---|---|---|
| Mean | 3.32 | 3.53 | 3.41 | 3.41 | 3.41 |
| Median | 3.25 | 3.63 | 3.38 | 3.25 | 3.45 |
| Standard deviation | 0.48 | 0.69 | 0.65 | 0.65 | 0.50 |
| Coefficient of variation | 14.55 % | 19.61 % | 19.28 % | 18.97 % | 14.79 % |
| (Cronbach's α) | 0.77 | 0.80 | 0.57 | 0.82 | 0.69 |

algebraic and analytical dimensions. (d) And a final category was associated with their goals as future teachers.

## Cognitive Processes and Emotions

Subjects' emotions when solving mathematical problems with a computer were analyzed to attempt to answer the question: what cognitive-emotional processes govern subjects' positive or negative appraisal of the use of GeoGebra to learn mathematics? Positive emotions were described by 52.6 %, negative emotions by 19.6 % and 27.8 % replied that they could not give a general answer, for their emotions were task-specific.

The frequency of emotional categories varied, partly depending on task typology. On the whole, the emotions that subjects reported as obstructions to learning were lack of self-confidence and apprehension, which respectively accounted for 15 and 25 % of all the responses. The reasons were mental blockage around the use of the tool and the application of mathematical knowledge with the software: "*At first you feel overwhelmed when you don't know how to apply your mathematical knowledge with the tools that perform these operations*" (A18-GA-E2). Frustration and disappointment were other negative emotions frequently reported in retrospect, attributed to an uncertain command of the technical language or the time that had to be devoted to solving the problem.

The characteristics of the cognitive-affective structure were explored by applying Statistical Implicative Analysis. Figure 1 shows the similarity analysis findings. The variables (listed in item 3.1.) grouped in two classes are underlined in red.

Group 1: this group was characterized by variables specifying a clear preference for visual reasoning and working on computers (EmoP VisualA) and positive attitudinal dimensions (MathconfA ((MathmotA MathengA) (CompmotA MathcompuintA))). One of the most prominent elements was cognitive difficulties and variables relating to the global control of geometric tasks that affect discursive and instrumental processes in technology-assisted work.

Group 2: this group was associated with the variable describing instrumental genesis cognition and problem visualization difficulties in the initial phase of problem solving and negative feelings toward computer-aided mathematics. Not significant.

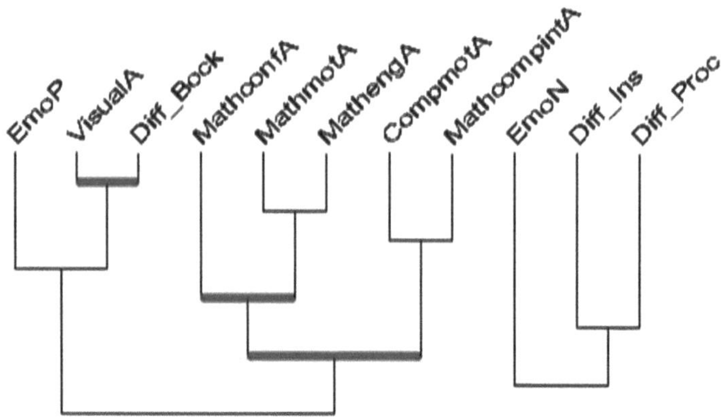


**Fig. 1** Similarity tree

The implicative graph revealed causal relationships with a reliability index of at least 84 % between: computer motivation and attitude toward the interaction between mathematics and computers (CompmotA$\rightarrow^{0.85}$ MathCompuint); non-preference for visual reasoning and instrumental difficulties (VisualD$\rightarrow^{0.84}$ Diff_Ins); and a positive preference for visual reasoning and variables relating to the global control of geometric tasks that affect visual, discursive and instrumental processes in technology-assisted work (VisualA$\rightarrow^{0.95}$ Diff_Block).

In a nutshell, motivation to use computers is the variable observed to have the heaviest impact on mathematics-technology interaction. The results also revealed the importance of a preference for visual reasoning in understanding and solving problems.

## *Study II: The Role of Emotion and Meta-emotion in Interactive Visualization*

The objective here was a more thorough review of the conceptual structure underlying visualization, along with productive pathways and meta-emotional intervention in locus problems.

### Beliefs and Emotions

In this study a similar response was received when the beliefs explored related to the use of dynamic geometry software as an aid to understanding and visualizing the geometric locus idea. All the subjects claimed to find it useful and 80 % expressed positive emotions based on its reliability, speedy execution and potential to develop their intuition and spatial vision. They added that the tool helped them surmount mental blocks and enhanced their confidence and motivation. As future teachers

**Table 4** Frequency of affective-cognitive and emotional pathways associated with visualization in the ladder exercise (N = 32)

| | R1 | R2 | R3 | EviP | EviN | EviM | EviInd |
|---|---|---|---|---|---|---|---|
| Problem 4 | 15 | 4 | 13 | 6 | 8 | 17 | 1 |

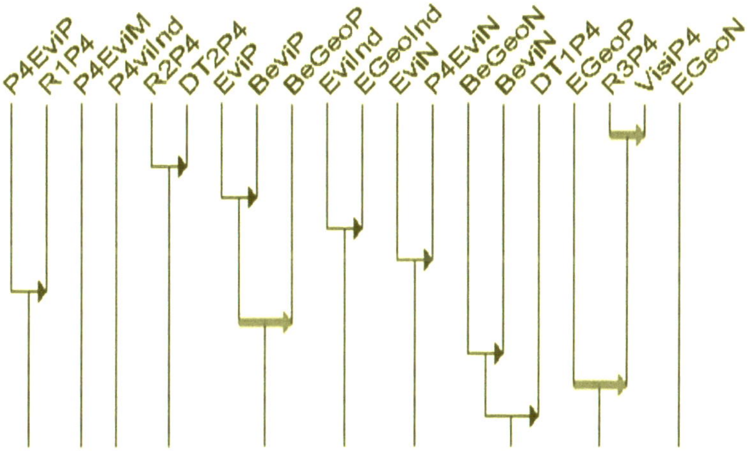


**Fig. 2** Hierarchy tree

they stressed that GeoGebra could favor not only visual thinking, but help maintain a productive affective pathway. They indicated that working with the tool induced positive beliefs towards mathematics itself and their own capacity and willingness to engage in mathematics learning (self-concept as a mathematics learner).

Table 4 summarizes the frequencies of pathways and emotions associated with visualization in the ladder problem. Mixed affective pathways were identified, with alternating negative and positive emotions and optimized self-control of emotions.

The question posed to study the mix of emotions and meta-emotion in greater detail was: what are the differences in a subject's choice of these three pathways? A preliminary analysis showed that pathway R3 was largely self-formulated and contained a much greater mix of emotions. In most cases, moreover, the trend was not as explicit as in R1 (positive) or R2 (negative). Rather, negative feelings (which were controlled) were attributed to certain stages of the visualization process, and positive feelings to success in representing the desired images. A hierarchy study of R3 yielded some significant affective-cognitive implications respecting visual processes, such as: R3P4$\rightarrow^{0.99}$ VisiP4 and R2P4$\rightarrow^{0.90}$ DT2P4 (Fig. 2).

Three of the nine nodes obtained in the hierarchy tree were significant and identified the following groups.

Group 1 (N (level 1, cohesion: 0.998) = (R3P4 VisiP4)), comprising over 40 % of the initially visualizing subjects (in problem 4) who indicated pathway R3 as the expression of their cognition-affect interaction. The most significant characteristic of these individuals was their positive feelings towards computers (use of GeoGebra (EGeoP) software).

Group 2 (N (level 7, cohesion: 0.276)=((EviP BeviP) BeGeoP))), where the most prominent finding was that a belief in the use of GeoGebra was attendant upon a belief in and a preference for visual reasoning.

## Meta-emotion and Visualization: Maintaining Productive Affective Pathways

The study subsequently focused on the Group 1 subjects to ascertain the characteristics of their R3 pathways and explore the implied relationships. The affective pathways they reported were compared to glean information on meta-emotion and visualization. The comparison revealed: (a) the use of visualization and associated emotion; and (b) their emotional self-control depended on their individual perception, which was influenced by style, disposition, type of activity or skill, instrumental command and belief systems around technology-aided mathematical learning.

By way of illustration of the foregoing, three case studies with varying characteristics are discussed below. The key characteristics of the case studies are given in Table 5.

The ladder problem was chosen for this analysis. According to the subjects' affective pathways for this problem given in Table 6, two identified with R3 and one with R2.

The first case was *S-19*, a visualizer. In the interview he said that the pleasure he derived from visualization was closely associated with his view of mathematics. He regarded visual reasoning as essential to problem solving to monitor, contribute to the intuitive dimension of knowledge and form mental images.

When he was asked whether his feelings were related to visualization and to specify the parts of the problem where they were, he replied: "*curiosity predominated in visualization. Since the problem was interesting. I was keen on finding the solution. I had a major mental block when it came to representing the problem and later, when I sought a strategy, I was unable to define a good strategy to find the answer. I was puzzled long enough to leave the problem unsolved and try again later. When I visualized the problem in a different way, I found a strategy: construct a circle with radius 5 to represent the ladder and another smaller circle to represent the point in question. When I reached that stage, I felt confident, happy and satisfied*" (S-19).

An in-depth analysis of the problem-solving protocol for this exercise and the affective-cognitive pathway reported showed that this subject was able to describe and control emotions and identify causes. First he tried to find the mathematical object, but reached a block (Table 7 (1)). He even attempted construction with physical objects. That led him to iconic visualization and from there to visual and semiotic exploitation within a dynamic geometry environment. For instance, in Table 7 (1), the subject's search for a "mental image", involved "implemented discovery" including empirical pursuit, as if his figures were objects of experimentation, anticipating

**Table 5** Three case studies: characteristics

| Case | Gender | Mathematical achievement | Visual style | Beliefs about computer learning | Feelings about computers | Beliefs about visual thinking | Feelings about visualization | Pathway | Global affect |
|---|---|---|---|---|---|---|---|---|---|
| Student 19 | Male | High | Visualizing subject | Positive | Likes | Positive | Likes | R3 | Positive self-concept |
| Student 20 | Female | Average | Non-visualizing subject | Positive | Dislikes | Positive | Dislikes | R3 | Positive self-concept |
| Student 6 | Female | Low | Style not clear | Positive | Dislikes | Positive | Likes | R2 | Negative self-concept |

**Table 6** Affective pathways and visual cognitive processes reported for this problem by three subjects

| Ladder exercise | Cognitive-emotional process | |
|---|---|---|
| Student 19 | Curiosity | Reading and understanding problem |
| Pathway R3 | Confusion | Drawing (patterns and lines/figure) |
| | | Analytical |
| | Puzzlement | (Search for mental image) (specific illustration and dynamic image) |
| | Mental block | |
| | Confidence | Search for mental image |
| | Perseverance-motivation | Search for mental image |
| | Exhilaration | Physical manipulation – kinetics |
| | | Kinesthetic learning |
| | | Mental image Identification mathematical object |
| | Confidence | Technological manipulation with the computer |
| | | Representing circle radius (specific illustrations) |
| | Confidence, joy | Interactive image generation, slider (analogical) |
| | Joy | Interactive image generation, slider (analogical) |
| | Perceived beauty | Specific illustration with interactivity (analogical) |
| | Satisfaction | Analytical-visual |
| | | Memorized formulaic typology |
| | Global affect | Positive self-concept |
| Student 20 | Curiosity | Reading problem |
| Pathway R3 | Frustration | Global visualization of problem |
| | | Pictorial image |
| | Confusion | Search for mental image |
| | | Inability to visualize the ladder as the radius of a circle |
| | Puzzlement | Search for mental image |
| | | Dynamic and interactive image with GeoGebra |
| | Stimulus, motivation | Technological manipulation with the computer |
| | | Pictorial representation with GeoGebra |
| | Satisfaction | Pictorial representation with the GeoGebra "trace on" function |
| | | Full construction from scratch |
| | | Obtaining a final solution |
| | Global affect | Positive self-concept |
| Student 6 | Curiosity | Reading problem |
| Pathway R2 | Puzzlement | Global visualization of problem |
| | | Pictorial image |
| | Bewilderment | Search for an instrumental image with GeoGebra |
| | Frustration | Computer skills |
| | Apprehension | Inability to visualize the ladder as the radius of a circle or to use the "trace on" function |
| | Fear/despair | Need for help to find the solution |
| | Global affect | Negative self-concept |

**Table 7** Analysis of S-19's solution to the ladder exercise and use of images as reported by the subject in his protocol

| Description of method, including visualization | Use of image | Emotions/affect |
|---|---|---|
| (1) First I sketched the problem on paper. I tried to find a way to solve it analytically but couldn't. I imagined the possible relationships between the triangles that the ladder would gradually generate as it slides downward against the wall to the ground, but that got me nowhere. | Drawing (patterns and lines/figure)<br>Analytical (search for mental image (specific figure/illustration and dynamic image) | Curiosity |
| (2) Then I tried to envisage the answer: would it be a straight line, an ellipse or a circle? | | Confusion |
| (3) I decided to come back to the problem another day, but I kept thinking about it meanwhile. I trusted my subconscious to help me. | Search for mental image | Puzzlement<br>Mental block<br>Confidence<br>Perseverance-motivation |
| (4) I came back to the problem with new energy. I experimented with a biro attached in the middle to a rubber band. It seemed to form the arc of a circle. At least I had something to go on. | Physical manipulation - kinetics (Kinesthetic manipulation)<br>Mental image – identification<br>Mathematical object | Exhilaration |
| (5) I started to work with GeoGebra. After trying a few straight line constructs, I noticed that the ladder could be viewed as a radius of a circle of length 5 running along the y-axis. | Technological manipulation with the computer<br>Representation of the radius of the circle (Specific illustrations) | Confidence |
| (6) I generated a slider that I called t and defined the center of the circle C = (0,t). The slider would shrink from 5 to 0 when the ladder is lying on the ground. Point B represented the intersection of the circle with the x-axis. | 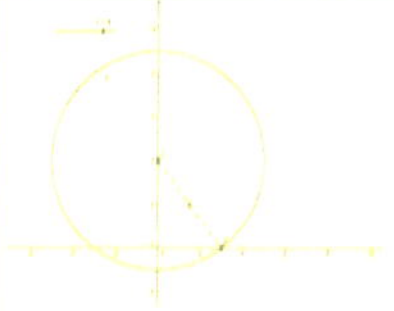<br>Interactive image generation, Slider (analogical) | Confidence, joy |
| (7) After drawing the ladder, I constructed another circle with a variable radius (using a second slider, p). This circle would be the path of the point constituting the object of the study. | 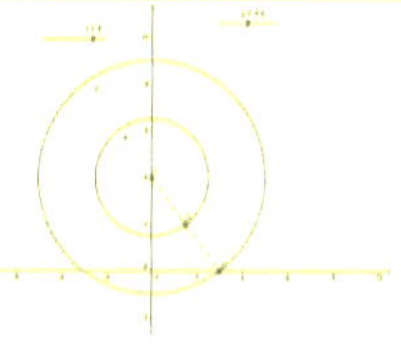 | Joy |
| In the first case, it is the midpoint of the line representing the ladder. | Interactive image generation Slider (analogical) | |

(continued)

**Table 7** (continued)

| Description of method, including visualization | Use of image | Emotions/affect |
|---|---|---|
| (8) I observed the path of the midpoint of the ladder when I activated the drawing. | 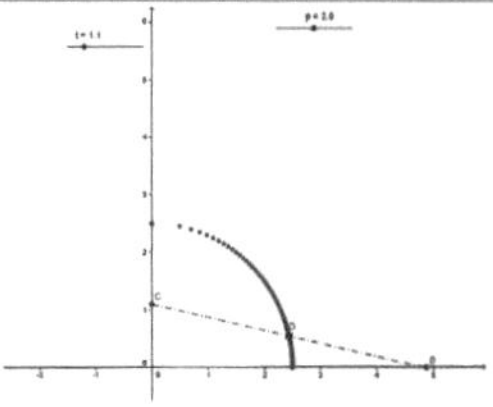<br>Specific illustration with interactivity (analogical) | Perceived beauty |
| (9) I found that I had an arc of a circle with a center C = (0,0) and radius r = 2.5. Then, I tried the same procedure with a point located 4 m from the bottom of the ladder in the initial position (on the y-axis). The result was an arc of an ellipse whose major vertical semi-axis was 4 and minor semi-axis 1. | 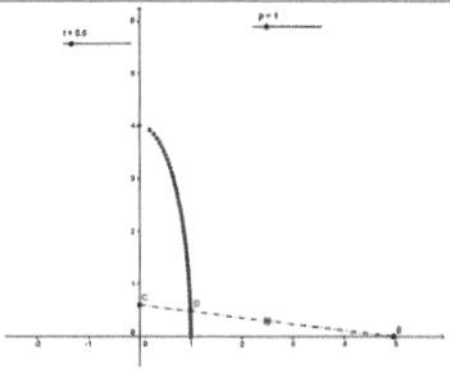<br>Specific illustration with interactivity (analogical) | Perceived beauty |
| (10) I calculated the geometric locus generated by point D to be an ellipse whose major semi-axis was max(h, 5-h), positioned vertically if h > (5-h) and horizontally if h < (5-h)); and whose minor semi-axis was min (h, 5-h) positioned horizontally if h > (5-h) and vertically if h < (5-h)), where h is the distance of the point from the base of the ladder when set vertically. | Analytical-visual<br>Formulation from knowledge of mathematics | Satisfaction |
| Where h = 2.5, the two semi-axes are equal, confirming that it is a circle. | | |

the right loci. Here, inductive reasoning could be seen to be supplemented with analytical reasoning in certain steps. In others, analytical reasoning was explicit, such as in Table 7 (10).

Three types of affective perspective were identified. First, S-19, always tried to find an answer even when in doubt or blocked. S-19 was continuously active, which is one way that many students cope with stress. Secondly, he was able to walk away from the problem, aware of the role of the subconscious in mathematics (Table 7 (3)). Thirdly, he struck a balance between the combination of graphic geometric thought and analytical task solving. These three behaviors were indicative of interaction between the cognitive-affective system and self-control. The description of emotions revealed that self-confidence, stimulus and joy were associated with the reproduction of physical forms and the visual/perceptive control implicit in a command of ancillary mathematical objects, from both a mathematical and a technical-instrumental perspective.

*S-20* was a non-visualizing thinker with positive beliefs about the importance of visual reasoning. However, she claimed that her preference for visualization depended on the problem and that she normally found visualization difficult. It was easier for her to visualize "real life" (such as in the ladder exercise) than more theoretical problems.

Her motivation and emotional reactions to the use of computers were not positive, although she claimed to have discovered the advantages of GeoGebra and found its environment friendly. She also found that working with GeoGebra afforded greater assurance than manual problem solving because the solution is dynamically visible. Convincing trainees such as S-20 that mathematical learning is important to teaching their future high school students helps them keep a positive self-concept, even if they don't always feel confident in problem-solving situations (Table 6).

*S-6*'s visual thinking style could not be clearly identified. She expressed a belief in the importance of positive visual reasoning ("because visual reasoning helps gain a better understanding of the problem and consequently the solution"). This confirmed a liking for visualization and representation because it made it easier to understand the problem and she found formalization helpful. She added, however, that she felt insecure applying technological software to mathematics, although she believed GeoGebra, specifically, to be useful. In her own words, "*I don't like it and never will. I feel a little nervous and unsure of myself, not because of GeoGebra but because computers intimidate me because I don't understand them completely. But when I managed to represent the problem with GeoGebra, I felt more satisfied with the result than when I solved it with paper and pencil*". Although S-6's pathway was essentially negative, she persisted until she found the solution. In some cases subjects were unaware of their mistakes and misunderstandings, however.

Comparing these three cases in terms of meta-emotion (self-awareness of the nature, cause and control of emotions) showed that the subjects exhibited meta-cognitive and meta-affective understanding in connection with the knowledge acquired and their own beliefs and cognitive processes. And further to Demetriou and Kazi (2001), this self-image makes a fairly effective contribution to learning, for it forms an integral part of mental control.

The analysis of the relationship between these three subjects' affective pathways (Table 6) and their cognitive visualization showed that negative feelings and interactions around visualization stemmed essentially from subjects' lack of familiarity with the tools. They were intimidated by their want of resources in their search for computer-transferable analogue images and their conversion from a paper and pencil to a computer environment in their interpretation of the mathematical object. The results nonetheless showed that although the construction of interactive images may be an obstacle, interactivity may be beneficial. Here, interactive computer programs provided feedback and clues to raise subjects' awareness of their cognitive and metacognitive processes. A review of S-19's pathways in the six problems revealed that the interaction between visual reasoning and negative feelings arose around the identification of interactive representation strategies and the formulation of certain images involving the identification of parametric variations. This subject's command of the use of concrete, kinesthetic and analogical images was very helpful and contributed to his global affect and his positive overall self-concept when engaging in computer-aided mathematics.

The data also revealed the relationship between beliefs, goals and emotional pathways. The analysis of S-20's responses showed that while she had no inclination to use computers, the importance she attached to mathematics and IT in specific objectives and the structuring of her overall objective kept her on a productive affective pathway (McCulloch 2011).

S-20's solution to the ladder exercise (Table 6), for instance, constitutes a good example of a productive pathway: despite negative feelings, she maintained a positive mathematical self-concept, which she reported when she explained her global affect. Questions designed to elicit the reasons for her positive mathematical self-concept in terms of technology showed that objectives, purposes and beliefs were clearly interrelated. Her own words were: "*I think that computers, not only the GeoGebra program, are an excellent tool for anyone studying mathematics. Nowadays, the two are closely linked: everyone who studies mathematics needs a computer at some point... mathematics is linked to computers and specifically to software like GeoGebra (if you want to teach high school mathematics, for instance. I at least am trying to learn more to be a math teacher)*" (S-20).

## Conclusions

Like the results of related research, the present findings revealed that in the usage schemes for technology-assisted mathematics learning, emotions interact with cognition, conation and instrumental dimensions that determine problem solving procedures and practice. The two studies described identified several emotional phenomena associated with technology-assisted learning: (a) an initially positive attitude toward computer-aided mathematics learning and a preference for visual reasoning; (b) instrumental genesis associated with social and contextual dimensions of emotion and cognition; and (c) the effect of meta-emotion on task performance and the development of visual processes.

Contrary to the evidence reported in earlier studies (Galbraith and Haines 2000), mathematics-computer interaction is determined not by these two dimensions separately: the relationship is more complex. While individuals' positive belief systems about visualization constitute a core value in mathematics-computer interaction, this does not confirm prior studies on preferences for visualization (e.g., Eisenberg 1994). Rather, different emotions are associated with such beliefs. This emotional plurality and the individual and social elements in visualization-related instrumental genesis were identified by analysis of artifact use from an instrumental and cognitive approach, focusing on instrumentalization and instrumentation processes.

For the group studied, instrumental mediation (DGS) favored visualization and revealed the existence of social usage processes in their beliefs. Emotions arose in connection with the appraisal of success. The value accorded the situation (importance of the goal, whether it be "to build an interactive image" or the longer term "to become a teacher") and its appraisal determined the emotional response (Pekrun 2006; Weiner 1985; Hannula 2002) and meta-emotion.

Evidence of positive instrumental mediation was observed in situation "controllabilty" (Weiner 1985) expressed as interaction with software that affords an immediate answer and lowers apprehension thanks, for instance, to the algebraic and geometric windows that contributed to analytical and geometric comprehension. Interaction with the medium was found to consist of both intrinsic and extrinsic components that favored emotional self-control.

Students exhibited a variety of cognitive-affective pathways in which emotional self-control was not always optimized. Pathway R3 was characteristic of subjects who optimized their emotions and learned how emotional activation could turn perplexity and confusion into success (or de-fuse apprehension, frustration or anger).

Other factors present in visual reasoning schemes included cognitive styles and beliefs about the technological context. The use of different theoretical frameworks favored the understanding of a preference for visualization. According to the data, subjects experienced positive feelings in the initial stage of problem solving, when seeking an image that would describe the structure sought; or of satisfaction and happiness when they were able to construct an interactive image. However, the attempted generation of interactive images, the use of analogical visualization and progressive schematization based on the analytical-algebraic analysis and geometric figures prompted confusion, mental blocks and frustration.

Finally, another source of information on performance in technological contexts was social persuasion (portrayal of computer use in education) and the individual's psychological and emotional appraisal. Summing up, in this group the two crucial elements that contributed to sustaining effective interaction between technology and mathematics learning were the subjects' self-perception or self-image and the meta-affective experiences involved in task performance, to which the instrument contributed.

## References


Arcavi, A. (2003). The role of visual representations in the learning of mathematics. *Educational Studies in Mathematics, 52*(3), 215–224.

Artigue, M. (2002). Learning mathematics in a CAS environment: The genesis of a reflection about instrumentation and the dialectics between technical and conceptual work. *International Journal of Computers for Mathematical Learning, 7*(3), 245–274.

Bodin, A., Coutourier, R., & Gras, R. (2000). *CHIC: Classification Hiérarchique Implicative et Cohésive-Version sous Windows–CHIC 1.2*. Association pour la Recherche en Didactique des Mathématiques Rennes.

Cobb, P., Cofrey, J., di Sessa, A., Lehrer, R., & Schauble, L. (2003). Design experiments in educational research. *Educational Researcher, 32*(1), 9–13.

De Corte, E., Depaepe, F., Op't Eynde, P., & Verschaffel, L. (2011). Students' self-regulation of emotions in mathematics: An analysis of meta-emotional knowledge and skills. *ZDM, 43*, 483–495.

de Guzmán, M. (2002). The role of visualization in the teaching and learning of mathematical analysis. In *Proceedings of the 2nd international conference on the teaching of mathematics (at the undergraduate level)*. University of Crete, Greece.

DeBellis, V. A., & Goldin, G. A. (1997). The affective domain in mathematical problem solving. In E. Pekhonen (Ed.), *Proceedings of the PME 21* (Vol. 2, pp. 209–216).

DeBellis, V. A., & Goldin, G. A. (2006). Affect and meta-affect in mathematical problem solving: A representational perspective. *Educational Studies in Mathematics, 63*(2), 131–147.

Demetriou, A., & Kazi, S. (2001). *Unity and modularity in the mind and the self.* London: Routledge.

Drijvers, P., Kieran, C., & Mariotti, M. A. (2010). Integrating technology into mathematics education: Theoretical perspective. In C. Hoyles & J. B. Lagrange (Eds.), *Mathematics education and technology-rethinking the terrain* (pp. 89–132). New York: Springer.

Duval, R. (2006). A cognitive analysis of problems of comprehension in a learning of mathematics. *Educational Studies in Mathematics, 61*, 103–131.

Eisenberg, T. (1994). On understanding the reluctance to visualize. *Zentralblatt für Didaktik der Mathematik, 26*, 109–113.

Evans, J. (2000). *Adults' mathematical thinking and emotions: A study of numerate practices.* London: Routledge Falmer.

Forgasz, H. J. (2006). Teachers, equity, and computers for secondary mathematics learning. *Journal for Mathematics Teacher Education, 9*(5), 437–469.

Galbraith, P., & Haines, C. (2000). *Mathematics-computing attitudes scales* (Monographs in continuing education). London: City University London.

Gliner, J. A., Morgan, G. A., & Leech, N. L. (2009). Sampling and introduction to external validity. In J. A. Gliner, G. A. Morgan, & N. L. Leech (Eds.), *Research methods in applied setting: An integrated approach to design and analysis* (pp. 115–133). New York: Routledge Taylor & Francis Group.

Goldin, G. A. (2000). Affective pathways and representation in mathematical problem solving. *Mathematical Thinking and Learning, 2*(3), 209–219.

Goldin, G. (2002). Affect, meta-affect, and mathematical belief structures. In G. Leder, R. Pehkonen, & G. Törner (Eds.), *Beliefs: A hidden variable in mathematics education?* (pp. 59–73). Dordrecht: Kluwer Academic Publishers.

Goldin, G. A. (2004). Characteristics of affect as a system of representation. In M. Johnsen Hoines & A. B. Fuglestad (Eds.), *Proceedings of the PME 28* (Vol. 1, pp. 109–114). Bergen: Bergen University College.

Goldin, G. A., Roesken, B., & Toerner, G. (2009). Beliefs – No longer a hidden variable in mathematics teaching and learning processes. In J. Maass & W. Schlöglmann (Eds.), *Beliefs and attitudes in mathematics education: New research results* (pp. 1–18). Rotterdam: Sense Publishers.

Gómez-Chacón, I. M. (1997). *Procesos de aprendizaje en matemáticas con poblaciones de fracaso escolar en contextos de exclusión social: Las influencias afectivas en el conocimiento de las matemáticas.* Doctoral dissertation of Universidad Complutense de Madrid (Published in electronic version in 2004).

Gómez-Chacón, I. M. (2000a). *Matemática emocional. Los afectos en el aprendizaje matemático* [Emotional mathematics. Affects in mathematics learning]. Madrid: Narcea.

Gómez-Chacón, I. M. (2000b). Affective influences in the knowledge of mathematics. *Educational Studies in Mathematics, 43*, 149–168.

Gómez-Chacón, I. M. (2008). *Suggesting practical advances in the research on affect in mathematical learning.* In ICME-11, 11th international congress on mathematical education, Monterrey, México.

Gómez-Chacón, I. M. (2011). Mathematics attitudes in computerized environments. A proposal using GeoGebra. In L. Bu & R. Schoen (Eds.), *Model-centered learning: Pathways to mathematical understanding using GeoGebra* (pp. 147–170). Rotterdam: Sense Publishers.

Gómez-Chacón, I. M., & Escribano, J. (2011). Teaching geometric locus using GeoGebra. An experience with pre-service teachers. *GeoGebra International Journal of Romania (GGIJRO), 2*(1), 209–224.

Gómez-Chacón, I. M., & Kuzniak, A. (2011). Les espaces de travail Géométrique de futurs professeurs en contexte de connaissances technologiques et professionnelles. *Annales de Didactique et de Sciences Cognitives, 16*, 187–216.

Gottman, J. M., Katz, L. F., & Hooven, C. (1997). Introduction to the concept of meta-emotion. In J. M. Gottman, L. F. Katz, & C. Hooven (Eds.), *Meta-emotion. New families communicate emotionally* (pp. 3–8). Mahwah: Lawrence Erlbaum Associates.

Gras, R., Peter, P., Briand, H., & Philippé, J. (1997). Implicative statistical analysis. In C. Hayashi, N. Ohsumi, K. Yajima, Y. Tanaka, H. Bock, & Y. Baba (Eds.), *Proceedings of the 5th conference of the International Federation of Classification Societies* (Vol. 2, pp. 412–419). New York: Springer.

Guin, D., Ruthven, K., & Trouche, L. (Eds.). (2004). *The didactical challenge of symbolic calculators: Turning a computational device into a mathematical instrument*. New York: Springer.

Hannula, M. S. (2002). Attitude towards mathematics: Emotions, expectations and values. *Educational Studies in Mathematics, 49*(1), 25–46.

Leder, G. C., Pehkonen, E., & Torner, G. (Eds.). (2002). *Beliefs: A hidden variable in mathematics education?* Dordrecht: Kluwer.

Liljedahl, P. (2005). Mathematical discovery and *affect*: The *effect* of AHA! experiences on undergraduate mathematics students. *International Journal of Mathematical Education in Science and Technology, 36*(2–3), 219–236.

Malmivuori, M. L. (2001). *The dynamics of affect, cognition, and social environment in the regulation of personal learning processes: The case of mathematics* (Research Report 172). Helsinki: Helsinki University Press.

Malmivuori, M. L. (2006). Affect and self-regulation. *Educational Studies in Mathematics, 63*, 149–164.

McCulloch, A. W. (2011). Affect and graphing calculator use. *Journal of Mathematical Behavior, 30*, 166–179.

McLeod, D. B. (1992). Research on affect in mathematics education: A reconceptualization. In D. A. Grouws (Ed.), *Handbook of research on mathematics learning and teaching* (pp. 575–596). New York: Macmillan.

McLeod, D. B. (1994). Research on affect and mathematics learning in the JRME: 1970 to the present. *Journal for Research in Mathematics Education, 25*(6), 637–647.

Monaghan, J. (2004). Teachers' activities in technology-based mathematics lessons. *International Journal of Computers for Mathematical Learning, 9*(3), 327–357.

Op't Eynde, P., & Turner, J. E. (2006). Focusing on the complexity of emotion-motivation issues in academic learning: A dynamical, component systems approach. *Educational Psychology Review, 18*, 361–376.

Pekrun, R. (2006). The control-value theory of achievement emotions: Assumptions, corollaries, and implications for educational research and practice. *Educational Psychology Review, 18*(4), 315–341.

Pierce, R., Stacey, K., & Brakatsas, A. (2007). A scale for monitoring students' attitudes to learning mathematics with technology. *Computer and Education, 48*, 285–300.

Presmeg, N. C. (2006). Research on visualization in learning and teaching mathematics. In *Handbook of research on the psychology of mathematics education: Past, present and future. PME 1976-2006* (pp. 205–235). Rotterdam: Sense Publishers.

Presmeg, N. C., & Bergsten, C. (1995). *Preference for visual methods: An international study*. In L. Meira, & D. Carraher (Eds.), Proceedings of PME 19 (Vol. 3, pp. 58–65), Recife.

Rabardel, P. (1995). *Les hommes et les technologies. Une approche cognitive des instruments contemporains*. Paris: Université Paris 8.

Schlöglmann, W. (2002). Affect and mathematics learning, In A. D. Cockburn, & E. Nardi (Eds.), *Proceedings of the 26th PME* (Vol. 3, pp. 185–192), Norwich, UK.

Schlöglmann, W. (2005). Affect and cognition – Two poles of a learning process. In C. Bergsten & B. Grevholm (Eds.), *Conceptions of mathematics. Proceedings of Norma 01* (pp. 215–222). Linköping: Svensk Förening för Matematikdidaktisk Forskning.

Stylianou, D. A. (2001). On the reluctance to visualize in mathematics: Is the picture changing? In M. van den Heuvel-Panhuizen (Ed.), *Proceedings of the 25th PME* (Vol. 4, pp. 225–232). Utrecht: Freudenthal Institute.

Weiner, B. (1985). An attributional theory of achievement motivation and emotion. *Psychological Review, 92*(4), 548–573.

Zan, R., Brown, L., Evans, J., & Hannula, M. S. (2006). Affect in mathematics education: An introduction. *Educational Studies in Mathematics, 63*(2), 113–121.

# Effects of Motivation on the Belief Systems of Future Mathematics Teachers from a Comparative Perspective

**Sigrid Blömeke and Gabriele Kaiser**

**Abstract** The paper examines the relationship between future teachers' professional motivation and their beliefs on the dynamic nature of mathematics as an academic discipline as well as on their transmission-oriented beliefs on the teaching and learning of mathematics. As motives, intrinsic-pedagogical, intrinsic-academic and extrinsic motives were examined. Based on IEA's "Teacher Education and Development Study in Mathematics" (TEDS-M), carried out in 2008, we analyzed data from two Western (Germany, Norway) and two East Asian countries (Singapore, Taiwan) which represent different educational cultures. Our results revealed that the level of the motivational facets and the beliefs facets differed between the four countries. The pattern of relationships between professional motivation and teacher beliefs was largely similar across countries though. This result indicates a generic effect of motivation but culturally shaped strength of the different characteristics.

**Keywords** Epistemological beliefs • Nature of mathematics • Transmission beliefs • Professional motivation • Intrinsic motives • Extrinsic motives

## Introduction

Motivation is often positively related to cognitive learning outcomes (Benware and Deci 1984; Grolnick and Ryan 1987), especially if *intrinsic motivation* – as one specific facet of motivation – is used as a predictor (Singh et al. 2002). This positive relationship applies not only to the K-12 student level but also to university

S. Blömeke (✉)
Centre for Educational Measurement at University of Oslo (CEMO), Oslo, Norway

Humboldt University of Berlin, Berlin

IPN–Leibniz Institute for Science and Mathematics Education at Kiel University, Kiel, Germany
e-mail: sigrid.bloemeke@cemo.uio.no

G. Kaiser
University of Hamburg, Hamburg, Germany
e-mail: gabriele.kaiser@uni-hamburg.de

 
B. Pepin, B. Roesken-Winter (eds.), *From beliefs to dynamic affect systems in mathematics education*, Advances in Mathematics Education,
DOI 10.1007/978-3-319-06808-4_11

students. Evidence exists that the level of their intrinsic motivation significantly predicts study success (Schiefele and Urhahne 2000). Also with respect to future teachers, studies revealed that intrinsic motivation is positively correlated with their professional knowledge (Blömeke et al. 2011; Brouwer and ten Brinke 1995; Keller-Schneider 2011; Watt et al. 2007). Longitudinal studies revealed correspondingly that future teachers' motivation influence their cognitive development during teacher education (König and Herzmann 2011; König and Rothland 2012; Mayr 2009).

However, how future teachers' motives to become teachers are related to their *beliefs*, is still an open question. The present paper intends to close this research gap with respect to future teachers from four countries. We selected these countries so that they represent different educational cultures, because we assume that the level of teacher motives and the relationship between motivation and beliefs differ across countries.

From a study on the relationship of future lower-secondary mathematics teachers' professional knowledge and their beliefs in Germany, Norway, Singapore and Taiwan based on data from the "Teacher Education and Development Study: Learning to Teach Mathematics" (TEDS-M, Blömeke 2012), we learned that these facets were correlated with each other in specific ways. The relationships pointed to belief *systems* which were largely the same across countries that belonged to different educational traditions – the two East Asian countries, Taiwan and Singapore, influenced by Confucian heritage, and the two Western European countries, Germany and Norway having a classical Greek-Roman background: Mathematics content knowledge (MCK) and mathematics pedagogical content knowledge (MPCK) were always positively or not related to a belief that the nature of mathematics is dynamic. In none of the countries, the relationship was negative. In turn, dynamic beliefs and transmission beliefs on the teaching of mathematics were always either negatively or not correlated with each other. In none of the countries, the relationship was positive. This pattern was thus more universal than we had expected, because teacher training programs had been described as being influenced by the context in which they are implemented (Leung et al. 2006; Even and Ball 2009).

In this paper, we intend to go beyond this relationship by including the motivational characteristics of the future teachers from Germany, Norway, Singapore and Taiwan and by examining how these influence their beliefs. The future teachers of these countries had different motives to enter mathematics teacher education: either intrinsic-pedagogical or intrinsic-academic motives or extrinsic motives. As it was with respect to knowledge and beliefs, the database of our study comes from the TEDS-M-study (see e.g. Blömeke et al. 2011; Blömeke et al. 2012).

These data point to varying levels of motivation why teacher education was entered across countries. So, the question arises, whether already at such an early point of the academic career the foundation of the future teachers' beliefs systems was developed. We assume that this in fact may be true and that the pattern of the relationships may once again be more universal than usually discussed in the literature. Furthermore, we assume that future teachers with an intrinsic pedagogical motive to enter the teaching profession may be more strongly convinced that mathematics is a dynamic discipline and that the teaching of mathematics should happen with strong participation of students and not in a teacher-directed way. In contrast, we assume that students with extrinsic motives to go into the teaching profession are more strongly convinced that a teacher-directed style may be the better way to go.

## Conceptual Framework and Research Questions

With respect to teachers, intrinsic motives to enter the profession can be distinguished into altruistic-pedagogical and subject-related motives (Brookhart and Freeman 1992; Watt and Richardson 2007). Altruistic-pedagogical motives include the motivation to work with children and to support their development whereas subject-related motives express the enjoyment of the content to be taught. Extrinsic motives represent another motivational facet. If this perspective is taken, somebody wants to become a teacher mainly because of the salary paid or other working conditions.

Comparative studies revealed different levels of these specific motives across countries (Watt et al. 2012). Whereas future teachers from Western countries typically stress altruistic-pedagogical motives only, those from some East Asian countries endorse extrinsic motives, too (Schmidt et al. 2011).

In TEDS-M, beliefs were defined as "understandings, premises or propositions about the world that are felt to be true" (Richardson 1996, p. 103). If beliefs are looked at alongside both the subject being taught and a professional task which needs to be mastered, evidence suggests that there is a link between teacher beliefs and pupil achievement (Bromme 2005). Beliefs are a crucial aspect of a teacher's perception of teaching situations and in their choice of teaching methods in the classroom (Leinhardt and Greeno 1986; Leder et al. 2002). Thus, they are also an indicator of the type of teaching methods future teachers will use in the classroom in the future (Nespor 1987).

Beliefs are, however, not a well-defined construct (Pajares 1992). Clear distinctions between beliefs and other conceptions such as attitudes, perceptions or conceptions are rare. Rodd (1997) points out, that beliefs rely on evaluative and affective components. At the same time, the distinction towards knowledge – in particular towards pedagogical content knowledge and general pedagogical knowledge – is more heuristic than that it can strictly be kept up (Furinghetti and Pehkonen 2002). Several efforts have been made to categorize the belief systems of teachers (Thompson 1992). TEDS-M distinguishes between beliefs about the nature of mathematics and beliefs about the teaching and learning of mathematics. Self-related beliefs were not covered in TEDS-M.

With respect to the interaction of teacher beliefs, Wehling and Charters (1969) discussed beliefs in terms of discrete sets of inter-related concepts. Kagan (1992) argued along the same line that teachers dispose of a highly personalized belief system that constrains their perception, judgment, and behavior. This system grows more coherently as a teacher's experience in classrooms grows. Pajares (1992) stresses the role of experience as well: Individuals develop a belief system which has an adaptive function in helping individuals define and understand the world and themselves. The different belief dimensions are connected to each other. However, since beliefs are generally contextualized and associated with a particular situation or circumstance (Kagan 1992), systems of beliefs may contradict each other.

Specifically with respect to mathematics teachers, Op't Eynde et al. (2002) intended to clarify empirically with their mathematics-related beliefs questionnaire (MRBQ) the structure of teachers' belief system and how its different facets relate to each other. Subsequently, Schommer-Aikins (2004) introduced the idea of an "embedded systemic model" of epistemological beliefs after she had already in the early 1990s – inspired by the seminal work of Perry (1968) and Schoenfeld (1983) – argued that these beliefs may form a system of discrete beliefs which are connected to each other. Epistemological beliefs are defined as those beliefs that are related to the nature and scope of knowledge. In her embedded model, Schommer-Aikins (2004) pointed to the need of including other cognitive and affective characteristics besides beliefs, because "epistemological beliefs do not function in a vacuum" (p. 23). Concerning the components of the belief systems Schommer (1990) included already in the early 1990s beliefs about learning to the range of beliefs about knowledge, on which her predecessors had limited their models. Boekaerts (1995), in particular, requested to take motivation into account in order to "bridge the gap between metacognitive and metamotivation theories".

Schommer-Aikins (2004) points also to the cultural context as an influential factor. In line with Schoenfeld (1998), affective-motivational characteristics can, in fact, be understood as culturally shaped mental constructs acquired in educational settings with traditions that vary across countries. Culture can be defined as the "shared motives, values, beliefs, identities, and interpretations or meanings of significant events that result from common experiences of members of collectives that are transmitted across generations" (House et al. 2004, p. 15). It is hypothesized that through socialization processes a country's culture has an impact on the preferred modes of learning (Hofstede 1986). In anthropology as well as cross-cultural psychology, several conceptualizations exist with which different dimensions of cultural differences can be described.

Referring to studies by Hofstede (2001), Triandis (1995) and others, Schommer-Aikins (2004) argues that the degree of *individualism* and *collectivism* may have consequences for the shape of a belief system, because goals and emotional attachment are directed differently in individualistic and collectivistic countries. In countries such as Germany and Norway, typically classified as individualistic countries on Hofstede's individualism-collectivism scale, individuals are regarded as largely independent and the needs and goals of the self are regarded more important than the needs and goals of the society as a whole (Felbrich et al. 2012). In educational processes, autonomy and emotional detachment are supported whereas in collectivistic countries such as Singapore and Taiwan group identity and emotional attachment are stressed. In individualistic countries, learners are then perceived as autonomous subjects acquiring knowledge mainly independently on their own (Triandis 1995). Lack of success in learning is often attributed to a misfit between the conditions of learning and the individual learner, i.e. in terms of composition of groups of learners or too demanding tasks, rather than to individual characteristics of the learner. In contrast, learners from collectivist countries engage in learning processes because of an obligation towards their teachers, their families and other societal entities, which in turn are seen as obliged to grant the learner the necessary support (Felbrich et al. 2012). School failure in these countries is attributed to a lack of effort by the learner. Hofstede also assumed that specific differences exist in both teacher–student and student–student interactions between individualistic and collectivistic countries. In individualistically oriented societies students expect to learn how to learn and to think, whereas in collectivistic societies they expect to learn how to do something. Whereas in the latter diploma certificates are of utmost importance, they have lower symbolic value in individualistic countries (Hofstede 1986).

As part of her embedded systemic model of epistemological beliefs, Schommer-Aikins (2004) assumes that such cultural relational views are key antecedents of beliefs on teaching and learning. Based on her ideas, it can be assumed that student teachers in collectivistic societies more strongly endorse a transmission view on teaching and learning and less a dynamic view on the nature of mathematics, because teachers and final examinations expect pupils to be proficient in the application of rules and formulae and that it is the responsibility of the teacher as the master of the content to deliver these. In contrast, student teachers from individualistic societies should feel less comfortable with a transmission view and should stress a dynamic view of mathematics which stresses individual approaches to mathematics.

The MT21 Study (Mathematics Teaching in the 21st century; Blömeke et al. 2008a; Schmidt et al. 2011) was the first study to compare future primary teachers' beliefs in six countries, namely Bulgaria, Germany, USA, Mexico, Taiwan and South Korea (Blömeke et al. 2008b). The results revealed country-specific differences in beliefs on the nature of mathematics. German, Mexican and US future teachers agreed more strongly with dynamic statements than Taiwanese, South Korean and Bulgarian future teachers (Schmidt et al. 2011). The results of the TALIS Study (Teaching and Learning International Survey, OECD 2009) which refer to teachers' epistemological beliefs on teaching and learning of mathematics, point to the same direction. In collectivistic countries, such as Malaysia and South American states, transmission views have more strongly been articulated by teachers than in Western countries (Klieme and Vieluf 2009; Vieluf and Klieme 2011). With this study cultural *patterns* of beliefs were identified for the first time.

In contrast to these studies, Andrews et al. (2011) report on a comparative study of mathematics-related beliefs on teaching and learning of students in England, Slovakia and Spain. They addressed specifically the question of structural equivalence of these beliefs across countries by applying the above mentioned instrument developed by Op't Eynde et al. (2002) to 220 students from England, 405 students from Spain and 250 students from Slovakia. The students were at ages 11–15. Andrews et al. (2011) found convincing evidence that the beliefs structure was structurally equivalent across these three countries. For each factor identified in one country they found an equivalent factor in the other two countries.

Against this state of research, we examine the following two research questions:

1. To what extent are motivational characteristics of future lower-secondary mathematics teachers related to their beliefs on the nature of mathematics and the teaching and learning of mathematics? Can we, in fact, assume an embedded systemic model as Schommer-Aikins points out?
2. To what extent is the structure of this relationship structurally equivalent across countries? Do we have to distinguish between individualistic and collectivistic countries as Schommer-Aikins points out or not as Andrews et al. describe?

## Study Design

### *Sampling*

The target group of the present study was defined as future teachers in their final year of teacher education who would receive a license to teach mathematics in lower secondary schools (Tatto et al. 2008). A teacher education program was identified as focused on lower secondary school level, if the qualification included to teach grade 8 (basic education, cycle 2; UNESCO 1997). In a two-stage process, random samples were drawn in each participating country. The TEDS-M quality standards required minimum participation rates for all target populations of the survey to ensure that bias resulting from non-response was kept within acceptable limits. The samples were organized according to important teacher education features such as the type of program (consecutive vs. concurrent programs), the school level to be taught (grade range included in the qualification, e.g. grades 5–10 vs. grades 7–12), the attention paid to learning opportunities (e.g., a major or minor in mathematics) and the region where the training was based (for example, federal states) in order to reflect accurately the future teachers' characteristics at the end of their education.

In 2008, more than 8,000 future lower-secondary teachers from 15 countries were surveyed. All countries had to meet the strong IEA quality requirements. These included controlling translation, monitoring test situations and meeting participation rates. If a country missed the participation benchmark only slightly, its results are reported with an annotation. The present study uses the samples from Germany, Norway, Singapore and Taiwan.

In Germany, the 771 future lower-secondary teachers (response rate: 81 %) came from the following three programs:

– "Primary and Lower Secondary Teachers", a 5.5 year consecutive model that trained teachers for grades 1 through 10 in two subjects (mathematics had to be one of these subjects to be included in TEDS-M); a 3.5 year university training was followed by a 2-year practical training with mathematics as one of two subjects
– "Lower-Secondary Teachers", a 5.5 year consecutive model that trained teachers for grades 5 through 10 in two subjects (mathematics had to be one of these sub-

jects to be included in TEDS-M); a 3.5 year university training was followed by a 2-year practical training with mathematics as one of two subjects
- "Lower and Upper Secondary Teachers", a 6.5 year consecutive model that trained teachers for grades 5 through 13 in two subjects (mathematics had to be one of these subjects to be included in TEDS-M); a 4.5 year university training was followed by a 2-year practical training with mathematics as one of two subjects

The 555 teachers in Norway (response rate below 60 %) came from the following three programs:

- "Allmennlærerutdanning", a 4-year concurrent model that trained generalist teachers in a broad range of subjects including mathematics for grades 1 through 10
- "Allmennlærerutdanning with extra mathematics", a 4-year concurrent model that trained generalist teachers in a broad range of subjects including mathematics for grades 1 through 10
- "Praktisk-Pedagogisk Utdanning", a 4-year consecutive model that trained teachers for grades 8 through 13 in two subjects; a 3 year university training was followed by a 1-year professional training with mathematics as one of two subjects

The sampling process for Norway was difficult, and the final country sample consisted of two subsamples that were likely to partly overlap. While information about the seriousness of this problem is not available, we realized that using only one subsample would lead to strongly biased country estimates. Combining both subsamples would lead to imprecise standard errors (for more details, see Tatto et al. 2012). After an extensive research of the Norwegian literature about teacher education, combining TEDS-M data with publicly availably evaluation data from Norway (NOKUT 2006), and recourse to expert reviews, we decided to combine the two subsamples in order to represent the future teachers' knowledge as appropriately as possible. However, the results should be regarded as a rough approximation only. The country's participation rate was below 60 % with respect to lower secondary teachers. In addition, the sample met only partly the TEDS-M definition of the target population (Tatto et al. 2008). Therefore, caution is necessary with respect to these results.

In Singapore, the 393 future lower-secondary mathematics teachers (response rate 91 %) came from the following two programs:

- "PGDE 7–10", a 5-year consecutive model that trained teachers for grades 7 through 10 in two subjects; a 4-year Bachelor program was followed by a 1-year Master program with mathematics as one of two subjects
- "PGDE 7–12", a 5-year consecutive model that trained teachers for grades 7 through 12 in two subjects; a 4-year Bachelor program was followed by a 1-year Master program with mathematics as one of two subjects

In Taiwan, the 365 future lower-secondary teachers (97 %) came from a concurrent model that trained mathematics specialists for grades 7 through 9 in one subject only.

## Data Sources

Regarding motives to go into the teaching profession, TEDS-M distinguished between intrinsic-pedagogical, intrinsic-academic and extrinsic motives. The scales were developed based on the study "Mathematics Teaching in the twenty-first century (MT21)" (Schmidt et al. 2011). Four items covered pedagogical motives, e.g. "I like working with young people." Academic motives were captured with two items, e.g. "I love mathematics". Three items covered extrinsic motives. "I seek the long-term security associated with being a teacher" is an example for this scale. All items had to be rated on 4-point Likert scales from "not a reason" to "a major reason". The means reported below represent the equally weighted means of each scale's items.

The future teachers' beliefs about the nature of mathematics were measured using an instrument developed by Grigutsch et al. (1998). This instrument originally consisted of 75 items, but due to time constraints it was reduced to 12 in TEDS-M based on the highest factor loadings in the original study and high-scale reliability in the TEDS-M pilot studies. The items' two-dimensional structure represented a static and a dynamic view on the nature of mathematics.

For the present paper we used the scale that represents a dynamic view which means that mathematics is seen as a process of enquiry. The scale's structure was confirmed through explorative and confirmatory factor analysis. The scale consists of six items which emphasize the process- and application-related character of mathematics, for example, "in mathematics you can discover and try out new things by yourself" or "many aspects of mathematics are of practical use". The future teachers had to express their agreement on a six-point Likert scale (1 = strongly disagree, 6 = strongly agree). The raw data were scaled using a partial credit IRT model (Tatto et al. 2012). For the sake of clarity, individual scores were transformed to a scale with a mean value of 10 based on the test characteristic curve. Conceptually, this mean represents the mean of the scale (corresponding to 3.5 on the initial scale) and thus a neutral view.

The future teachers' beliefs about the teaching and learning of mathematics were measured with another well-established scale from instructional research (Peterson et al. 1989). One scale represented a transmission view. Teachers who agreed strongly with its four items tended to see mathematics learning as teacher-centered with the pupil's role being to follow instructions given. Two examples of its items are: "The best way to do well in mathematics is to memorize all the formulae"; and "Pupils need to be taught exact procedures for solving mathematical problems".

Since our intention is to get pure regression parameters for professional motivation predicting teachers' epistemological beliefs on the nature of mathematics, we control for the future teachers' professional knowledge. Mathematics content knowledge (MCK) and mathematics pedagogical content knowledge (MPCK) were assessed in a 60-min paper-and-pencil test that had to be completed during a standardized and monitored session. The items were supposed to depict classroom performance as closely as possible. Many of them therefore represent problems and situations constitutive for mathematics teaching (NCTM 2000). In order to capture the desired breadth and depth of teacher knowledge, a matrix design was applied. Three test booklets were developed that had rotated blocks of items ("Balanced Incomplete Block Design"). The items of the mathematics test covered number, algebra, geometry and to a small extent data. The items of the mathematics pedagogy test covered pre-active curricular and planning knowledge which is necessary before a teacher enters the classroom and interactive knowledge about how to enact mathematics for teaching and learning. Three item formats were used: multiple-choice, complex multiple-choice, and constructed-response items.

All parameter estimations for this paper (in particular means and regression parameters) were carried out using the International Database Analyzer provided by IEA. As a consequence all results are based on weighted data (taking unequal selection probabilities into account as well as non-response adjustments) and using appropriate estimations of standard errors (taking the complex sample design into account by applying the balanced repeated replication technique).

Whether systematic mean differences between countries exist was estimated by taking the confidence intervals into account. As a rough "thumb of rule", the approach can be explained as follows: If intervals around the means of two countries which have a size of 1.96*Standard Error do not overlap, a 95 % probability exists that the differences are significant. We provide the standard error of the means in each table.

## Results

### *Descriptive Results: Job Motivation and Professional Beliefs of Teachers*

Regarding their motivation to become a teacher, the future teachers from all four countries agree the most with intrinsic-pedagogical motives (see Table 1). However, we can distinguish between Germany and Norway on the one hand and Singapore and Taiwan on the other hand. Whereas intrinsic-pedagogical motives are an important reason to go into the teaching profession in Germany and Norway, in particular teachers from Taiwan are more neutral in this respect. Taking the confidence

**Table 1** Intrinsic-pedagogical motives in Germany, Norway, Singapore and Taiwan

| Country | Mean | Standard error | Standard deviation | min–max |
|---|---|---|---|---|
| Germany | 3.23 | 0.03 | 0.48 | 1.25–4.00 |
| Norway | 3.26 | 0.02 | 0.44 | 1.75–4.00 |
| Singapore | 2.99 | 0.03 | 0.61 | 1.00–4.00 |
| Taiwan | 2.76 | 0.04 | 0.73 | 1.00–4.00 |

**Table 2** Intrinsic-academic motives in Germany, Norway, Singapore and Taiwan

| Country | Mean | Standard error | Standard deviation | min–max |
|---|---|---|---|---|
| Germany | 2.73 | 0.04 | 0.65 | 1.00–4.00 |
| Norway | 2.22 | 0.03 | 0.72 | 1.00–4.00 |
| Singapore | 2.51 | 0.03 | 0.75 | 1.00–4.00 |
| Taiwan | 2.27 | 0.03 | 0.61 | 1.00–4.00 |

**Table 3** Extrinsic motives in Germany, Norway, Singapore and Taiwan

| Country | Mean | Standard error | Standard deviation | min–max |
|---|---|---|---|---|
| Germany | 2.43 | 0.04 | 0.68 | 1.00–4.00 |
| Norway | 1.99 | 0.02 | 0.56 | 1.00–3.33 |
| Singapore | 2.01 | 0.03 | 0.68 | 1.00–3.67 |
| Taiwan | 2.26 | 0.03 | 0.66 | 1.00–4.00 |

**Table 4** Dynamic beliefs of future teachers in Germany, Norway, Singapore and Taiwan

| Country | Mean | Standard error | Standard deviation | min–max |
|---|---|---|---|---|
| Germany | 11.98 | 0.08 | 1.4 | 7.95–15.48 |
| Norway | 11.67 | 0.08 | 1.4 | 8.54–15.48 |
| Singapore | 11.75 | 0.06 | 1.3 | 9.10–15.48 |
| Taiwan | 12.08 | 0.07 | 1.4 | 9.10–15.48 |

intervals into account, the pedagogical motivation is significantly higher in the first two countries than in the latter two. The variation is much smaller in the first two countries, too, compared to the latter countries.

In all four countries, the view on intrinsic-academic motives is more neutral. In this case, the difference is the largest between Germany and Norway. Whereas German future teachers take a slightly positive view, Norwegian teachers reject this view to some extent. The difference between these countries is significant, if one takes the confidence intervals into account (Table 2).

The future mathematics teachers from Norway and Singapore reject in addition extrinsic motives whereas future teachers from Taiwan and particularly those from Germany have a more neutral view. The differences between Norway and Singapore on the one hand and Taiwan on the other hand and again between Taiwan and Germany are significant. In addition, the variation in Norway is lower than in the other countries, meaning that future teachers are more homogenous in their rejection of extrinsic motives than those from the other countries (Table 3).

When it comes to professional beliefs, there is also variation between the four countries examined. Future teachers from all four countries react positively to statements stressing that mathematics is creative and useful. However, the support of such a view is significantly stronger in Taiwan than in Norway if one takes the confidence intervals into account (see Table 4).

**Table 5** Transmission beliefs on the teaching and learning of mathematics

| Country | Mean | Standard error | Standard deviation | min–max |
|---|---|---|---|---|
| Germany | 8.90 | 0.06 | 0.8 | 4.98–14.80 |
| Norway | 8.96 | 0.03 | 0.7 | 6.21–10.72 |
| Singapore | 9.50 | 0.04 | 0.8 | 4.98–14.80 |
| Taiwan | 9.02 | 0.04 | 0.8 | 4.98–11.08 |

When considering transmission beliefs on the teaching and learning of mathematics, the variation between the four countries is even larger (see Table 5). Lower-secondary teachers in Germany, Norway and Taiwan reject teacher-led learning significantly more strongly than those in Singapore. The mean of the future teachers from Singapore is still below a neutral position though.

## *Effects of Motivation on Teacher Beliefs*

We examined the relationship between the different beliefs with two regression models. In these models served the future mathematics teachers' dynamic beliefs on the nature of mathematics or their transmission-oriented beliefs on the teaching and learning of mathematics as dependent variables and the motivational characteristics as predictors while we controlled for the future teachers' knowledge (not displayed in Tables 6 and 7). The variance explained by the predictors was generally higher in the first case than in the latter. If we use the adjusted $R^2$ as a measure of the effect size in order to take the sample size and the number of predictors into account, we can point out that – according to Cohen's (1988) classification of effect sizes – the practical relevance of job motivation as a predictor was between medium and large with respect to dynamic beliefs whereas it was between small and medium with respect to transmission-oriented beliefs.

With respect to our first research question, the data revealed that significant relationships between motivation to become a teacher and teacher beliefs exist if the knowledge is controlled for. The relationships vary by facets of the predictors and by the dependent variables. With respect to our second research question the data revealed that the patterns of these relationships were surprisingly homogenous across countries in the case of epistemological beliefs on the nature of mathematics but less so in the case of beliefs on the teaching and learning of mathematics.

The more strongly future teachers were intrinsically motivated to go into the teaching profession, the more strongly they believed in the dynamic nature of mathematics. This result applies to intrinsic-academic professional motives as well as to intrinsic-pedagogical motives and it applies to teachers from Taiwan and Singapore as well as from Germany and Norway. Although the practical relevance of the standardized beta coefficients was large in all cases (Roussos and Stout 1996), the strength in the relationships varied: If one controls for the professional knowledge of the future teachers, academic motives were particularly important in Norway and

**Table 6** Regression parameters for professional motivation predicting teachers' dynamic beliefs on the nature of mathematics (while controlling for knowledge; not displayed)

| | Taiwan | | Singapore | | Germany | | Norway | |
|---|---|---|---|---|---|---|---|---|
| | b | β | b | β | b | β | b | β |
| MOT_ACA | .26 | .12† | .33 | .20*** | .48 | .23*** | .56 | .29*** |
| MOT_PED | .53 | .28*** | .27 | .13* | .49 | .18*** | .34 | .11** |
| MOT_EXT | ns | ns | ns | ns | ns | ns | −.16 | −.06† |
| $R^2_{adj}$ | .10 | | .06 | | .12 | | .10 | |

*MOT_ACA* intrinsic-academic professional motive, *MOT_PED* intrinsic-pedagogical professional motive, *MOT_EXT* extrinsic professional motive, $R^2_{adj}$ determination coefficient adjusted for sample size and number of predictors (MCK and MPCK were controlled for), *ns* not significant, † = $p<.10$, * = $p<.05$, ** = $p<.01$, *** $p<.001$

**Table 7** Regression parameters for professional motivation predicting teachers' transmission beliefs on the teaching and learning of mathematics (while controlling for knowledge; not displayed)

| | Taiwan | | Singapore | | Germany | | Norway | |
|---|---|---|---|---|---|---|---|---|
| | b | β | b | β | b | β | b | β |
| MOT_ACA | | .14** | | ns | | −.17** | | −.07† |
| MOT_PED | | ns | | ns | | ns | | ns |
| MOT_EXT | | .11† | | .15** | | .16* | | .12** |
| $R^2_{adj}$ | .07 | | .04 | | .06 | | .02 | |

*MOT_ACA* intrinsic-academic professional motive, *MOT_PED* intrinsic-pedagogical professional motive, *MOT_EXT* extrinsic professional motive, $R^2_{adj}$ determination coefficient adjusted for sample size and number of predictors (MCK and MPCK were controlled for), *ns* not significant, † = $p<.10$, * = $p<.05$, ** = $p<.01$, *** $p<.001$

still more important in Germany than in Singapore and Taiwan. In contrast, pedagogical motives were particularly important in Taiwan but less important in the other three countries (although still significant). Extrinsic motives tended to have a negative relationship with dynamic beliefs on the nature of mathematics. However, only in Norway the correlation became marginally significant ($p<.10$).

The patterns were different with respect to beliefs on the teaching and learning of mathematics. The more strongly future teachers were extrinsically motivated to go into the teaching profession, the more strongly they believed in a transmission-oriented teaching style. This result applied to teachers from all four countries, although it was only marginally significant in Taiwan. The practical relevance of the beta coefficients was large in all cases (Roussos and Stout 1996). In contrast, pedagogical motives were not systematically related to these beliefs.

Interestingly, intrinsic-academic professional motives had differential effects on the future teachers' beliefs. In the two Western countries, the data revealed the following relationship: The more strongly future lower-secondary mathematics teachers from Germany and (marginally) Norway were intrinsic-academically motivated to go into the teaching profession, the less strongly they believed into a transmission-oriented teaching style. This result applied although we controlled for the mathematics-related knowledge level of the teachers. In contrast, future teachers from Taiwan (and the tendency was the same in Singapore) believed more strongly in a transmission-oriented teaching style, if they were more strongly intrinsic-academically motivated. Also in this case, the result applied although we controlled for mathematics-related knowledge and although the relationship between this knowledge and the transmission beliefs were negative. It has to be taken into account, that these differential effects were quite small and not always significant.

## Summary and Conclusions

In this paper, we extended the state of research on (future) teachers' beliefs system by including motivational characteristics as demanded by the discussion on teacher education and its development. Since our prior research had pointed to an interrelationship of knowledge and beliefs we carried out our study while controlling for mathematical content knowledge and mathematics pedagogical content knowledge. With respect to epistemological beliefs on the nature of mathematics a homogenous pattern across the four countries examined emerged: The more strongly future teachers were intrinsically motivated to go into the teaching profession, either from a pedagogical or an academic point of view, the more strongly they believed in the dynamic nature of mathematics in Taiwan and Singapore as well as in Germany and Norway. Regarding beliefs on the teaching and learning of mathematics, we found another homogenous pattern with respect to extrinsic motives to go into the teaching profession. The more strongly future teachers from these four countries were motivated this way, the more strongly they believed in a transmission-oriented teaching style.

In contrast, intrinsic-academic professional motives had differential effects on the future teachers' beliefs, although these effects have to be treated cautiously, because they were quite small and not in all countries significant. In the two Western countries, future lower-secondary mathematics teachers from Germany and (marginally) Norway more strongly motivated in an intrinsic-academic way believed less strongly in a transmission-oriented teaching style. In contrast, future teachers from Taiwan (and the tendency was the same in Singapore) believed more strongly in a transmission-oriented teaching style, if they were more strongly academically motivated.

Thus, our data support Schommer-Aikins' (2004) "embedded systemic model" of beliefs by including not only epistemological beliefs but also beliefs about the teaching and learning of mathematics. In addition, our data support Boekaerts' (1995) request to take motivation into account to "bridge the gap between metacognitive and metamotivation theories". In contrast, the study supports Schommer-Aikins' (2004) assumption of the cultural context as an influential factor only to a very limited extent and this is the small differential effect of one of the three motives (intrinsic-academic) on one of the two outcomes (transmission beliefs). Mainly, the relationships are the same across individualistic and collectivist countries. With this study we therefore confirm Andrews' et al. (2011) findings that belief systems are structurally equivalent across countries.

To close we see the following three conclusions, which can be drawn from this study:

At first place the results point out, that the differences between East Asian and Western countries concerning their beliefs on the teaching and learning of mathematics and the epistemological beliefs on the nature of mathematics are increasingly vanishing, the two cultures do apparently not exist as monolithic blocks anymore. However, in this respect it must be considered that Singapore and Taiwan are countries which are strongly influenced by Western culture. Taken together, these results indicate that a dynamic perspective onto mathematics, which traditionally had been identified with Western countries, nowadays is more and more accepted in East Asian countries too. Therefore, this trend might be summarized by the statement that internationalization leads to blurring of cultural differences in education within a globalized world.

Secondly, the data point to the fact – which might be valid worldwide or at least for East Asian and Western countries –, that those future teachers who are less interested in the subject mathematics, show more transmission oriented beliefs, which is a traditional model worldwide no longer accepted. It can be expected that internationally seen these groups of future teachers may become a problem for the future development of schools and educational systems, which expect openness for innovations and the willingness to accept change in contrast to these future teachers, who tend to stick to traditional beliefs on the organization of learning as transmission-oriented processes. As a consequence, it can be stated that self-assessments at study entry containing motivational elements will become increasingly important in the future and might be applied wide-spread in order to identify potential risk groups already at the beginning of their university studies.

Thirdly, we can note, that there are still enough culturally significant differences within the belief systems of future teachers. In East Asian countries teachers with a more learner-oriented motivation in their studies can be clearly distinguished from those teachers, who are more guided by subject-oriented beliefs regarding mathematics as science. For those more subject-oriented teachers transmission-oriented conceptions play a greater role even though we controlled for knowledge. According to this position, learners do not need such constructivistic-oriented learning environments, but want to be fostered rapidly and effectively. This indicates that the greater need of time for constructivist learning is regarded quite sceptically by East Asian teachers, and that these kinds of approaches are seen to be more adequate for low performing learners. On the contrary, in Western countries constructivist attempts are deeply rooted so that irrespective to learning groups and learning environments, transmission-oriented attempts are generally rejected.

Overall it is obvious how much the different education systems have come closer together, while some identifiable differences between East Asian and Western countries still exist.

## References


Andrews, P., Diego-Mantecón, J., Vankúš, P., & Op't Eynde, P. (2011). Construct consistency in the assessment of students' mathematics-related beliefs: A three-way cross-sectional pilot comparative study. *Acta Didactica Universitatis Comenianae – Mathematics, 11*, 1–25.

Benware, C., & Deci, E. L. (1984). Quality of learning with an active versus passive motivational set. *American Educational Research Journal, 21*, 755–765.

Blömeke, S. (2012). Content, professional preparation and teaching methods: How diverse is teacher education across countries? *Comparative Education Review, 56*, 684–714.

Blömeke, S., Kaiser, G., & Lehmann, R. (Eds.). (2008a). *Professionelle Kompetenz angehender Lehrerinnen und Lehrer: Wissen, Überzeugungen und Lerngelegenheiten deutscher Mathematikstudierender und -referendare – Erste Ergebnisse zur Wirksamkeit der Lehrerausbildung*. Münster: Waxmann.

Blömeke, S., Müller, C., Felbrich, A., & Kaiser, G. (2008b). Epistemologische Überzeugungen zur Natur der Mathematik. In S. Blömeke, G. Kaiser, & R. Lehmann (Eds.), *Professionelle Kompetenz angehender Lehrerinnen und Lehrer: Wissen, Überzeugungen und Lerngelegenheiten deutscher Mathematik-Studierender und -referendare – Erste Ergebnisse zur Wirksamkeit der Lehrerausbildung* (pp. 219–246). Münster: Waxmann.

Blömeke, S., Suhl, U., & Kaiser, G. (2011). Teacher education effectiveness: Quality and equity of future primary teachers' mathematics and mathematics pedagogical content knowledge. *Journal of Teacher Education, 62*, 154–171.

Blömeke, S., Suhl, U., Kaiser, G., & Döhrmann, M. (2012). Family background, entry selectivity and opportunities to learn: What matters in primary teacher education? An international comparison of fifteen countries. *Teaching and Teacher Education, 28*, 44–55.

Boekaerts, M. (1995). Self-regulated learning: Bridging the gap between metacognitive and metamotivation theories. *Educational Psychologist, 30*, 195–200.

Bromme, R. (2005). Thinking and knowing about knowledge: A plea for and critical remarks on psychological research programs on epistemological beliefs. In M. Hoffmann, J. Lenhard, & F. Seeger (Eds.), *Activity and sign – Grounding mathematics education* (pp. 191–201). New York: Springer.

Brookhart, S. M., & Freeman, D. J. (1992). Characteristics of entering teacher candidates. *Review of Educational Research, 62*, 37–60.

Brouwer, N., & ten Brinke, S. (1995). Der Einfluss integrativer Lehrerausbildung auf die Unterrichtskompetenz (II). *Empirische Pädagogik, 9*, 289–330.

Cohen, J. (1988). *Statistical power analysis for the behavioral sciences* (2nd ed.). Hillsdale: Lawrence Erlbaum Associates.

Even, R., & Ball, D. (Eds.). (2009). *The professional education and development of teachers of mathematics*. New York: Springer.

Felbrich, A., Kaiser, G., & Schmotz, C. (2012). The cultural dimension of beliefs: An investigation of future primary teachers' epistemological beliefs concerning the nature of mathematics in 15 countries. *ZDM – The International Journal on Mathematics Education, 44*(2), 355–366.

Furinghetti, F., & Pehkonen, E. (2002). Rethinking characterizations of beliefs. In G. C. Leder, E. Pehkonen, & G. Törner (Eds.), *Beliefs: A hidden variable in mathematics education?* (pp. 39–57). Dordrecht: Kluwer Academic Publishers.

Grigutsch, S., Raatz, U., & Törner, G. (1998). Einstellungen gegenüber Mathematik bei Mathematiklehrern. *Journal für Mathematik-Didaktik, 19*, 3–45.

Grolnick, W. S., & Ryan, R. M. (1987). Autonomy in children's learning: An experimental and individual difference investigation. *Journal of Personality and Social Psychology, 52*, 890–898.

Hofstede, G. (1986). Cultural differences in teaching and learning. *International Journal of Intercultural Relations, 10*, 301–320.

Hofstede, G. (2001). *Culture's consequences – Comparing values, behaviors, institutions and organizations across nations*. London/Thousand Oaks: Sage.

House, R. J., Hanges, P. J., Javidan, M., Dorfman, P. W., & Gupta, V. (Eds.). (2004). *Culture, leadership, and organizations*. Thousand Oaks: Sage.

Kagan, D. M. (1992). Implications of research on teacher belief. *Educational Psychologist, 27*(1), 65–90.

Keller-Schneider, M. (2011). Die Bedeutung von Berufswahlmotiven von Lehrpersonen in der Bewältigung beruflicher Anforderungen in der Berufseingangsphase. *Lehrerbildung auf dem Prüfstand, 4*, 157–185.

Klieme, E., & Vieluf, S. (2009). Teaching practices, teachers' beliefs and attitudes. In OECD (Ed.), *Creating effective teaching and learning environments: First results from TALIS* (pp. 87–135). Paris: OECD.

König, J., & Herzmann, P. (2011). Lernvoraussetzungen angehender Lehrkräfte am Anfang ihrer Ausbildung: Erste Ergebnisse aus der wissenschaftlichen Begleitung des Kölner Modellkollegs Bildungswissenschaften. *Lehrerbildung auf dem Prüfstand, 4*, 186–210.

König, J., & Rothland, M. (2012). Motivations for choosing teaching as a career: Effects on general pedagogical knowledge during initial teacher education. *Asia-Pacific Journal of Teacher Education, 40*, 289–315.

Leder, C., Pehkonen, E., & Törner, G. (Eds.). (2002). *Beliefs: A hidden variable in mathematics education?* Dordrecht: Kluwer Academic Publishers.

Leinhardt, G., & Greeno, J. G. (1986). The cognitive skill of teaching. *Journal of Educational Psychology, 78*, 75–95.

Leung, F. K. S., Graf, K.-D., & Lopez-Real, F. J. (Eds.). (2006). *Mathematics education in different cultural traditions – A comparative study of the East Asia and the West: The 13th ICMI study*. New York: Springer.

Mayr, J. (2009). LehrerIn werden in Österreich: Empirische Befunde zum Lehramtsstudium. *Erziehung und Unterricht, 159*, 14–33.

NCTM (National Council of Teachers of Mathematics). (2000). *Professional standards for teaching mathematics* (ENC-001451). Reston: NCTM.

Nespor, J. (1987). The role of beliefs in the practice of teaching. *Journal of Curriculum Studies, 19*, 317–328.

NOKUT = Nasjonalt Organ for Kvalitet i Utdanningen. (2006). *Evaluering av allmennlærerutdanningen i Norge 2006: Hovedrapport*. Retrieved from http://www.nokut.no/Documents/NOKUT/Artikkelbibliotek/Norsk_utdanning/SK/alueva/ALUEVA_Hovedrapport.pdf.

OECD. (2009). *Creating effective teaching and learning environments: First results from TALIS*. Paris: OECD Publishing.

Op't Eynde, P., De Corte, E., & Verschaffel, L. (2002). Framing students' mathematics-related beliefs. A quest for conceptual clarity and a comprehensive categorization. In G. C. Leder, E. Pehkonen, & G. Törner (Eds.), *Beliefs: A hidden variable in mathematics education?* (pp. 13–37). Dordrecht: Kluwer Academic Publishers.

Pajares, M. F. (1992). Teachers' beliefs and educational research: Cleaning up a messy construct. *Review of Educational Research, 62*, 307–332.

Perry, W. G., Jr. (1968). *Patterns of development in thought and values of students in a liberal arts college: A validation of a scheme*. Cambridge: Harvard University.

Peterson, P. L., Fennema, E., Carpenter, T., & Loef, M. (1989). Teachers' pedagogical content beliefs in mathematics. *Cognition and Instruction, 6*, 1–40.

Richardson, V. (1996). The role of attitudes and beliefs in learning to teach. In *Handbook of research on teacher education* (pp. 102–119). New York: Macmillan.

Rodd, M. M. (1997). Beliefs and their warrants in mathematics learning. In E. Pehkonen (Ed.), *Proceedings of the 21st international conference of the international group for the Psychology of Mathematics Education (PME)* (Vol. 4, pp. 64–65). Helsinki: University of Helsinki, Lahti Research and Training Center.

Roussos, L., & Stout, W. (1996). A multidimensionality-based DIF analysis paradigm. *Applied Psychological Measurement, 20*, 355–371.

Schiefele, U., & Urhahne, D. (2000). Motivationale und volitionale Bedingungen der Studienleistung. In U. Schiefele & K.-P. Wild (Eds.), *Interesse und Lernmotivation. Untersuchungen zur Entwicklung, Förderung und Wirkung* (pp. 183–205). Münster: Waxmann.

Schmidt, W. H., Blömeke, S., & Tatto, M. T. (2011). *Teacher education matters: A study of the mathematics teacher preparation from six countries*. New York: Teacher College Press.

Schoenfeld, A. H. (1983). Beyond the purely cognitive: Belief systems, social cognitions, and metacognitions as driving focuses in intellectual performance. *Cognitive Science, 7*, 329–363.

Schoenfeld, A. H. (1998). Toward a theory of teaching-in-context. *Issues in Education, 4*, 1–94.

Schommer, M. (1990). Effects of beliefs about the nature of knowledge on comprehension. *Journal of Educational Psychology, 82*, 498–504.

Schommer-Aikins, M. (2004). Explaining the epistemological belief system: Introducing the embedded systemic model and coordinated research approach. *Educational Psychologist, 39*, 19–29.

Singh, K., Granville, M., & Dika, S. (2002). Mathematics and science achievement: Effects of motivation, interest and academic engagement. *Journal of Educational Research, 95*, 323–332.

Tatto, M. T., Schwille, J., Senk, S., Ingvarson, L., Peck, R., & Rowley, G. (2008). *Teacher Education and Development Study in Mathematics (TEDS-M): Policy, practice, and readiness to teach primary and secondary mathematics. Conceptual framework*. East Lansing: Teacher Education and Development International Study Center, College of Education, Michigan State University.

Tatto, M. T., Schwille, J., Senk, S. L., Ingvarson, L., Rowley, G., Peck, R., et al. (2012). *Policy, practice, and readiness to teach primary and secondary mathematics in 17 countries: Findings from the IEA Teacher Education and Development Study in Mathematics (TEDS-M)*. Amsterdam: IEA.

Thompson, A. G. (1992). Teachers' beliefs and conceptions: A synthesis of research. In D. A. Grouws (Ed.), *Handbook of research on mathematics teaching and learning* (pp. 127–146). New York: Macmillan.

Triandis, H. C. (1995). *Individualism and collectivism*. San Francisco: Westview Press.

UNESCO. (1997). *International standard classification of education*. http://www.unesco.org/education/information/nfsunesco/doc/isced_1997.htm

Vieluf, S., & Klieme, E. (2011). Cross-nationally comparative results on teachers' qualification, beliefs, and practices. In Y. Li & G. Kaiser (Eds.), *Expertise in mathematics instruction* (pp. 295–325). New York: Springer.

Watt, H. M. G., & Richardson, P. W. (2007). Motivational factors influencing teaching as a career choice: Development and validation of the FIT-choice scale. *Journal of Experimental Education, 75*, 167–202.

Watt, H. M. G., Richardson, P. W., & Tysvaer, N. M. (2007). Profiles of beginning teachers' professional engagement and career development aspirations. In A. Berry, A. Clemans, & A. Kostogriz (Eds.), *Dimensions of professional learning: Professionalism, practice and identity* (pp. 155–176). Rotterdam: Sense Publishers.

Watt, H. M. G., Richardson, P. W., Klusmann, U., Kunter, M., Beyer, B., Trautwein, U., & Baumert, J. (2012). Motivations for choosing teaching as a career: An international comparison using the FIT-choice scale. *Teaching and Teacher Education, 28*, 791–805.

Wehling, L., & Charters, W. (1969). Dimensions of teacher beliefs about the teaching process. *American Educational Research Journal, 6*, 7–29.

# Affect and Gender

**Helen Forgasz, Gilah Leder, David Mittelberg, Hazel Tan, and Adelino Murimo**

**Abstract** In this chapter the authors present findings from recent research studies, conducted in different contexts, in which gender issues associated with a range of affective variables included in explanatory models for gender differences in mathematics learning outcomes – achievement and participation – were explored. The studies encompass data gathered from various different groups of students (Indigenous, primary, secondary), parents, mathematics teachers, and the general public. What emerges is an international profile of gender-related affective measures, with varying levels of agreement, which highlight the significance of contextual factors in this field of research. The authors explore the implications of their findings on classroom practice, policy, and future research.

**Keywords** Gender • Indigenous • Stereotyping • Occupational aspirations • Parents • Attributions • Calculators • Computers • International comparisons

## Introduction

Mathematics is an enabling discipline for STEM-based [Science, Technology, Engineering and Mathematics-based] studies at university and related careers and, around the world, males continue to dominate these fields (Blickenstaff 2005; United Nations Educational, Scientific and Cultural Organization [UNESCO] 2012). The contemporary relevance and importance of overcoming this gender imbalance internationally was highlighted at the United Nation's 55th Commission on the Status of Women, held in New York in 2011. It was noted that "… quality education and full and equal access and participation in science and technology for women of all ages are imperative for achieving gender equality and the

H. Forgasz (✉) • G. Leder • H. Tan • A. Murimo
Faculty of Education, Monash University, Melbourne, Australia
e-mail: Helen.Forgasz@monash.edu; gilah.leder@monash.edu

D. Mittelberg
Oranim Academic College of Education, Kiryat Tivon, Israel

 
B. Pepin, B. Roesken-Winter (eds.), *From beliefs to dynamic affect systems in mathematics education*, Advances in Mathematics Education,
DOI 10.1007/978-3-319-06808-4_12

empowerment of women, and an economic necessity…" (Commission on the Status of Women 2011, p. 2).

Serious study of affect within the field of mathematics education stemmed from the early work of those focussing on understanding the observed and consistent patterns of gender differences favouring males in mathematics achievement and in participation in challenging mathematics courses and mathematics-related careers. This work began in English-speaking western nations in the mid 1970s (e.g., in the USA: Fennema and Sherman 1977) and has spread around the globe (see Burton 1990; Forgasz et al. 2010).

Early research efforts resulted in a range of factors being identified as contributors to gender differences in mathematics learning outcomes. Various explanatory models for these gender differences began to emerge in the 1980s (e.g., Eccles et al. 1983; Fennema and Peterson 1985). At the time, it was also postulated that biological factors were implicated in males' superior mathematical performance. These explanations, however, lacked scientific evidence. The implied inevitability of change provoked debate and disquiet among many in the research community. More recent international performance comparisons in the Trends in International Mathematics and Science Studies [TIMSS] (see Mullis et al. 2012) continue to challenge biological explanations, as it is now apparent that in some countries females outperform males.

Other than refining and adding to the psycho-social and socio-cultural variables included in the early explanatory models for gender differences in mathematics performance and participation in challenging mathematics subjects, in participation in higher level tertiary studies and in careers involving mathematics, there has been little new theoretical work in the field. Leder (1990) provided an explanatory framework which incorporated the many common elements encompassed within the postulated models. These were described by Leder (1992) as follows:

> …the emphasis on the social environment, the influence of other significant people in that environment, students' reactions to the cultural and more immediate context in which learning takes place, the cultural and personal values placed on that learning and the inclusion of learner-related affective, as well as cognitive, variables. (p. 609)

Among the variables included in Leder's (1990) explanatory framework were the following learner-related affective variables: confidence; sex-role congruity; perceived usefulness of mathematics; and motivational variables including attributional style, learned helplessness, mastery orientation, and fear of success. Previous research into these variables reported by Leder (1992) had identified gender differences favouring males. Compared to females, males had been found to be more confident about their mathematical capabilities, more comfortable with the view of mathematics and related fields as male domains, and to perceive mathematics as useful; they were also more likely to have functional (leading to future success) attributional styles, that is, they were more likely than females to be mastery oriented and less likely than females to be learned helpless or believe that there was a price to pay for being successful in a male domain (fear of success). To this day, these elements remain useful in designing research studies in mathematics education in which gender equity considerations are relevant (Forgasz 2008).

In this chapter we present findings from five contemporary research studies in mathematics education in which gender and affective variables were of interest. Each study had unique dimensions related to the social context and geographic location, and the participants of interest. The affective variables central to each study varied, although there was some overlap. All were directly related to the range of learner-related variables included in Leder's (1990) explanatory framework for gender differences in mathematics learning outcomes.

Study 1. Primary-aged Australian Indigenous students were the focus. Perceptions of their own mathematical achievement and that of others, their beliefs about the relevance of mathematics for the future, and the students' feelings as they learn mathematics were of interest.

Study 2. Comparisons between Israeli and Australian secondary mathematics teachers' beliefs about the gendering of mathematics as a male, female, or neutral domain, and their beliefs about the reasons girls and boys would give for their successes and failures in mathematics were explored. Whether male and female teachers in each country viewed these issues differently was also of interest.

Study 3. Mozambique was the context for this study. Grade 7 boys' and girls' beliefs about and attitudes towards mathematics, including self-perceptions of achievement and the usefulness of mathematics, the gendering of mathematics, and career aspirations. The students' views were also compared with parents' beliefs for their sons and daughters.

Study 4. In Singapore and Australia, advanced calculators (graphics in Singapore, and Computer Algebra Systems [CAS] in Victoria) are mandated in the high stakes examinations at the end of secondary schooling. Comparisons between Singaporean and Victorian (Australia) male and female senior high school students' self-perceptions of self-competence with mathematics and with the advanced calculators, as well as their enjoyment and confidence with the calculators for mathematics learning were examined.

Study 5. In this study, an online survey, using Facebook as the means to recruit participants from around the world, was conducted to gauge the beliefs of the general public about the gender-stereotyping of mathematics and technology capability. The responses of participants from nine countries were compared.

The background, aims, instruments, and results for each study are presented in turn. The chapter ends with a general discussion of the findings from the five studies, overall conclusions, implications, and reflections on the direction for future work in this field.

## Study 1: Australian Indigenous Students, Gender, and Attitudes to Mathematics

A new phase in the assessment of student achievement in mathematics began in Australia in 2008 with the introduction of the National Assessment Program – Literacy and Numeracy [NAPLAN]. Each year since then the NAPLAN tests have been administered to students throughout Australia in Years 3, 5, 7, and 9. Although participation in NAPLAN testing is not compulsory, compliance is high. For example, in 2012 approximately 95 % of the Australian Year 3 cohort and 92 % of the Year 9 cohort completed NAPLAN tests. NAPLAN numeracy tests contain both multiple choice and open-ended items. Some student background information is also gathered: age, gender, Indigenous status, language background status (English/non-English), state/Territory, geolocation (metropolitan, provincial, remote, and very remote), parental educational background, and parental occupation. NAPLAN data are reported overall as well as separately by these different categories. However, no affective data are gathered.

## *The "Make It Count" Project*

One disturbing finding that has emerged consistently from the NAPLAN Numeracy tests, as well as from other data sources (e.g., TIMSS and PISA [Programme for International Student Assessment]) is the lower performance of Australian Indigenous students compared with their non- Indigenous counterparts.

Over the years, various interventions have been introduced to improve the mathematics performance of Indigenous students, including the recently concluded national *Make it Count* [MiC] project.[1] Around 40 primary and secondary schools in eight metropolitan and provincial localities participated in all, or some, components of the MIC project. Attention to affect was part of the project. A survey suitable for Year 3–6 students was devised and administered in participating schools. Responses were examined for gender differences. Findings for Indigenous students were of particular interest. Some schools elected not to participate in the survey.

## *The Survey, Sample, and Selected Findings*

The survey comprised some items with closed response formats and some with open-response formats to explore students' attitudes towards and beliefs about mathematics; similar questions were included about reading. Most items were modelled on those used in the TIMSS and PISA tests (see Thomson et al. 2008; de Bortoli and Thomson 2010). Among the survey items, students were shown three mathematics problems (Q1: calculate arrival time given time of departure and length

[1] *Make it Count: Numeracy, mathematics and Indigenous learners* is a national project that seeks to develop an evidence base of practices to improve the learning outcomes of Aboriginal and Torres Strait Islander (that is, Indigenous) students in mathematics. Eight clusters of schools across urban and regional Australia have been provided with various sources of support "to develop responsive mathematics pedagogy that will engage and inspire Aboriginal and Torres Strait Islander students and contribute to improved learning outcomes." For more details see http://makeitcount.aamt.edu.au/

**Table 1** Response frequencies to selected items by indigeneity and sex

| Item | Response | Indigenous | | Non-indigenous | | Comments |
|---|---|---|---|---|---|---|
| | | Males (%) | Females (%) | Males (%) | Females (%) | |
| How good were you at mathematics last year? | Above average | 67.8 | 58.5 | 66.7 | 68.2 | Indigenous females (F) lowest % |
| How much do you like mathematics this year? | Yes (A little/ very much) | 70.0 | 68.8 | 71.7 | 73.3 | Little between group variation but Indigenous F lowest % |
| How much do you like reading? | Yes (A little/ very much) | 68.3 | 83.1 | 70.3 | 84.1 | All students, higher % of F than M like reading, $\chi^2_2$,=27.09, p<.001 |
| How much does your best friend like mathematics? | Yes (A little/ very much) | 71.7 | 58.5 | 66.7 | 67.2 | Indigenous F lowest % |
| Is mathematics important for grown-ups? | Yes | 86.0 | 89.1 | 86.4 | 87.7 | Little between group variation but Indigenous F highest % |
| How sure are you that you can do this question Q1? | Sure | 82.5 | 76.2 | 81.5 | 80.7 | Indigenous F lowest % |
| How sure are you that you can do Q2? | Sure | 87.5 | 71.4 | 81.8 | 85.0 | Indigenous F lowest % and significantly lower than non-Indigenous F, $\chi^2_2$,=7.854, p<.05 |
| How sure are you that you can do Q3? | Sure | 72.2 | 67.7 | 80.3 | 73.3 | Indigenous F lowest % |

of trip; Q2: select a 3-dimensional object after rotation; and Q3: add 17 to a list of three numbers). They were asked to indicate how certain they were that they could answer each correctly; they were NOT asked to find the answers to the problems.

Over 1,200 students completed the survey. Of these, 125 (60 male, 65 female) were Indigenous; 1,108 (561 M, 547 F) were non-Indigenous. Responses to a selection of readily codable items are shown in Table 1.

As can be seen in Table 1, a subtle but unmistakable trend emerged. As a group, Indigenous females were more negative about mathematics, but not about reading, and were somewhat less confident about their mathematics proficiency. Additional information was gleaned from responses to particular open ended items. Detailed responses to two of these open-ended items are shown below.

**ITEM 1: Circle the words that show how you feel when you do mathematics (students were shown six positive and six negative adjectives)**

Of the adjectives circled by Indigenous males and females, 74 % and 63 % respectively were positive. For the non-Indigenous students, responses were virtually identical: of the adjectives circled by males 69 % were positive, compared with 70 % for females. Thus as a group, non-Indigenous females circled fewer positive adjectives than the other groups. For four adjectives there were notable differences in the proportions circled by female and male Indigenous students: '*Worried*' (M = 5 %; F = 17 %); '*Don't understand*' (M = 12 %; F = 28 %), '*Unhappy*' (M = 8 %; F = 18 %); and '*Not worried*' (M = 37 %; F = 22 %). These responses may help in explaining the Indigenous females' lower confidence in the likely correctness of their answers to the three mathematics problems (see Table 1).

**ITEM 2: How will you use mathematics as a grown up?**

All groups, it can be seen from Table 1, expected to use mathematics as adults. How much credence should be given to this uniformly high response with respect to the long term utility of mathematics? And do Indigenous and non-Indigenous females have the same expectations?

For both Indigenous and non-Indigenous students, using mathematics for shopping or working out the cost of items featured prominently. So did using mathematics at work, although the nature of that work was often not stated. Some roles were specified by the Indigenous group: shop assistant, teacher, dance teacher, and cook. The only roles nominated by the non-Indigenous group were working in a shop – as a sales person or a cashier (several students) and working in the air force (one student). Usage of mathematics in daily in-home activities was mentioned more frequently by the Indigenous females than by the non-Indigenous females, though again few definite activities were mentioned. A notable exception was the entry from one of the Indigenous students who expected to use mathematics "for counting lots of birds when you are sighting things". Given the age of the sample, students in Years 3–6, the lack of specificity with respect to the relevance of mathematics for their long term life goals is probably to be expected. Realistic expectations of adult life are still elusive.

## *Summary*

The sample comprised students in Years 3–6 at schools involved in the MiC project. For this group it was found that the attitudes of Indigenous females towards mathematics and their expectations of success in that subject were generally less functional (less likely to lead to success) than those of Indigenous males and non-Indigenous students. The scope of the MiC project did not allow a direct link to be made between the students' affective responses and their mathematics performance as measured by the NAPLAN tests. Others, however, have pointed to the impact of affective factors on academic performance. For example, writing about the 2011 TIMSS results for Australia, Thomson et al. (2012, p. 159) noted that:

> Developing positive attitudes towards reading, mathematics and science are important goals of the curriculum, particularly in primary school. Within Australia, students who expressed more positive attitudes and reported a higher level of self-confidence in reading, mathematics and science scored higher in the cognitive assessments than those who expressed less positive attitudes.

To what extent the findings of the MiC sample can be generalised beyond the group is debatable. The sample was drawn from schools directly involved in a project aimed at improving the mathematics performance of Indigenous students at metropolitan and provincial schools. How might findings from schools not involved in such a project differ from those presented here? Should we expect the less functional responses of Indigenous females in other school settings to be lessened or exacerbated? The attitudinal survey used was designed to cater for a wide age range of students attending geographically dispersed schools. The findings are sufficiently provocative to warrant a more in-depth investigation with Australian Indigenous students of different ages and in different settings. How representative these findings are with respect to gender and affect for other Indigenous groups around the globe is also worthy of further study.

## Study 2: Australian and Israeli Teachers' Gendered Beliefs About Mathematics

Cross-national comparisons of students' mathematics achievements, and the factors contributing to them, provide valuable insights into the complexities associated with finding explanations for observed patterns of gender difference in cognitive and affective measures. Cultural, ethnic, and societal factors have been found to interact with gender in many such explorations (e.g., Barkatsas et al. 2002; Forgasz and Mittelberg 2008).

This study was conducted in Israel and in Australia. Both societies are built of large waves of immigration and are comprised of a diverse number of ethnic/cultural groups. In addition, Israeli society is further divided into two major groups: Jews and Arabs. Israeli Arabs are the largest minority in Israel, comprising 19.7 % of the overall population in 2006 (Central Bureau of Statistics, Israel 2007). The educational systems for Arabs and Jews are segregated, but both offer the same curriculum and are run by the Ministry of Education. In Australia, state and territory governments are responsible for the provision and major funding of schooling. There is a national curriculum but states/territories implement and conduct the assessment of it. In both countries more females than males successfully complete secondary schooling.

## *Aims, Sample, and Instrument*

The aims of the study were to examine Australian and Israeli mathematics teachers' gendered perceptions of mathematics and to compare the findings by country (Israel and Australia) and by gender within country. A quantitative cross-sectional research design was adopted. Using pre-existing instruments when possible, a survey questionnaire was developed and encompassed a number of affective variables previously identified as contributors to gender differences in mathematics outcomes (see Leder 1992) including: the gendering of mathematics, and attributions for mathematics success and failure. The survey was administered online to voluntary samples of secondary mathematics teachers in both countries.

The bi-national sample was comprised of 181 secondary mathematics teachers: 47 % Israeli (85: 28 M, 57 F), and 53 % Australian (96: 36 M, 60 F). It should be noted that not all teachers responded to every item on the survey, reducing sample sizes for particular analyses. The length of the survey is likely to have contributed to this attrition.

Several demographic and biographical items were included in the survey. These differed slightly by country and, except for respondent gender, are irrelevant to the discussion here. With respect to the affective variables tapped, items from the following pre-existing instruments were used or modified: *Mathematics as a Gendered Domain* (Leder and Forgasz 2002); *Mathematics Attribution Scales* (Fennema et al. 1979). Other previously used items tapping perceptions of mathematics achievement, and differences between boys and girls, were employed.

The affective elements of the online survey included the following:

(i) *Mathematics as a Gendered Domain instrument* (Leder and Forgasz 2002)
Following Mittelberg and Forgasz (2009), only eight items from each of the three subscales – Mathematics as a Male Domain [MD], Mathematics as a Female [FD], and Mathematics as a Neutral Domain [ND] – were used. As in the original scales, five-point Likert-type response formats (Strongly Disagree = 1 to Strongly Agree = 5) were adopted. Sample items:

MD: Boys understand mathematics better than girls do
FD: Girls are more suited than boys to a career in a mathematically-related area
ND: Boys are just as likely as girls to help friends with their mathematics

(ii) *Mathematics Attribution Scales* [MAtS]. The original MAtS are comprised of eight items (four success, and four failure). Each has a stem statement, and four "causes" for the success/failure are listed, one corresponding to each of: ability, effort, task difficulty, and luck/environment; a response to each cause is required (5-point Likert: SD to SA). In this study, a simplified version of MAtS was used to determine teachers' views on the attributions for mathematical success/failure they believed girls and boys would provide. There were four items which were paired to explore for stereotyped beliefs. The teachers were asked to select the most likely reason that a boy/girl would give for success/failure. The items were:

If a boy/girl is successful/unsuccessful at mathematics, which is the most likely reason he/she would give? [Select one of four statements related to each of: ability, effort, task, environment]

## Findings

### Mathematics as a Gendered Domain: Differences by Gender and Country

Independent groups t-tests were conducted to test for differences in the mean scores (range 1–5) on each the three subscales (MD, FD, and ND) by country and then by gender within country. The t-tests revealed no statistically significant differences by country for each of the three subscales:

| | | |
|---|---|---|
| MD: | Israel = 2.53 | Australia = 2.62 |
| FD: | Israel = 2.61 | Australia = 2.44 |
| ND: | Israel = 3.83 | Australia = 4.00 |

These data indicate that on average, and to the same extent, respondents in both countries disagreed that mathematics was a male domain (mean scores < 3), disagreed that mathematics was a female domain, and agreed that it was a neutral domain (mean scores > 3). The results of the independent t-tests by gender within country are shown in Table 2.

The data in Table 2 clearly reveal that the male and female teachers in Australia were of one mind with respect to the gendering of mathematics. In Israel, on the other hand, the male teachers held more gender-stereotyped views than the females: they disagreed less strongly that mathematics was a male domain or a female domain, and agreed less strongly that mathematics was a neutral domain.

### Attributions for Success and Failure

The frequencies and percentages of the mathematics teachers in each country selecting each of the four attributions (ability, effort, task, environment) about boys' and girls' attributions for success and for failure are shown in Table 3.

The findings in Table 3 are consistent with the literature on gender differences in boys' and girls' own attributions for success: *ability* was the most frequent response provided by the Israeli (60.8 %) and the Australian teachers (40.7 %) for boys' most likely success attribution; for girls, *effort* was the most frequently provided response by teachers in both countries (Israeli: 53.8 %, Australian: 62.3 %).

**Table 2** MD, FD, and ND: sample size, means and T-test results by gender (within country)

| | | Male | | Female | | | |
|---|---|---|---|---|---|---|---|
| | Country | *N* | Mean | *N* | Mean | *t* | *p*-value |
| MD | Israel | 19 | 2.95 | 39 | 2.31 | 3.88 | <.001 |
| | Australia | 23 | 2.55 | 41 | 2.65 | | *ns* |
| FD | Israel | 19 | 2.92 | 40 | 2.45 | 2.97 | <.01 |
| | Australia | 19 | 2.45 | 37 | 2.44 | | *ns* |
| ND | Israel | 20 | 3.61 | 39 | 3.93 | 2.21 | <.05 |
| | Australia | 20 | 4.08 | 41 | 3.96 | | *ns* |

**Table 3** Teachers' beliefs about the success and failure attributions of boys and girls: frequencies and percentages of Israeli and Australian teachers' responses

| | Success | | | Failure | |
|---|---|---|---|---|---|
| | Boys' attributions | | | Boys' attributions | |
| Attribution | Israel | Australia | Attribution | Israel | Australia |
| Ability | 31 (60.8 %) | 22 (40.7 %) | Lack of ability | 2 (4.0 %) | 13 (24.5 %) |
| Effort | 4 (7.8 %) | 14 (25.9 %) | Lack of effort | 13 (26.0 %) | 19 (35.8 %) |
| Task | 9 (17.6 %) | 9 (16.7 %) | Task difficulty | 20 (40.0 %) | 11 (20.8 %) |
| Environment | 7 (13.7 %) | 9 (16.7 %) | Poor environment | 15 (30.0 %) | 10 (18.9 %) |
| Total *N* | 51 | 54 | | 50 | 53 |

| | Success | | | Failure | |
|---|---|---|---|---|---|
| | Girls' attributions | | | Girls' attributions | |
| Attribution | Israel | Australia | Attribution | Israel | Australia |
| Ability | 8 (15.4 %) | 3 (5.7 %) | Lack of ability | 12 (22.6 %) | 23 (42.6 %) |
| Effort | 28 (53.8 %) | 33 (62.3 %) | Lack of effort | 20 (37.7 %) | 21 (38.9 %) |
| Task | 5 (9.6 %) | 4 (7.5 %) | Task difficulty | 10 (18.9 %) | 7 (13.0 %) |
| Environment | 11 (21.2 %) | 13 (24.5 %) | Poor environment | 11 (20.8 %) | 3 (5.6 %) |
| Total *N* | 52 | 53 | | 53 | 54 |

Australian teachers' views, but not those of Israeli teachers, were also consistent with the literature on gender differences in boys' and girls' own attributions for failure. For boys, the most frequent response was *lack of effort* (35.8 %), and for girls it was *lack of ability* (38.9 %).

Interestingly, a majority of both the Israeli and the Australian teachers believed that girls would attribute failure to internal factors (lack of ability and lack of effort) – total Israeli 60.3 %; total Australian 81.5 %. Also, a majority of Israeli, but not Australian, teachers believed that boys would attribute failure to external factors (task difficulty and poor environment) – Total 70.0 %. Attributing success to internal factors and failure to external factors is consistent with a *mastery orientation* and considered a functional (likely to lead to future success) attribution pattern (Kloosterman 1990), while attributing failure to internal factors and success to external factors characterises the *learned helpless* individual and is not considered a functional attribution pattern (Leder 1992). The patterns of attribution from teachers in both countries suggest that they would consider boys more likely than girls to have a mastery orientation.

## Summary

The cross-national comparisons revealed that in both countries the teachers generally believed that mathematics was a neutral domain. Within country, there were no differences in the views of Australian male and female teachers. Among the Israelis, however, the males appeared to hold more traditionally gender-stereotyped views than their female counterparts. For teachers from both countries, there was consistency with earlier literature (e.g., see Leder 1992) with respect to most likely success attributions: boys to "ability" and girls to "effort". Also echoing previous research, the Australian teachers believed that girls were most likely to attribute failure to "lack of ability" and boys to "lack of effort". Based on response patterns, teachers from both countries are more likely to view boys than girls as having a mastery orientation (Kloosterman 1990), leading to future success in mathematics.

## Study 3: Grade 7 Students' and Parents' Gendered Views About Mathematics Learning in Mozambique

Reported here are findings from affective variables incorporated in a larger study aimed at understanding factors influencing mathematics learning outcomes in primary schools in Mozambique. The study was inspired by on-going concerns about persistent patterns of gender differences in mathematics achievement favouring boys among grade 6 students in Mozambique (Saito 2010).

A convenience sample of 300 Grade 7 pupils (134 boys and 166 girls) and 225 parents of these children (109 fathers and 116 mothers) participated in the study. The children attended schools in urban, rural, and remote areas of the Sofala province of Mozambique. The ages of the children varied from 11 to 16, with a mean of 13. The affective variables examined included: perceived achievement in mathematics; perceived usefulness of mathematics; perceptions of boys' and girls' behaviours and dispositions towards mathematics learning; and occupational aspirations. These variables have been identified as influencing students' task-choices, effort expended at tasks, and performance, and are implicated in explanations for gender differences in mathematics learning outcomes (Eccles et al. 1983). The data were gathered from students and their parents using pencil-and-paper surveys. Items related to the affective variables described above were drawn from pre-existing instruments, and other data gathered included biographical and background information.

### *Findings*

#### Perceived Achievement in Mathematics

Both boys and girls viewed mathematics as the most difficult subject they learnt at school, and physical education was identified as their best subject. Parents also believed mathematics was the most challenging school subject for themselves and for their children. Gender differences in self-perceived achievement were found only in regard to moral and civic education, and favoured girls. No statistically significant differences in these views were found by parent gender, or between parents of sons and parents of daughters.

## Perceived Usefulness of Mathematics

All parents believed mathematics was important for their children. However, they did not associate mathematics with technology or jobs. Parents tended to hold utilitarian views about mathematics, that is, that it is about counting, calculating, reasoning, and developing memory skills. The students viewed mathematics as more useful for them than did parents for their children.

The youngest children, children studying in urban schools, children having more educated parents, and children having fewer than three siblings held more positive views about mathematics than the other children. No influence of child gender was evident for perceived usefulness of mathematics. Having electricity at home, piped water, a television set, a computer, internet access, and school uniforms were positively associated with perceived usefulness of mathematics, but not with perceived achievement. Having calculators, cell phones, textbooks for reading and for mathematics, and the language spoken at home were not statistically significantly associated with the children's responses.

## Perceptions of Boys' and Girls' Behaviours and Dispositions to Learn Mathematics

To determine the extent to which the students and their parents gender-stereotyped mathematics learning as a male domain, the 30 items from the *Who and mathematics* [W&M] instrument (Leder and Forgasz 2002) were used. Leder and Forgasz (2002) claimed that the 30 items were worded to reflect typical beliefs and/or behaviours towards mathematics learning that had been identified in previous research to be associated with the gendering of mathematics. Consider the item, "Think mathematics will be important in their adult life". Respondents are asked to decide whether the belief or behaviour consistent with the wording of the item was more likely to be true for boys (B), girls (G), or whether there was no difference (ND). For this item, previous research indicated that respondents tended to match the wording of the item with 'boys' (Leder and Forgasz 2002), a response aligned with the traditional view that mathematics is a male domain.

For 17 of the 30 items, the responses of the Grade 7 males were consistent with Leder and Forgasz' (2002) predictions based on previous research. For the Grade 7 females there were only five items: the female students agreed that boys "are asked more questions by the mathematics teacher", "distract other students from their mathematics work", "like using computers to work on mathematics problem", "tease girls if they are good at mathematics", and that girls "get on with their work in class". The grade 7 females' responses to most other items reflected the view of mathematics as a gender-neutral domain.

Parents responded similarly to most items and reflected the traditional view that boys are more suited than girls to learn mathematics (Fathers: 21 items, Mothers: 21 items). Both believed girls would "give up when they find the mathematics problem is too difficult" and that girls would not be good at mathematics. In contrast, they indicated that boys would "enjoy mathematics", "think mathematics will be important in their adult life", and "need mathematics to maximize future employment opportunities". The parents' views were consistent with the view that mathematics is a male domain (see Leder and Forgasz 2002).

#### Occupational Aspirations

The students were asked which job they would like to have when they are about 30 years of age. Parents were also asked to indicate the job that they would like their children to have in the future. Both students and parents were asked to justify their choices. A higher proportion of girls than boys indicated a preference to engage in people-oriented occupations (e.g., teachers, lawyers, and physicians). In contrast, a higher proportion of boys than girls wanted to be engineers. Previous studies have also reported that females tend to select 'people-oriented' occupations and males 'things-oriented' occupations (Eccles 2007). Interestingly, a higher proportion of mothers wanted their daughters to be engineers than did fathers.

### *Summary and Recommendations*

Both students and parents viewed mathematics as the most difficult school subject. Mothers, fathers, and Grade 7 males tended to view mathematics as more suited to males than to females, while Grade 7 females believed mathematics learning was equally appropriate for boys and girls. Occupational aspirations also reflected a traditional gender divide. The sample involved in this study had both strengths and limitations. The diverse settings in which the data gathering took place (urban, rural, and remote settings) added to the robustness of the findings. The sample size of 300 students at only one grade level from only one province in Mozambique was realistically feasible, yet served as an inevitable constraint on the generalizability of the findings. Nevertheless, the findings have implications for government policy and mathematics teaching in Mozambique. Efforts are necessary to show the relevance and importance of mathematics to parents and students in Mozambique and that mathematics is a field of study that suits every person, regardless of gender or other affiliation.

## Study 4: Male and Female Students' Attitudes Towards Mathematics and Calculators

This study was part of a larger Ph.D. research study in which students' learning preferences, beliefs about and attitudes towards mathematics and advanced calculators were explored. The aim of the study reported in this chapter was to investigate if there were any gender differences in the students' attitudes towards

**Table 4** Items from Online Survey and Response Formats

| Variables and items | Response formats (and coding) |
|---|---|
| Mathematics competency self-rating (MSR): Currently for mathematics, I consider myself… | 1 = Weak, 2 = Below average, 3 = Average, 4 = Good, 5 = Excellent |
| Calculator competency self-rating (CalSR): In terms of GC/CAS calculator[a] skills, I consider myself… | 1 = Weak, 2 = Below average, 3 = Average, 4 = Good, 5 = Excellent |
| Calculator enjoyment (Cal_Enj): I enjoy using calculators to learn mathematics. | 1 = Strongly disagree, 2 = Disagree, 3 = Neutral, 4 = Agree, 5 = Strongly agree |
| Calculator confidence (Cal_Conf): I feel confident doing mathematics using calculators. | 1 = Strongly disagree, 2 = Disagree, 3 = Neutral, 4 = Agree, 5 = Strongly agree |

[a]"GC" for Singaporean students and "CAS calculator" for Victorian students

mathematics and advanced calculators, for example graphics calculators (GCs) and calculators with computer algebra system (CAS). Senior secondary (grades 11 and 12) students from Singapore and Victoria, Australia were surveyed. In Singapore, GCs are used in all of the senior secondary pre-university mathematics examinations; in Victoria, CAS calculators are used in some of the senior secondary high-stakes mathematics examinations. An anonymous online survey was designed using SurveyMonkey (http://www.surveymonkey.com) and used in the data collection. A summary of the relevant questionnaire items and the response formats is presented in Table 4.

These items were included in the questionnaire developed for the PhD study. A total of 964 Singaporean senior secondary students (63 % females, 37 % males) from four schools responded. Various recruitment strategies were adopted in Victoria because of the initial poor response from a stratified sample of schools (government and non-government) invited to participate. In all, 176 Victorian senior secondary students (69 % females, 31 % males) participated, with 19 % from government schools and 81 % from non-government schools. The small number of Victorian students and the high proportion of students from non-government schools limited the generalisability of the Victorian data. While it is unclear why there were higher proportions of female respondents in both regions, this seems consistent with other survey studies (e.g., Sax et al. 2003).

## Results

The Statistical Package for Social Sciences (SPSS) was used to analyse the data. T-tests were used to investigate gender differences in the Singaporean sample, and Mann-Whitney U tests were performed on the Victorian data due to the small sample size. Findings from the items noted in Table 4, including effect sizes, are shown in Table 5.

**Table 5** Mean scores and results of statistical comparisons by gender in Singapore and Victoria

| | | Singapore | | | | Victoria | | | |
|---|---|---|---|---|---|---|---|---|---|
| | Gender | N | Mean | t | Effect size $r^a$ | N | Mean | z (Mann-Whitney U test) | Effect size $r^a$ |
| MSR (M>F) | F | 605 | 2.79 | −4.01***[c] | 0.13 | 118 | 3.53 | NS | |
| | M | 358 | **3.08**[b] | | | 48 | **3.67** | | |
| CalSR (M>F) | F | 604 | 2.87 | −3.59*** | 0.12 | 118 | 3.53 | NS | |
| | M | 358 | **3.07** | | | 48 | **3.79** | | |
| Cal_Enj (M>F) | F | 589 | 3.28 | −3.39** | 0.11 | 99 | 3.33 | NS | |
| | M | 349 | **3.49** | | | 23 | **3.65** | | |
| Cal_Conf (M>F) | F | 588 | 3.08 | −4.64*** | 0.18 | 99 | 3.49 | 3.07** | 0.28 |
| | M | 349 | **3.40** | | | 23 | **4.22** | | |

[a]Effect size for t-test: $r = \sqrt{\frac{t^2}{t^2+df}}$ ; for Mann-Whitney U test, $r = \frac{z}{\sqrt{N}}$ (Field 2005)

[b]For each variable the higher of the two scores (male and female mean scores) is in bold

[c]p-values: *p<.05, **p<.01, ***p<.001

It can be seen in Table 5 that there were differences by region in the patterns of gender differences. In the Singaporean sample there were statistically significant gender differences in students' attitudes towards mathematics and calculators (GCs), and their confidence and enjoyment of calculators. In terms of mathematics and calculator competencies, males, on average, rated themselves slightly above average ($MSR_M = 3.08$, $CalSR_M = 3.07$), whereas females rated themselves slightly below average ($MSR_F = 2.79$, $CalSR_F = 2.87$). Both males and females indicated that they enjoyed using calculators and were confident doing mathematics using calculators, but males had higher mean scores than females ($Cal\_Enj_M = 3.49$, $Cal\_Enj_F = 3.28$, $Cal\_Conf_M = 3.40$, $Cal\_Conf_F = 3.08$).

For the Victorian sample, students generally scored above average in mathematics and calculator competencies, and agreed to the statements about calculator enjoyment and confidence. There was a statistically significant gender difference in students' confidence with doing mathematics using CAS calculators, with males having higher mean score than females ($Cal\_Conf_M = 4.22$, $Cal\_Conf_F = 3.49$). Males had higher mean scores than females for the other variables, but the differences were not statistically significant.

It is interesting to note that although the effect sizes were small for all the gender differences found, the largest effect sizes were for calculator confidence. Also, the effect size for calculator confidence was larger for the Victorian than Singaporean students. The findings that males were more confident than females with using technology are consistent with past research on gender and technology (e.g. Pierce et al. 2007). The findings in the Singaporean data that males had higher mean scores than females for mathematics and calculator competencies are consistent with past studies on Singaporean national examinations (Kaur 1995).

### *Summary*

Overall, these findings reveal that males were more confident users of calculator technologies than females. Since the high stake examinations in both regions have mandatory calculator components, the greater confidence in calculators might be translated into better examination performance. In the Victorian context there were suggestions that males might be advantaged by CAS calculator use in mathematics (Forgasz and Tan 2010). Further studies are needed to examine student outcomes in relation to their attitudes in order to ascertain if this is indeed the case.

## Study 5: International Comparisons of Gendered Beliefs About Mathematics and Technology

As noted earlier in this chapter, much research indicates that "negative stereotypes about girls' and women's abilities in mathematics and science persist despite girls' and women's considerable gains in participation and performance in these areas during the last few decades" (Hill et al. 2010, p. 38). Parents and teachers have also been found to hold gender-stereotyped beliefs about and expectations of children's mathematical capabilities (e.g., Tiedemann 2000). Explanatory models for gender differences in mathematics learning outcomes include the views of society at large (see Leder 1992). Yet views from the general public are gathered less often than from stakeholders such as parents and teachers.

### *The Study: Aim, Method, Instrument, and Sample*

The aims of the study were to gather the views of the general public in a variety of countries around the world, to explore whether mathematics continues to be viewed as a male domain, and to make comparisons by country. Facebook was used as the means of participant recruitment (see Tan et al. 2012 for details on how this was done). To maximise response rates, the survey was limited to 15 items. Reported here are findings from six questions related to the gendering of mathematics and technology competence:

Q1. Who are better at mathematics, girls or boys?
Q2. Is it more important for girls or boys to study mathematics?
Q3. Who do parents think are better at mathematics, girls or boys?
Q4. Who do teachers think are better at mathematics, girls or boys?
Q5. Who are better at using calculators, girls or boys?
Q6. Who are better at using computers, girls or boys?

**Table 6** Numbers of responses and valid percentages by country

| Country | n | Valid (%) | Country | n | Valid (%) | Country | n | Valid (%) |
|---|---|---|---|---|---|---|---|---|
| Canada | 35 | 6.4 | India | 66 | 12.0 | UAE | 46 | 8.4 |
| China | 76 | 13.8 | Israel | 31 | 5.6 | UK | 58 | 10.5 |
| Egypt | 84 | 15.3 | Singapore | 35 | 6.4 | Australia | 119 | 21.6 |

For each item, participants selected: Boys/Girls/Same/Don't know/depends. Responses were received from 784 participants representing 81 countries. There were nine countries – Canada, China, Egypt, India, Israel, Singapore, UAE, UK, and Australia – with at least 30 responses from each; there were 505 responses from these nine countries, representing 70.2 % of all responses. The response frequencies from the nine countries, and the valid percentages represented, are shown in Table 6.

Using SPSS, the responses to the six questions were analysed by country, and chi-square tests were conducted to determine if there were statistically significant differences in the response frequency distributions by country.

## *Findings*

### Questions About the Gendering of Mathematics: Q1, Q2, Q3, and Q4

The frequency distributions of responses by country for Q1 (Who are better at mathematics, girls or boys?), Q2 (Is it more important for girls or boys to study mathematics?), Q3 (Who do parents think are better at mathematics?) and Q4 (Who do teachers think are better at mathematics?) are shown in Figs. 1, 2, 3, and 4 respectively. Statistically significant differences were found in the frequency distributions of responses by country to Q1 ($\chi^2=56.0$, df=24, p<.001), Q3 ($\chi^2=8.49$, df=24, p<.001), and Q4 ($\chi^2=88.23$, df=24, p<.001).

While "same" was the most frequent response to Q1 in five countries (Canada, Egypt, Israel, UAE, and UK), it is clear from Fig. 1 that more participants in each country considered "boys" than considered "girls" to be better at mathematics. Respondents from China held this view, the traditional gender-stereotyped view, more strongly than in other countries. Figure 2 shows that an overwhelming majority in each country considered it equally important for girls and boys to study mathematics (Q2). Interestingly, among the minorities in each country who held gendered views, slightly higher proportions felt that it was more important for boys than for girls to study mathematics.

The pattern of responses to Q3 and Q4 were similar. With respect to beliefs about who parents (Q3) and teachers (Q4) would consider to be better at mathematics, more respondents in each country indicated "boys" than "girls" - see Fig. 3 (Q3) and Fig. 4 (Q4). In response to both questions, respondents from China held to the traditional gender-stereotype more strongly than elsewhere.

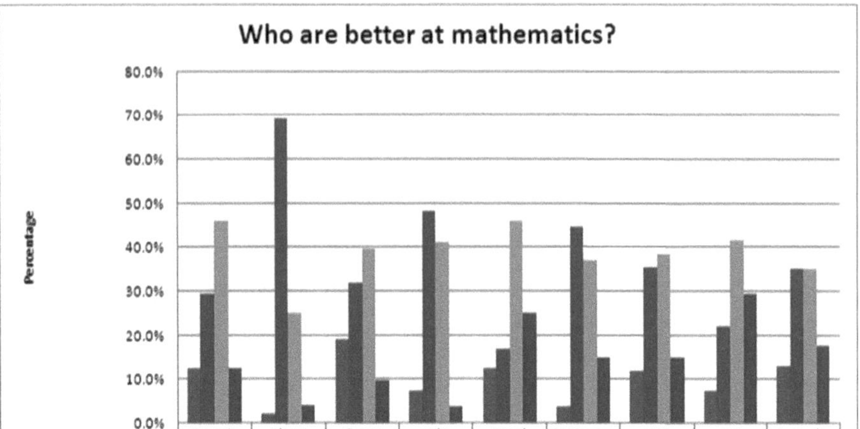


| | Canada | China | Egypt | India | Israel | Singapore | UAE | UK | Australia |
|---|---|---|---|---|---|---|---|---|---|
| Girls | 12.5% | 1.9% | 19.0% | 7.1% | 12.5% | 3.7% | 11.8% | 7.3% | 12.8% |
| Boys | 29.2% | 69.2% | 31.7% | 48.2% | 16.7% | 44.4% | 35.3% | 22.0% | 34.9% |
| Same | 45.8% | 25.0% | 39.7% | 41.1% | 45.8% | 37.0% | 38.2% | 41.5% | 34.9% |
| Don't know/depends | 12.5% | 3.8% | 9.5% | 3.6% | 25.0% | 14.8% | 14.7% | 29.3% | 17.4% |

**Fig. 1** Frequency responses by country: who are better at mathematics?

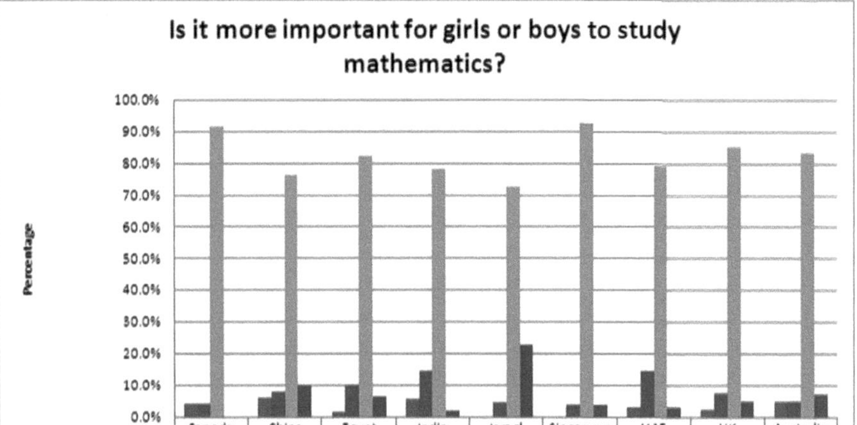


| | Canada | China | Egypt | India | Israel | Singapore | UAE | UK | Australia |
|---|---|---|---|---|---|---|---|---|---|
| Girls | 4.2% | 5.9% | 1.6% | 5.5% | 0.0% | 0.0% | 2.9% | 2.4% | 4.7% |
| Boys | 4.2% | 7.8% | 9.7% | 14.5% | 4.5% | 3.7% | 14.7% | 7.3% | 4.7% |
| Same | 91.7% | 76.5% | 82.3% | 78.2% | 72.7% | 92.6% | 79.4% | 85.4% | 83.5% |
| Don't know/depends | 0.0% | 9.8% | 6.5% | 1.8% | 22.7% | 3.7% | 2.9% | 4.9% | 7.1% |

**Fig. 2** Frequency responses by country: is it more important for girls or boys to study mathematics?

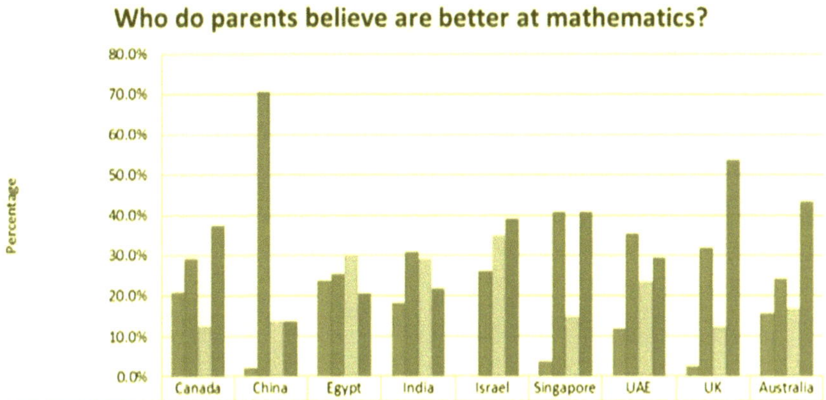


| | Canada | China | Egypt | India | Israel | Singapore | UAE | UK | Australia |
|---|---|---|---|---|---|---|---|---|---|
| Girls | 20.8% | 2.0% | 23.8% | 18.2% | 0.0% | 3.7% | 11.8% | 2.4% | 15.7% |
| Boys | 29.2% | 70.6% | 25.4% | 30.9% | 26.1% | 40.7% | 35.3% | 31.7% | 24.1% |
| Same | 12.5% | 13.7% | 30.2% | 29.1% | 34.8% | 14.8% | 23.5% | 12.2% | 16.9% |
| Don't know/depends | 37.5% | 13.7% | 20.6% | 21.8% | 39.1% | 40.7% | 29.4% | 53.7% | 43.4% |

**Fig. 3** Frequency responses by country: who do parents think are better at mathematics?

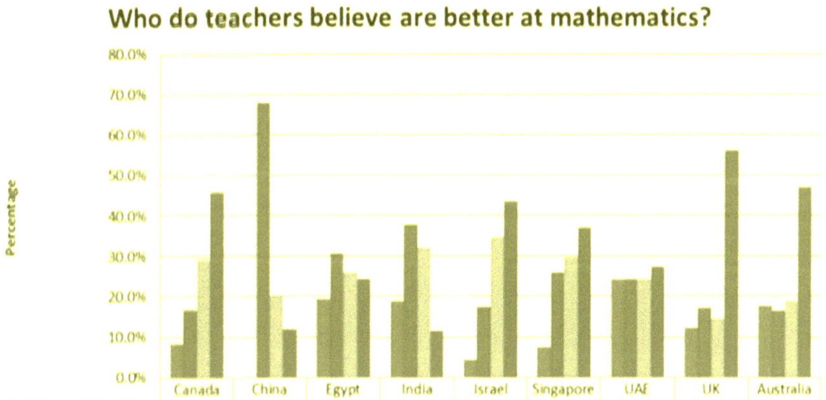


| | Canada | China | Egypt | India | Israel | Singapore | UAE | UK | Australia |
|---|---|---|---|---|---|---|---|---|---|
| Girls | 8.3% | 0.0% | 19.4% | 18.9% | 4.3% | 7.4% | 24.2% | 12.2% | 17.6% |
| Boys | 16.7% | 68.0% | 30.6% | 37.7% | 17.4% | 25.9% | 24.2% | 17.1% | 16.5% |
| Same | 29.2% | 20.0% | 25.8% | 32.1% | 34.8% | 29.6% | 24.2% | 14.6% | 18.8% |
| Don't know/depends | 45.8% | 12.0% | 24.2% | 11.3% | 43.5% | 37.0% | 27.3% | 56.1% | 47.1% |

**Fig. 4** Frequency responses by country: who do teachers think are better at mathematics?

## Competence with Technology: Q5 and Q6

The response frequency distributions by country for Q5 (Who are better at using calculators, girls or boys?) and Q6 (Who are better at using computers, girls or boys?) are shown in Figs. 5 and 6 respectively. The response distributions to both

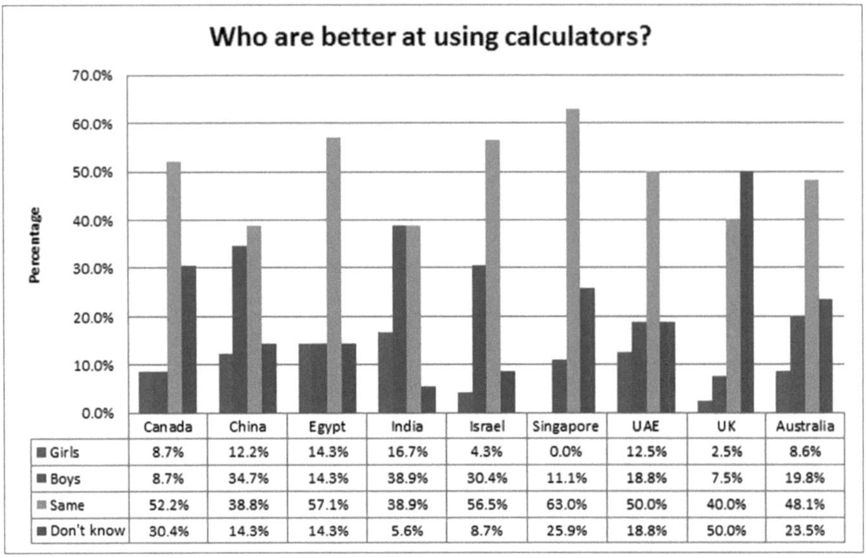


**Fig. 5** Frequency responses by country: who are better at using calculators, girls or boys?

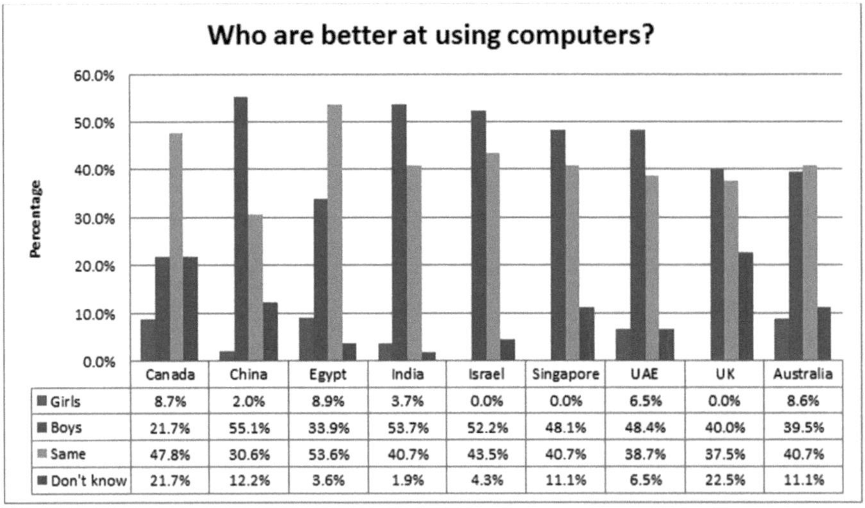


**Fig. 6** Frequency responses by country: who are better at using computers, girls or boys?

questions were statistically significantly different: Q5 ($\chi^2 = 63.9$, df = 24, $p < .001$) and Q6 ($\chi^2 = 38.7$, df = 24, $p < .05$).

Interestingly, the patterns of response to Q5 and Q6 about technology were different. Boys were considered more capable than girls with respect to both calculator and computer capability, but this gender-stereotyped view was more strongly held with respect to computers than calculators.

It can be seen in Fig. 5 that in seven countries (Canada, Egypt, China, Israel, Singapore, UAE and Australia), the most frequent response was "same", that is, that boys and girls were considered equally capable with calculators. In Fig. 6, considerably more respondents in each of the nine countries indicated that "boys" rather than "girls" were better at using computers. In fact, in six countries (China, India, Israel, Singapore, UAE and UK) "boys" was the most frequent response. In China, India, and Israel, a majority (ie. over 50 %) of respondents said "boys" were better with computers, indicating that the traditional gender-stereotyped view was very strongly held by respondents from these countries.

## *Summary*

One positive outcome of this study was that in all nine countries there was strong endorsement of mathematics as an important study for all students irrespective of gender. Differences by country were evident, however, when it came to perceptions of boys' and girls' capabilities with mathematics and with technology (calculators and computers). While many people in the nine countries did not distinguish between boys and girls with respect to mathematics or technology use, it was evident that in all countries, to varying extents, it was more likely that the traditional gender-stereotyped view of mathematics as a male domain (that is, that mathematics is more suited to boys than to girls) prevailed. It was noteworthy that there was no item for which the response "girls" had a higher percentage response rate than the response "boys" in any of the nine countries. Participants from China appeared to hold the strongest traditional beliefs about mathematics as a male domain. Participants from English-speaking countries appeared to be more likely than participants from non-English speaking countries to hold gender-neutral beliefs (ie. more likely to respond "same"). Despite the study's limitations – English as the language used in the survey, and the potential age and socio-economic bias inherent to Facebook users – the consistency in the direction of the findings in support of the traditional male stereotype provides strong evidence that gendered perceptions of mathematics are still evident in many parts of the world.

## Final Words

Over time, there has been an overwhelming volume of evidence indicating that affect is critical in understanding gender differences in mathematics achievement and participation.

In this chapter we have provided contemporary evidence from five separate research studies covering a range of affective variables with participants of different ages and from different cultural and national contexts to illustrate this point. Collectively, the data presented reveal that the attitudes towards mathematics learning of primary-aged Indigenous female students (Study 1) and of grade 7 female students in Mozambique (Study 3) were less functional (likely to lead to future success) than those of their male counterparts. The parents of the Mozambican students (Study 3), Israeli and Australian mathematics teachers (Study 2), and members of the general public around the world (Study 5) also held views about girls and boys that imply expectations of boys as more likely than girls to have functional dispositions towards mathematics learning. When it came to technology capability, now integral to mathematics learning and science-related occupations, Singaporean and Australian high school boys (Study 4) were found to be more confident in using sophisticated calculators and the general public around the world (Study 5) viewed males as more competent with calculators, although to a lesser extent than they viewed males to be superior with computers.

In summary, beliefs and attitudes that mathematics is still a male domain, that is, that males are more suited than females to pursue studies in mathematics and to follow associated career paths, persist. It also appears that the distribution of these gendered beliefs varies across nations and may be stronger than the differences between males and females within countries. All the data point to the constructed nature of gendered beliefs, cultural and societal drivers and, by implication, the potential for the malleability of these views. What is needed and what can be done?

In our view it is incumbent on mathematics education researchers, whether focussing on affect or not, to consider including gender as an independent variable when designing research studies and when gathering and analysing data. It is also critical to recognize the context in which the research is being conducted and take into consideration the societal and cultural practices that may be influencing affective measures and which, in turn, impact on children's mathematics learning experiences, teachers' pedagogical practices, and students' subsequent mathematics learning outcomes.

It is clear, and endorsed by the United Nations (Commission on the Status of Women 2011), that the *gender problem* has not been overcome, despite evidence of more equitable achievement outcomes and greater levels of educational participation more generally. Why do the attitudes and beliefs of girls, and those of the significant others around them, remain less optimistic in respect of mathematics? There is an imperative for gender and affect to remain on the research agenda and for a re-examination of the means to address and overcome females' persistent mathematical disadvantage.

**Acknowledgments** Study 1: The research was funded by the Australian Government and was administered by the Australian Association of Teachers of Mathematics.

Study 2: The research in Israel was supported by The MOFET Institute and the Department of Teacher Education at the Israeli Ministry of Education.

Study 5: The research was funded by the Faculty of Education, Monash University.

## References

Barkatsas, A. N., Forgasz, H. J., & Leder, G. C. (2002). The stereotyping of mathematics: Gender and cultural factors. *Themes in Education, 3*(2), 91–216.

Blickenstaff, J. C. (2005). Women and science careers: Leaky pipeline or gender filter? *Gender and Education, 17*(4), 369–386. doi:10.1080/09540250500145072.

Burton, L. (Ed.). (1990). *Gender and mathematics: An international perspective*. London: Cassell.
Central Bureau of Statistics, Israel. (2007). *Israel in figures. 2007*. Retrieved from http://www1.cbs.gov.il/www/publications/isr_in_n07e.pdf
Commission on the Status of Women. (2011). *Report on the fifty fifth session* (Economic and Social Council Official Records, 2011, Supplement No. 7). New York: United Nations. Retrieved from http://www.un.org/ga/search/view_doc.asp?symbol=E/2011/27
De Bortoli, L. J., & Thomson, S. (2010). *Contextual factors that influence the achievement of Australia's indigenous students: Results from PISA 2000–2006*. Camberwell: Australian Council for Educational Research Ltd.
Eccles, J. S. (2007). Where are all the women? Gender differences in participation in physical science and engineering. In S. J. Ceci & W. M. Williams (Eds.), *Why aren't more women in science? Top researchers debate the evidence* (pp. 199–210). Washington, DC: American Psychological Association.
Eccles, J. S., Adler, T. F., Futterman, R., Goff, S. B., Kaczala, C. M., Meece, J. L., et al. (1983). Expectancies, values, and academic behaviors. In J. T. Spence (Ed.), *Achievement and achievement motivation*. San Francisco: W. H. Freeman.
Fennema, E., & Peterson, P. (1985). Autonomous learning behavior: A possible explanation of gender-related differences in mathematics. In L. C. Wilkinson & C. B. Marrett (Eds.), *Gender related differences in classroom interaction* (pp. 17–35). New York: Academic.
Fennema, E., & Sherman, J. (1977). Sex-related differences in mathematics achievement, spatial visualization and affective factors. *American Educational Research Journal, 14*(1), 51–71.
Fennema, E., Wolleat, P., & Pedro, J. D. (1979). Mathematics attribution scale. *JSAS: Catalog of Selected Documents in Psychology, 9*(5), 26 (Ms. No. 1837).
Field, A. (2005). *Discovering statistics using SPSS* (2nd ed.). London: Sage.
Forgasz, H. (2008). Stars, compass, and GPS: Navigating currents and charting directions for mathematics education research on gender issues. In M. Goos, R. Brown, & K. Makar (Eds.), *Conference proceeding MERGA 31* (pp. 5–15). Brisbane: MERGA.
Forgasz, H. J., & Mittelberg, D. (2008). Israeli Jewish and Arab students' gendering of mathematics. *ZDM – The International Journal on Mathematics Education, 40*(4), 545–558. doi:10.1007/s11858-008-0139-3.
Forgasz, H., & Tan, H. (2010). Does CAS use disadvantage girls in VCE mathematics? *Australian Senior Mathematics Journal, 24*(1), 25–36.
Forgasz, H. J., Becker, J. R., Lee, K., & Steinthorsdottir, O. (Eds.). (2010). *International perspectives on gender and mathematics education*. Charlotte: Information Age Publishing.
Hill, C., Corbett, C., & St. Rose, A. (2010). *Why so few? Women in science, technology, engineering and mathematics*. Washington, DC: AAUW. Retrieved from http://eric.ed.gov/PDFS/ED509653.pdf
Kaur, B. (1995). Gender and mathematics: The Singapore perspective. In P. Rogers & G. Kaiser (Eds.), *Equity in mathematics education: Influences of feminism and culture* (pp. 129–134). London: Falmer Press.
Kloosterman, P. (1990). Attribution, performance following failure, and motivation in mathematics. In E. Fennema & G. C. Leder (Eds.), *Mathematics and gender* (pp. 96–127). New York: Teachers College Press.
Leder, G. C. (1990). Gender differences in mathematics: An overview. In E. Fennema & G. C. Leder (Eds.), *Mathematics and gender* (pp. 10–26). New York: Teachers College Press.
Leder, G. C. (1992). Mathematics and gender: Changing perspectives. In D. A. Grouws (Ed.), *Handbook of research in mathematics teaching and learning* (pp. 597–622). New York: Macmillan.
Leder, G. C., & Forgasz, H. J. (2002). Two new instruments to probe attitudes about gender and mathematics. *ERIC, resources in education (RIE)* (ERIC document number: ED463312). Retrieved from http://files.eric.ed.gov/fulltext/ED463312.pdf
Mittelberg, D., & Forgasz, H. (2009). *Israeli Jewish and Arab pre-service students' gendering of mathematics*. Unpublished report to Oranim College of Advanced Education.

Mullis, I. V. S., Martin, M. O., Foy, P., & Arora, A. (2012). *TIMSS 2011 international results in mathematics*. Boston/Amsterdam: TIMSS & PIRLS International Study Center/IEA.

Pierce, R., Stacey, K., & Barkatsas, A. (2007). A scale for monitoring students' attitudes to learning mathematics with technology. *Computers and Education, 48*(2), 285–300. doi:10.1016/j.compedu.2005.01.006.

Saito, M. (2010). The gender gap in learning – Back to square one. A need to change the focus from access and participation to learning achievement. *UNESCO International Institute for Educational Planning [IIEP] Newsletter, 28*(3), 6–7. Retrieved from http://www.sacmeq.org/research.htm

Sax, L. J., Gilmartin, S. K., & Bryant, A. N. (2003). Assessing response rates and nonresponse bias in web and paper surveys. *Research in Higher Education, 44*(4), 409–432.

Tan, H., Forgasz, H., Leder, G., & McLeod, A. (2012, August 17–19). Survey recruitment using Facebook: Three studies. In *Proceedings of the 2012 international conference on internet studies (NETs2012), Bangkok, Thailand.* Retrieved from http://www.internet-studies.org/

Thomson, S., Wernert, N., Underwood, C., & Nicholas, M. (2008). *TIMSS 2007: Taking a closer look at mathematics and science in Australia*. Camberwell: Australian Council for Educational Research. Retrieved from http://www.acer.edu.au/documents/timss_2007-australiafullreport.pdf

Thomson, S., Hillman, K., Wernert, N., Schmid, M., Buckley, S., & Munene, A. (2012). *Monitoring Australian year 4 student achievement internationally: TIMSS & PIRLS 2011*. Camberwell: Australian Council for Educational Research. Retrieved from http://www.acer.edu.au/documents/TIMSS-PIRLS_Monitoring-Australian-Year-4-Student-Achievement.pdf

Tiedemann, J. (2000). Parents' gender stereotypes and teachers' beliefs as predictors of children's concept of their mathematical ability in elementary school. *Journal of Educational Psychology, 92*(1), 144–151.

United Nations Educational, Scientific and Cultural Organization [UNESCO]. (2012). *World atlas of gender equality in education*. Paris: UNESCO. Retrieved from http://unesdoc.unesco.org/images/0021/002155/215522E.pdf

# Reaction to Section 2: The Relevance of Affective Systems and Social Factors: A Commentary

**Markku S. Hannula**

**Abstract** This commentary of the six chapters in this section will address three issues. First, a metatheoretical framework for research on mathematics-related affect will be presented. It consists of three dimensions: (1) emotional, motivational and cognitive components of affect, (2) state and trait aspects of affect, and (3) theories of affect as a social, psychological and physiological phenomenon. Secondly, there will be a discussion on the structure of affect. Lastly, there will be a reflection on social influences on individual affect.

**Keywords** Emotion • Motivation • Beliefs • Affect • State and trait • Social influences • Structure of affect • Mathematics

## Relevance

The first part of this section's title, "Relevance in the field", relates in my mind to two things. Firstly, research in mathematics-related affect is relevant when it identifies which of the many affective components of the individual are the most important cornerstones of their view of mathematics. As an educator, I would love to know which set of affective components is such that, when challenged and changed, would cause a chain reaction of permanent changes throughout the person's view of mathematics. Secondly, it would be important to identify which of the affective components can be changed and how. Knowledge of cornerstones of affect is not helpful, unless there are ways to shake them.

My commentary will consist of three sections. The first part focuses on paving the way through establishing a metatheoretical framework and a vocabulary to discuss the different chapters. The second part of this commentary looks at what the six chapters reveal about the systemic nature of affect. The last part this commentary, will focus on what the chapters are able to tell about the social aspects influencing affect.

M.S. Hannula (✉)
University of Helsinki, Helsinki, Finland
e-mail: markku.hannula@helsinki.fi

 
B. Pepin, B. Roesken-Winter (eds.), *From beliefs to dynamic affect systems in mathematics education*, Advances in Mathematics Education,
DOI 10.1007/978-3-319-06808-4_13

## A Metatheory of Affect

Ambiguous terminology is a known problem in research of mathematics-related affect (Furinghetti and Pehkonen 2002; Hannula 2011, 2012; Goldin 2004). Most notably, some researchers use attitude as the umbrella concept consisting of cognitive (beliefs), affective (emotions) and conative (behaviour) dimensions (e.g. Hart 1989) while several others define attitudes, beliefs and emotions as three dimensions of affect (e.g. McLeod 1992). This problem is not severe in this section, as all chapters give sufficient theoretical background and define their concepts (norm, motivation, goal, view, attitude, affective pathway, local and global beliefs, and belief system) clearly. Yet, the reader may find it difficult to relate these studies to some others using a different terminology.

In this commentary, we will be using the terminology by Hannula (2011, 2012). The terminology is related to a metatheory that aims at linking and contextualizing theories for mathematics related affect (Fig. 1). This terminology distinguishes not only the cognitive, emotional and motivational aspects of affect, but also separates the relatively stable traits and the more dynamically changing states in all three. The *cognitive traits* include beliefs and other mental representations to which it makes sense to attribute a truth value (c.f. Goldin 2002). The *emotional traits* include emotional dispositions, i.e. tendencies to feel joy, anxiety or other emotions in relation to certain objects or situations (such as mathematics). The third category, *motivational traits* include personal preferences. The distinction from the cognitive aspect is that preferences are subjective and it is not possible to attribute truth value to them. Respectively, the dynamically changing mental states, such as thoughts, feelings and motives, refer to the respective cognitive, emotional and motivational states. In addition, the metatheory identifies three levels of theorizing affect; one focusing on affect as a social phenomenon, the second looking at affect as part of individual psychology and the third that looks at affect as a biological/physiological phenomenon.

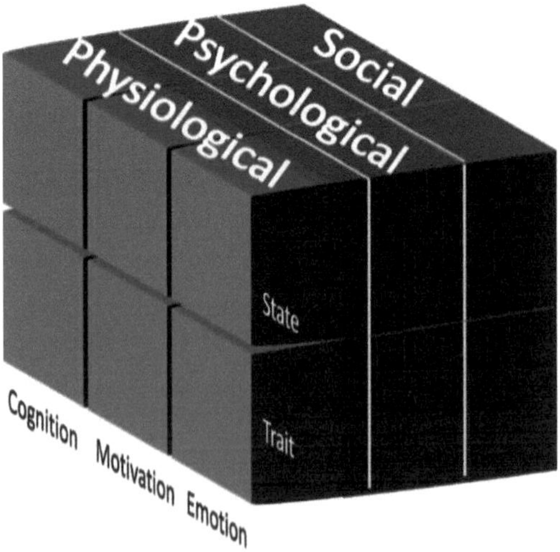


**Fig. 1** Illustrating the three dimensions for a metatheory of mathematics-related affect (Hannula 2011, 2012)

When we examine the six chapters of this section using the metatheoretical framework for affect, we see some areas that are hardly, or not at all, touched in them. Firstly, none of the chapters discusses affect as a biological/physiological phenomenon. Secondly, only Gómez-Chacón discusses the dynamics of affective states. Hence, this commentary will focus mainly on two aspects of affective systems. First there will be an analysis of what the six chapters tell about the structure of different cognitive, motivational and emotional traits, i.e. how different aspects of these traits relate to each other. Secondly there will be an analysis of how different aspects of the social level may influence student and teacher affect.

## The Structure of Student Affect

The unsurprising conclusion from the affect-related research is that those students who do well in mathematics tend to have a more positive affect towards mathematics than those who do less well. However, it has been more problematic to establish the direction of causality (see Hannula 2011 for a detailed discussion). Taken together, studies suggest a reciprocal rather than unidirectional causality between achievement and affect (Ma and Kishor 1997a, b; Ma 1999; Ma and Xu 2004; Williams and Williams 2010; Minato and Kamada 1996).

The tendency for positive variables to correlate positively seems to hold true also for relations between different affective variables, but this is only a tendency. In their chapter, Ding, Pepin and Jones observed that, in Shanghai, Chinese students' liking of mathematics (an emotional trait) and their perceived competence (a cognitive trait) were positively related. However, cases "I like it but can't do it" and "I can do it but I dislike it" were much more frequent (over 20 % of the responding students) than previously observed in Italy (Zan and Di Martino 2007). Ding et al. provide also an interesting preliminary glimpse of the reasons these Chinese students provide for their attitudes. Some of these reasons seem to relate to an internal motivation (e.g. it's interesting or they like the teacher), while some seem to relate more to an external motivation (e.g. it is a key subject in examination or it is important for future career). While the positive relation between internal motivation and positive emotions is a well known phenomenon, motivation theories and related research do not suggest a general positive relationship between external motivation and positive emotions (e.g. Deci et al. 1991; Pekrun et al. 2006).

One reason to be interested in affect is its assumed relation with metacognition and self-regulation (McLeod 1992; Goldin 2002). The chapter of Gómez-Chacón explores affect in the context of teacher students solving geometry problems using a dynamic geometry software. In addition to analysing the emotional, cognitive and motivational traits in this situation, she also explores how these traits and student meta-emotions (awareness and control of emotions) influence their cognitive-affective pathways (including emotional states) during the instrumental genesis. Her analysis exemplifies how specific can the relations between affective traits and states be: students who preferred visualizing (motivational trait) tended to encounter different types of difficulties than students who had a preference for non-visual approaches. Moreover, her report illustrates the highly individual patterns related to dynamics of affect.

The systemic nature of beliefs was first discussed by Green (1971), who identified three characteristics of belief systems. Firstly, primary beliefs are used as arguments to reason for derivative beliefs (quasi-logical ordering). Secondly, psychologically central beliefs are held more strongly than peripheral beliefs. And thirdly, beliefs are situated, i.e., belief clusters relate to specific situations and contexts. The contextual nature of beliefs has been frequently acknowledged in the elaborations of beliefs specific to content areas. Also in this section there are chapters that pay attention to the contextual nature of beliefs. Depaepe, De Corte and Verschaffel focus on teacher and students beliefs in the specific area of word problems. The way they focus on beliefs in this very specific context, they are implicitly acknowledging the contextual nature of beliefs. Gómez-Chacón addresses both general beliefs (e.g. self-confidence in mathematics) and more specific beliefs (e.g. emotions regarding use of GeoGebra). The contextual nature of beliefs is the explicit focus of the Eichler and Erens chapter, where they explore teacher beliefs in context of calculus, statistics and geometry, analysing how teachers' beliefs are different in these different contexts.

In a similar way that direction of causality is problematic between affect and achievement, it is also problematic between different affective variables. There is often little empirical evidence to assume a direction of causality between any two affective variables, and yet, any advanced theorizing or developed methodology forces us to assume causality. For example, McLeod's theory (1992) assumed a direction of causal effects, where student beliefs have their origin in individual experiences and the social context, these beliefs would influence the onset of emotions, which, when repeated, would lead to attitudes. However, Bandura's self-efficacy theory (1978) assumes a reciprocal relationship between the individual and the social level; emotions are known to influence the interpretation of experiences through directing attention and biasing memory (Power and Dalgleish 1997; Linnenbrink and Pintrich 2004) and the narrow definition of attitude used by McLeod defines attitudes as a tendency to feel certain emotions. Hence, all of the causal directions McLeod suggest, are likely to be reciprocal and we should be cautious with any assumed causalities. While Gómez-Chacón explores the possible causalities empirically using a data mining method and Depaepe et al. are explicit that they are not making any "hard causal statements", Blömeke and Kaiser have assumed a direction of causality from motivation to beliefs.

## How the Social Influences the Individual

The main contribution of these six chapters is in the richness of how different social aspects influence student and teacher affect. The importance of the social influences on affect has been long recognized, for example with respect to gender differences (McLeod 1992; Frost et al. 1994) and social norms of the classroom (Cobb et al. 1989).

In their chapters, Depaepe et al., and Forgasz et al. report cross-national differences in the strength of different affective traits, while Ding et al. and Blömeke and Kaiser also report differences in the structure of affect across countries studied. We already mentioned above that in the Ding et al. study the connection between liking mathematics and self-confidence was weaker than in previous study in Italy. The authors discuss the specific features of Chinese culture for learning, where education is the key to mobility and success and schools produce high level learning outcomes although they fail to fulfil features identified in (Western) research as characteristics of good learning environment. In fact, the response pattern where students say that they like mathematics although they have low self-confidence, fits together with previous results where Chinese students in Hong Kong and Macao were found to perceive their competencies (a cognitive trait) to be low and yet their levels of anxiety (emotional trait) were only of medium level (Lee 2009). This pattern is interestingly different from the patterns in Japan and Korea, where perceived competences were even lower than in China and student anxiety was very high (Lee 2009).

In addition to cross-national comparisons, the influence of the teacher is a repeated issue in this section. Most thoroughly the influence of teacher beliefs is discussed in the chapter by Dapaepe et al., who also observe a number of other influencing factors. They reviewed evidence on how certain students' beliefs explain their poor performance in realistic word problems and where these beliefs might originate from. The studies show that students interpret mathematical tasks in the context of schooling, suspending real-life information when it conflicts the "culture and practices of school mathematics". Student beliefs about word problems are influenced by the unrealistic tasks they encounter in textbooks and by the way teachers treat word problems in class. They observe that mathematics classrooms have shifted towards more realistic tasks and that most of the word problems in textbooks are still "stereotyped, easy and non-challenging". Their study exemplifies how their teacher's beliefs and other features of the learning environment can promote student beliefs that are counter to the explicit aims of the curriculum but also how educational initiatives do have an influence on educational practice.

Also Ding et al. mention that some of the observed effects may be influenced by features of the 11 schools in their study. The unusual combinations of high perception of confidence and negative affective relation with mathematics in their study were more frequent in high-achieving schools. Moreover, the response patterns across grade levels varied a lot between the schools, responses being quite uniform in some schools (e.g. in school 18 there were strikingly few students disliking mathematics on all grades) and very varied in some schools (e.g. in school 15 one third of grade 6 students disliked mathematics and on grade 8 nobody disliked mathematics). These suggest that both the school level and the level of classroom/teacher might be influential to student responses.

As teacher's affect seems to be highly relevant for the instructional choices they make, it is important that attention is paid also to teacher affect. Two of the chapters focus on teacher affect, both examining the relationship between teacher motivation and their beliefs about the nature of mathematics (cognitive trait). These two studies exemplify how qualitative and quantitative studies have different strengths.

Eichler and Erens explore teacher motivation and beliefs qualitatively within the framework of intended curriculum. Their study explores the structure of teacher affect with respect to context, the quasi-logical ordering of teachers' beliefs, and also the psychological centrality of different beliefs (c.f. Green 1971). Such a rich and detailed description would not have been possible using a quantitative approach. However, we can not know how generalizable these observations are. On the other hand, Blömeke and Kaiser used a large international data set (TEDS-M) to test the causal relationships from teachers' intrinsic professional motivations to a more dynamic view of mathematics and from their extrinsic professional motivation to more transmission oriented teaching style. Moreover, they chose two Western and two East Asian countries. The choice of four different countries and the statistical power of the large data provide strong evidence for the findings that are similar across all four countries: intrinsic motivation to become a teacher is related to an epistemological belief that mathematics is a dynamically developing science, while extrinsic professional motivation is related to a belief that transmission-oriented teaching style is efficient.

While the social structures of nations, schools, and classrooms are neatly nested, there are also social identities that cut across all these social groups, such as social class, gender and ethnicity. While Depaepe et al. and Ding et al. provide examples for how different features of learning environment promote certain affective traits in students, the study by Forgasz et al. reminds us that these influences are not the same for all students. In the same class, boys and girls are facing different expectations and interpretations by their teachers. Their article discusses also ethnicity in relation to gender and how students are affected by the beliefs of their parents and other members of society. However, none of the chapters in this section has studied the influences of social class, although Ding et al. discuss school socio-economic status as a variable to pay attention in their future analysis.

Although the six chapters in this section address a variety of social influences from textbooks and teachers to gender, ethnicity and nationality, they fail to address the agency of the student and their role in negotiating the social norms of the classroom. Discussion of classroom norms is reduced to norms of teaching. Hannula (2012) discusses the importance of agency, not only in the context of school, but also in all the multiple social roles that students and teachers take as family members, friends, citizens and as members of different social groups.

> Each group and each role requires building interpersonal relations and negotiating about shared norms, values and understandings, i.e., learning in the community of practice (Wenger 1998). For this negotiation, it is not necessary to explicate values and norms. Rather, norms and values become established as participants enact them. In this process of negotiation, both the individual and the social system change (Bandura 1978). Even a passive adaptation to existing rules and norms influences the system, validating the status quo. (Hannula 2012, p. 151)

Mathematics education has already made the social turn (Lerman 2000) and many researchers have taken the strong social position (Lerman 2006) to study the discourse, classroom climate, and other social phenomena emerging in classrooms, schools and more broadly in society. Such studies have observed how school culture and broader socio-cultural situation penetrate to the classroom microculture (e.g. Cobb and Yackel 1996; Partanen 2011), and how the microculture of the classroom may also build resilience against overall educational policy (e.g. Ciani et al. 2010). Although the chapters in this section were rich in their discussion of the social influences of affect, we should also note that the strong social position was not present.

## Conclusions

In the introduction, I framed the relevance of research on mathematics-related affect to consist of two components: to identify the most important aspects of affect, and to identify how to influence them. What I observed in the six chapters, was that it seemed to be more important for the students and teachers what they want (motivational trait) than what they believe to be true (cognitive trait) or what they tend to feel (emotional trait). When the importance of sense making is being emphasized, the teachers shift towards this way of teaching word problems (Depaepe et al.). When students see mathematics to be important for personal future, they continue to like mathematics even if they lose their confidence (Ding et al.). Students' preference for visual or non-visual approaches predicts what kind of problems they will encounter when solving problems in a dynamic geometry environment (Gómez-Chacón). Eichler and Erens discuss the psychological centrality of beliefs explicitly, and in their study these beliefs relate mostly to teachers' instructional goals (motivational trait). Although this is mainly based on personal impression, I would conclude that motivational traits are in the centre of mathematics-related affect.

However, when it comes to changing the affect, the picture becomes much more complex. We see a number of social factors that influence student and teacher affective traits. Teacher beliefs and practices influence student approaches to and beliefs about non-realistic word problems (Depaepe et al.). The student gender, age, school and class all seem to influence how student emotional and cognitive traits develop (Ding et al.). The subject to teach influences the instructional goals of the teacher (Eichler and Erens). And also the country where you are influences how student and teacher affect is developing (Ding et al., Blömeke and Kaiser). However, the big variation in student affect across the grade levels of the same school (Ding et al.) suggests that the classroom level is an important factor in this development. Depaepe et al., Eichler and Erens, Blömeke and Kaiser and Forgasz et al. show that there is great variation between teachers and their teaching across a number of factors. There is temptation to conclude that the different national educational policies are reflected in teacher affect, and they would implement in their classrooms teaching that reflects their beliefs and instructional goals. This would highlight the importance of top-down interventions. However, that would ignore the agency of teachers and students. Although none of the chapters in this section addressed the complex processes of negotiating new norms in the classrooms, this certainly is an issue to be taken into account.

# References

Bandura, A. (1978). The self-system in reciprocal determinism. *American Psychologist, 33*, 344–358.

Ciani, K. D., Middleton, M. J., Summers, J. J., & Sheldon, K. M. (2010). Buffering against performance classroom goal structures: The importance of autonomy support and classroom community. *Contemporary Educational Psychology, 35*, 88–99.

Cobb, P., & Yackel, E. (1996). Constructivist, emergent and sociocultural perspectives in the context of developmental research. *Educational Psychologist, 31*(3/4), 175–190.

Cobb, P., Yackel, E., & Wood, T. (1989). Young children's emotional acts during mathematical problem solving. In D. B. McLeod & V. M. Adams (Eds.), *Affect and mathematical problem solving: A new perspective* (pp. 117–148). New York: Springer.

Deci, E. L., Vallerand, R. J., Pelletier, L. G., & Ryan, R. M. (1991). Motivation and education: The self-determination perspective. *Educational Psychologist, 26*(3–4), 325–346.

Frost, L. A., Hyde, J. S., & Fennema, E. (1994). Gender, mathematics performance, and mathematics related attitudes and affect: A meta-analytic synthesis. *International Journal of Educational Research, 21*(4), 373–385.

Furinghetti, F., & Pehkonen, E. (2002). Rethinking characterizations of beliefs. In G. C. Leder, E. Pehkonen, & G. Törner (Eds.), *Beliefs: A hidden variable in mathematics education* (pp. 39–58). Dordrecht: Kluwer.

Goldin, G. (2002). Affect, meta-affect, and mathematical belief structures. In G. C. Leder, E. Pehkonen, & G. Törner (Eds.), *Beliefs: A hidden variable in mathematics education* (pp. 59–72). Dordrecht: Kluwer.

Goldin, G. (2004). Characteristics of affect as a system of representation. In M. J. Høines, & A. B. Fuglestad (Eds.), *Proceedings of the 28th conference of the IGPME* (Vol. 1., pp. 109–114), Bergen University Collage.

Green, T. F. (1971). *The activities of teaching*. New York: McGraw-Hill.

Hannula, M. S. (2011). *The structure and dynamics of affect in mathematical thinking and learning*. In M. Pytlak, E. Swoboda, & T. Rowland (Eds.), Proceedings of the seventh congress of the European Society for Research in Mathematics Education (pp. 34–60), University of Rzesów.

Hannula, M. S. (2012). Exploring new dimensions of mathematics-related affect: Embodied and social theories. *Research in Mathematics Education, 14*(2), 137–161.

Hart, L. (1989). Describing the affective domain: Saying what we mean. In D. McLeod & V. Adams (Eds.), *Affect and mathematical problem solving* (pp. 37–45). New York: Springer.

Lee, Y. (2009). Universals and specifics of math self-concept, math self-efficacy, and math anxiety across 41 PISA 2003 participating countries. *Learning and Individual Differences, 19*, 355–365.

Lerman, S. (2000). The social turn in mathematics education research. In J. Boaler (Ed.), *Multiple perspectives on mathematics teaching and learning* (pp. 19–44). Westport: Ablex.

Lerman, S. (2006). Cultural psychology, anthropology and sociology: The developing 'strong' social turn. In J. Maasz & W. Schloeglmann (Eds.), *New mathematics education research and practice* (pp. 171–188). Rotterdam: Sense.

Linnenbrink, E. A., & Pintrich, P. R. (2004). Role of affect in cognitive processing in academic contexts. In D. Y. Dai & R. J. Sternberg (Eds.), *Motivation, emotion, and cognition: Integrative perspectives on intellectual functioning and development* (pp. 57–88). Mahwah: Lawrence Erlbaum.

Ma, X. (1999). A meta-analysis of the relationship between anxiety toward mathematics and achievement in mathematics. *Journal for Research in Mathematics Education, 30*, 520–541.

Ma, X., & Kishor, N. (1997a). Assessing the relationship between attitude toward mathematics and achievement in mathematics: A meta-analyses. *Journal for Research in Mathematics Education, 28*(1), 26–47.

Ma, X., & Kishor, N. (1997b). Attitude toward self, social factors, and achievement in mathematics: A meta-analytic review. *Educational Psychology Review, 9*, 89–120.

Ma, X., & Xu, J. (2004). Determining the causal ordering between attitude toward mathematics and achievement in mathematics. *American Journal of Education, 110*(May), 56–280.

McLeod, D. B. (1992). Research on affect in mathematics education: A reconceptualization. In D. A. Grouws (Ed.), *Handbook of research on mathematics learning and teaching* (pp. 575–596). New York: Macmillan.

Minato, S., & Kamada, T. (1996). Results on research studies on causal predominance between achievement and attitude in junior high school mathematics of Japan. *Journal for Research in Mathematics Education, 27*, 96–99.

Partanen, A. M. (2011). *Challenging the school mathematics culture: An investigative small-group approach: Ethnographic teacher research on social and sociomathematical norms* (Acta Universitatis Lapponiensis 206), University of Lapland, Rovaniemi, Finland.

Pekrun, R., Elliot, A. J., & Maier, M. A. (2006). Achievement goals and discrete achievement emotions: A theoretical model and prospective test. *Journal of Educational Psychology, 98*(3), 583–597.

Power, M., & Dalgleish, T. (1997). *Cognition and emotion: From order to disorder*. Hove: Psychology Press.

Williams, T., & Williams, K. (2010). Self-efficacy and performance in mathematics: Reciprocal determinism in 33 nations. *Journal of Educational Psychology, 102*(2), 453–466. doi:10.1037/a0017271.

Zan, R., & Di Martino, P. (2007). Attitude toward mathematics: Overcoming the positive/negative dichotomy. *The Montana Mathematics Enthusiast, Monograph, 3*, 157–168.

# Part III
# Methodological Issues in Affect Research

The third part of the book focusses on the methodological aspects of affect research. It raises the question of how to study beliefs and affect of teachers and students, and how affect is linked to knowledge and self-efficacy. The methodological discussion encompasses the different research paradigms of quantitative and qualitative approaches, and reflects how the constructs of beliefs and affect require diverse operationalizations with respect to the research questions.

# Analyzing Data and Drawing Conclusion on Teachers' Beliefs

**Qian Chen and Frederick Koon Shing Leung**

**Abstract** This chapter discusses two major methodological issues in studying teachers' beliefs. The two issues are analyzing data on teachers' beliefs and drawing conclusion on teachers' beliefs. Furthermore, the authors use a cross-cultural study of teachers' mathematics beliefs to illustrate the two issues. Suggestions are provided at the end of this chapter.

**Keywords** Teacher beliefs • Mathematics • Methodological issues • Curriculum reform • China

## Introduction

Over the past two decades, affect has become one of most popular research areas in the field of mathematics education. The driving force behind research on affect is the conviction that emotional and cognitive aspects deeply interact, and are thus important factors in mathematics learning processes (Martino and Zan 2011). McLeod (1992) divided the affective domain into three subdomains: emotions, attitudes, and beliefs; and regarded them as ranged along a dimension of increasing stability and decreasing intensity, with emotions as most intense/least stable, beliefs as most stable/least intense, and attitudes in between.

Zan et al. (2006) suggest that two different foci apparent in 1960s and 1970s mathematics education research on affect are 'mathematics anxiety' (emotion) and 'attitude toward mathematics'. It is only until 1980s that 'mathematics beliefs' have aroused attention from the researchers (Dobson and Dobson 1983; Munby 1982; Thompson 1982). Researchers' interest in beliefs arises from a conviction about the existence of a

Q. Chen (✉)
School of Mathematics and Software Science, Sichuan Normal University, Chengdu, China
e-mail: chenqianhku@gmail.com

F.K.S. Leung
Faculty of Education, The University of Hong Kong, Hong Kong, SAR, China
e-mail: frederickleung@hku.hk



B. Pepin, B. Roesken-Winter (eds.), *From beliefs to dynamic affect systems in mathematics education*, Advances in Mathematics Education,
DOI 10.1007/978-3-319-06808-4_14

relationship between beliefs and behavior (Martino and Zan 2011). Since Thompson's (1992) holistic review of research on teachers' beliefs and conceptions, belief research within mathematics education have gained remarkable development, as exemplified by the publication of the seminal work *Beliefs: A hidden variable in mathematics education* (Leder et al. 2002). However, "belief research is notorious for its conceptual and methodological problems" (Skott 2013, p. 548). As an attempt to overcome these problems, this chapter is largely concerned with the methodological issues in studying teachers' beliefs. Theoretical discussions about the properties of beliefs are beyond the scope of this chapter, and can be found elsewhere (e.g. Green 1971; Rokeach 1968). In addition, given that classic methodological issues like defining the concept 'belief' and measuring beliefs have been expatiated by many researchers (e.g. Leder and Forgasz 2002; Pajares 1992; Speer 2005; Thompson 1992), this chapter will mainly talk about two underexplored methodological issues: analyzing data on teachers' beliefs and drawing conclusion on teachers' beliefs. Particularly, a cross-cultural study of teachers' mathematics beliefs conducted by the first author of this chapter will be used to illustrate the two issues.

## Two Methodological Issues

In this section, the two above-mentioned issues are discussed separately for the sake of clarity. However, it should be noted that they are actually related to each other. It is not hard to understand that in a study of teachers' beliefs, how to analyze data on teachers' beliefs will have a direct impact on the conclusion to be drawn for the study.

### *Analyzing Data on Teachers' Beliefs*

In the broad research literature on teachers' beliefs, there seems to be a correspondence between the way of measuring beliefs and the way of analyzing the belief data. Roughly speaking, data collected through quantitative approaches (e.g. Likert scale survey) tend to be analyzed in quantitative or statistical manner while data elicited from qualitative approaches (e.g. interview, observation) usually go through qualitative thematic analysis, typically following the basic principles of grounded theory approach (Strauss and Corbin 1998). Nevertheless, we must stress here that although data analysis methods are affected or constrained by data collection methods to a certain extent, decisions regarding data analysis methods are actually more dependent on the specific research purpose and research questions in studies. That is, regardless of how belief data are collected, both quantitative and qualitative methods can be used to analyze the data, as long as the chosen methods can serve the research purpose and help address the research questions in studies. In fact, in many studies, especially those using a mixed methodology, quantitative and qualitative analysis are well integrated so as to answer the research questions.

Quantitative analysis of belief data is perhaps relatively straightforward because the type of research questions to be answered is sort of pre-determined or predictable. For example, Wilkins (2008) investigated 481 in-service elementary teachers' level of mathematical content knowledge, attitudes toward mathematics, beliefs about the effectiveness of inquiry-based instruction, use of inquiry-based instruction and modeled the relationship among these variables through path analysis. One major finding from his study was that teachers' beliefs had the strongest effect on teachers' practice.

As compared to quantitative analysis, qualitative analysis of belief data tends to be more complex and demand substantial judgment by researchers. Very often, operable analytical schemes are essential to such analysis, because they enable effective organization, presentation and understanding of the data. With the help of analytical schemes, data from different sources can be analyzed in a consistent and comparable way, which can boost the validity of research. Take Cross's (2009) case study as an example, where the relationship between beliefs and classroom practices of five high school mathematics teachers was investigated, the process of analyzing data from semi-structured interviews and classroom observation is described as below:

> Thematic analysis was employed for analysis of the data. Specifically, using Strauss and Corbin's (1990) open coding technique, the participants' narratives from the transcribed interviews were examined for statements relevant to their beliefs about mathematics. From the open coding, I observed certain patterns among the codes from which categories were developed. The development of categories and refining of the categories was an ongoing iterative process that was repeatedly re-evaluated to ensure they reflected the participant's descriptions of their experiences. Each transcript was read multiple times to verify that for the codes and categories developed the "empirical reality fit the emerging theoretical framework" (Charmaz 2000, p. 514). These themes will be described illustrating how the teachers conceptualized and talked about their mathematics-related beliefs. The field notes from the classroom observations and lesson plans were analyzed and placed in categories that described the teachers' practices in three areas: (a) organizing the classroom environment, (b) role in teacher–student and student–student discourse and interactions, and (c) types and use of assessments. Within these categories, the coding scheme developed from the interviews was applied. Descriptions of particular classroom behaviors and practices that were reflective of these beliefs are also discussed. (Cross 2009, p. 331)

## *Drawing Conclusion on Teachers' Beliefs*

Drawing conclusion on belief is not simple, especially for researchers who adopt a mixed methodology in their studies. Often, researchers employ specific frameworks to characterize individual teachers' beliefs and differentiate one type of beliefs from another. Such frameworks can either be conceptualized from certain theories, or emerge from empirical data. However, even if the same specific framework is used in a particular study, the analysis of belief data from different sources may result in quite different findings. In this case, drawing a conclusion on what type of beliefs individual teacher actually holds becomes a challenging task.

Some explanations for the disparities in findings have been suggested from theoretical and methodological perspectives (Beswick 2003, 2005, 2007; Lerman 2001; Speer 2005; Thompson 1992), although they are intended to explain inconsistencies between teachers' beliefs and practice. For example, Speer (2005) argues from a theoretical perspective that reported discrepancies between the beliefs that teachers profess and those that are inferred from their behaviors are likely to be artifacts of the research methods employed. She maintains that all beliefs are to some degree inferred (from either the teacher's words or behaviors) and thus the dichotomy is false. Apparent conflicts can also be caused by a lack of shared understanding of the same words between teachers and researchers.

Another important reason is concerned with the contextual nature of beliefs. Beswick (2003, 2005, 2007) argues that the apparent inconsistencies between beliefs and practice may be due to the mismatches between aspects of the contexts in which teachers provided data concerning their beliefs, and the contexts in which their practices were observed or described. Teachers' beliefs and practices are contextualized, as suggested by Lerman (2001):

> Teachers' actions and utterances, that is to say their classroom practices and the response they give to interviews, questionnaires, or other research methods, are contextualized and cannot be interpreted outside of a consideration of the social situation (p. 45).

Thus, consistency cannot be expected when the contexts in which the teachers' beliefs are considered and their practices observed are not closely matched (Beswick 2003). As noted by Beswick (2005), "it is unreasonable to expect consistency between broad collections of beliefs that are not closely linked with a specific context, and practice that is not described in equally broad, contextually independent terms" (p. 42). Agreeing with Beswick on the context-dependent nature of beliefs, the authors of this chapter further hypothesize that the discrepancies in findings regarding teachers' beliefs may be due to the potential influence of larger organizational, societal or cultural contexts. For example, in the context of curriculum reform, teachers within a 'collectivist' culture may profess more reform-oriented beliefs than those within an 'individualist' culture, because the former tend to report what the society expects rather than their own true thoughts. On the other hand, when teachers' enacted beliefs (beliefs as reflected in classroom practices) are examined, traditional approaches may be identified in both kinds of culture as teaching is more affected by individual's ideas and values.

To sum up, for whatever reasons, the discrepancies among findings concerning teachers' beliefs as elicited in multiple ways deserve due attention. We argue that when drawing a conclusion on teachers' beliefs, it is important for researchers to report these discrepancies so that a more comprehensive understanding of teachers' beliefs can be reached.

## An Illustrative Study

In this section, Chen's (2010) study is used to illustrate the two issues discussed above. Before that, important background information of her study is provided. It is worth noting that apart from the two issues, Chen's (2010) study considered some classic methodological issues.

### *Background of Chen's (2010) Study*

At the turn of twenty-first century, there was a wave of mathematics curriculum reform around the world. Research literature shows that for curriculum reform to be successful, teachers must change their beliefs in a reform-oriented way (Battista 1994; Wilson and Goldenberg 1998; Xu 2003). Chen (2010) argued that the level of consistency between teachers' mathematics beliefs and the underlying philosophy of the reform-oriented curriculum can be an important, albeit not the only, indicator of the success of mathematics curriculum reform. She further investigated the level of consistency between junior secondary school teachers' mathematics beliefs and the underlying philosophy of the reform-oriented curriculum, and then explored the factors influencing the level of consistency in two different Chinese cities: Hong Kong and Chongqing (in mainland China). Given that the two cities have some significant similarities as well as differences, especially in terms of culture, it was expected that a comparative study could allow the role of culture in changing teachers' mathematics beliefs to be revealed.

In her study, Chen (2010) adopted Raymond's (1997) definition of 'mathematics beliefs' as "personal judgments about mathematics formulated from experiences in mathematics, including beliefs about the nature of mathematics, learning mathematics, and teaching mathematics" (p. 552). Therefore, teacher's mathematics beliefs consisted of three dimensions: beliefs about the nature of mathematics, beliefs about learning mathematics, and beliefs about teaching mathematics. In line with this definition, a uniform theoretical framework (Table 1) was proposed based on the influential work of Ernest (1989), Kuhs and Ball (1986) so as to characterize the teachers' mathematics beliefs (espoused as well as enacted) and the underlying philosophy of the reform-oriented mathematics curriculum. In doing so, the level of consistency between the two aspects was made obvious.

According to Ernest (1989), there are three different conceptions of the nature of mathematics. Firstly, the Platonist view regards mathematics as a static but unified body of certain knowledge. Mathematics is discovered, not created. Secondly, the instrumentalist view describes mathematics as an accumulation of facts, rules and skills to be used in the pursuance of some external end. Thus mathematics is a set of unrelated but utilitarian rules and facts. Thirdly, the problem solving (or social constructivist) view takes mathematics as a dynamic, continually expanding field of

**Table 1** Theoretical framework for characterizing the underlying philosophy of the reform-oriented mathematics curriculum and teachers' mathematics beliefs

| View of the nature of mathematics (Ernest 1989) | View of learning mathematics (Ernest 1989) | View of teaching mathematics (Kuhs and Ball 1986) |
|---|---|---|
| Platonism (absolutism) | Reception view, i.e. learning as reception of knowledge (behaviorism) | Content-focused with an emphasis on conceptual understanding (teacher-centered) |
| Instrumentalism (absolutism) | Learning as mastery of skills (behaviorism) | Content-focused with an emphasis on performance (teacher-centered) |
| Problem solving, or social constructivism (fallibilism) | (social) constructivist view, i.e. learning as active (social) construction of understanding (constructivism or social constructivism) | Learner-focused (student-centered) |

human creation and invention, a cultural product. Mathematics is a process of inquiry and coming to know, not a finished product, for its results remain open to revision. The former two views lie within the domain of absolutism while the latter one within the domain of fallibilism (Roulet 1998). Associated with the three views of mathematics respectively are three views of learning mathematics (Ernest 1989) and three views of teaching mathematics (Kuhs and Ball 1986).

As far as research methodology was concerned, Chen (2010) employed a mixed-method approach involving the use of both quantitative and qualitative methods. Specifically, data about teachers' mathematics beliefs were collected through a combination of questionnaire survey, videotaped classroom observation and semi-structured interview. Teachers' espoused mathematics beliefs were inferred from their self-report in the questionnaires and interviews, and their enacted mathematics beliefs were inferred based on their videotaped classroom teaching practices. Particularly, two key research instruments were developed for use. One was the teacher questionnaire, and the other was the semi-structured interview schedule.

The teacher questionnaire was composed of two parts: (1) Collier's (1972) two Scales, i.e. Beliefs About Mathematics Scale (BAMS) and Beliefs About Mathematics Instruction Scale (BAMIS), aimed at measuring individual teacher's beliefs about mathematics and its teaching and learning; (2) personal particulars, aimed at recording the demographic characteristics of individual teacher, including gender, educational level, specialty, mathematics teaching experience, new mathematics curriculum experience, medium of instruction, and number of times of attending in-service training on the new curriculum etc. BAMS and BAMIS were initially developed by Collier (1969) to understand beliefs that would help teachers "shift from an authoritarian, teacher-dominated classroom, to a child-centered classroom" and also "shift from a program emphasizing formal mathematical content to a program emphasizing the creative, investigative nature of mathematics". Seaman et al. (2005) regarded the two Scales as a reasonable measure of constructivist philosophy and ideas about instruction that follow from that philosophy. Thus, the two Scales were considered as suitable to measure teachers' beliefs about mathematics and mathematics instruction and thereby to a certain degree examine the level of consistency between their mathematics beliefs and the underlying philosophy of the new curriculum in Hong Kong and Chongqing.

BAMS contained 20 items that were devoted to distinguishing "the degree to which an individual viewed mathematics as (a) including elements of originality and creativity and characterized by the existence of choices as opposed to the view (b) that mathematics is based on fixed, established forms and requires scrupulous adherence to rule" (Collier 1969, p. 1). For example, on one of the items in this scale, students were asked to rate their strength of agreement with the statement, "The laws and rules of mathematics severely limit the manner in which problems can be solved". BAMIS contained 20 items that were designed to "give an indication of the degree to which individuals viewed the mathematics teacher as one who (a) encourages self-discovery and independence from memorized rules, as opposed to the view of the mathematics teacher as one who (b) defines and explains procedures for students" (Collier 1969, p. 1). On this scale, respondents were asked to rate their strength of agreement with statements such as, "all students should be required to memorize the procedures that the text uses to solve problems".

All the 40 items had 6-point Likert scale response options (from strongly disagree to strongly agree). Half of the items were phrased in a positive manner (advocating informal/constructivist ideas) and half in negative manner (advocating formal approaches to mathematics). Positive items received the scale value checked as their score, but negative items received 7 minus the scale value checked as their score. The possible range on each scale was 20 (achieved if a respondent scores a 1 on each of 20 items on that scale) to 120 (achieved if a respondent scores a 6 on each of 20 items on that scale), with 70 being a neutral score. Collier (1972) described both scales as lying along a formal-informal dimension. A score higher than 70 was in the "informal" direction and a score less than 70 was in the "formal" direction.

Chen (2010) noted that Collier's two Scales were not specially developed for her study, thus they could only provide a rough measure of teachers' mathematics beliefs and the level of consistency between their beliefs and the underlying philosophy of the reform-oriented mathematics curriculum. Specifically, in both Scales, if individual teacher's beliefs were in the "informal" direction, then his or her beliefs were generally considered as being consistent with the underlying philosophy of the reform-oriented curriculum, but to varying extent. The closer his or her score was to 120, the higher the level of consistency was. On the other hand, if individual teacher's beliefs were in the "formal" direction, then his or her beliefs were generally considered as being inconsistent with the underlying philosophy of the reform-oriented curriculum, but to varying extent. The closer his or her score was to 20, the lower the level of consistency was.

The semi-structured interview schedule was framed based on an extensive review of literature regarding curriculum reform, mathematics beliefs and belief (conceptual) change. All interview questions were mainly divided into three types: (1) mathematics beliefs; (2) factors influencing teachers' mathematics beliefs or belief

**Table 2** Categorization of interview questions

| Question type | Sample questions |
|---|---|
| Mathematics beliefs | What is mathematics all about? |
| | How should students learn mathematics? |
| | How should teachers teach mathematics? |
| Factors influencing teachers' mathematics beliefs or belief change | Do you think your beliefs about teaching mathematics have changed over years? |
| | If yes, what factors do you think promote your belief change? |
| | What factors do you think inhibit your belief change? |
| Teachers' opinions about the new mathematics curriculum and its implementation | What are the underlying ideas of the new mathematics curriculum? |
| | What are the differences between them and those of the old curriculum? |
| | Do you agree with these ideas? Why? |
| | Is your current teaching approach different from the reform-oriented approach? If yes, what are the differences? |

change; (3) teachers' opinions about the new curriculum and its implementation. The first type of questions were aimed at eliciting the teachers' mathematics beliefs, including beliefs about the nature of mathematics, beliefs about learning mathematics, and beliefs about teaching mathematics; the second type of questions were aimed at exploring the factors influencing the level of consistency between teachers' mathematics beliefs and the underlying philosophy of the reform-oriented curriculum; the last type of questions were aimed at inferring the extent to which the conditions for teachers' belief change were met. For each type of questions, the sample questions are displayed in Table 2.

In Chen's (2010) study, a total of 113 and 114 junior secondary school mathematics teachers in Hong Kong and Chongqing respectively participated in the questionnaire survey, and three case teachers were selected from the survey sample in each place for subsequent research. The case teachers were required to arrange their schedule for the data collection, including videotaped classroom observation and interview. It was basically up to the teachers themselves to select the topic, time and venue. Nevertheless, two basic requirements were conveyed to them. Firstly, the lesson(s) to be observed should deal with a completely new mathematical topic, instead of being review or exercise-oriented. It was believed that through analyzing how a teacher developed teaching around a new topic, his or her beliefs were easier to be captured. Secondly, the interviews need to be carried out as soon as possible after the classroom observation. It was expected that both the teacher and the researcher could have a vivid memory of the videotaped lesson so that they could make points based on the events just happened in the classroom teaching, if necessary.

For each case teacher in the two places, one lesson which dealt with a completely new topic was videotaped for analysis. The topics were not the same due to their different schedules. Before the observation, the teachers were required to submit a set of teaching materials concerning the topic, including textbook pages, worksheets if any. During the process of classroom observation, a digital video camera focusing on teacher was located in the back of classroom to record the teacher's teaching practice. After classroom observation, each case teacher was interviewed within the same day. The semi-structured interview schedule was used as major instrument for data collection. In addition, the case teachers' views inferred from their responses in the questionnaires were checked during the interviews. Some interesting points that arose during the interviews were also probed into further. All interviews were audio-taped with the permission from participants.

## *Analyzing Data on Teachers' Mathematics Beliefs*

In Chen's (2010) study, both quantitative and qualitative techniques were used for data analysis. Specifically, the data from the questionnaire survey were analyzed in quantitative manner. In addition to descriptive statistics, Pearson correlation tests were used to examine the relationship between beliefs about mathematics and beliefs about mathematics instruction, and independent-samples T-test was used to find the difference between Hong Kong and Chongqing teachers in terms of both kinds of beliefs.

After the classroom teaching videotaping was done, the videos were transferred from the digital camera to the computer. All the videos were transcribed verbatim for analysis. During the process of data analysis, the transcripts composed the major data, although the videos were also referred back to from time to time to ensure that the description represented the reality as closely as possible. All videos were analyzed by the researcher herself. The teaching suggestions in the key curriculum documents for both Hong Kong and mainland China were used as important reference for data analysis, but they were too general. In order to analyze the teachers' behavior and infer their enacted beliefs, a more operable scheme (Table 3) was developed based on the curriculum documents and relevant literature (Artzt and Armour-Thomas 2002; Henningsen and Stein 1997; Hiebert et al. 1997; Stein et al. 1996; Stein and Smith 1998).

Furthermore, in analyzing the teaching videos, both quantitative and qualitative measures, as summarized in Table 4, were used in the hope that the analysis could best reflect the classroom reality in a meaningful way. Finally, the case teachers' enacted mathematics beliefs were inferred and then characterized in reference to the theoretical framework (Table 1).

Like the videos, all audio-taped data from the interviews were transcribed verbatim for the subsequent analysis. Following the principles of thematic analysis, the transcribed data were organized and sorted according to three themes: (1) mathematics beliefs; (2) the factors influencing the level of consistency between teachers' mathe-

**Table 3** Lesson dimensions and dimension Indicators of a reform-oriented mathematics classroom

| Dimension | Indicator | Description of dimension indicators |
|---|---|---|
| Mathematical tasks | Task features | *Contexts*: The tasks have "real-life" contexts. |
| | | *Solution strategies*: The tasks are solved in multiple ways. |
| | | *Representations*: The tasks include the use of multiple representations, e.g. words, diagrams, manipulatives, computers, or calculators. |
| | | *Communication*: The tasks enable students to produce mathematical explanations or justifications. |
| | | *Collaboration*: The tasks enable students work in a collaborative way. |
| | Cognitive demand | The tasks demand the students to engage in high-level cognitive process—either the active "doing of mathematics" or the use of procedures with connection to concepts, meaning, or understanding. |
| Learning environment | Social and intellectual climate | The teacher establishes and maintains a positive rapport with and among students by showing respect for and valuing students' ideas and ways of thinking. The teacher does not play a role of authority of knowledge. |
| | Modes of instruction | The teacher uses instructional strategies that encourage and support student involvement as well as facilitate goal attainment. He or she provides time necessary for students to express themselves and explore mathematical ideas and problems. |
| Classroom discourse | Teacher-student interactions | The teacher communicates with students in a nonjudgmental manner and encourages the participation of each student. He or she requires students to give full explanations and justifications or demonstrations orally and/or in writing. He or she listens carefully to students' ideas and makes appropriate decisions regarding when to offer information, provide clarification, model, lead, and let students grapple with difficulties. |
| | Student-student interactions | The teacher encourages students to listen to, respond to, and question each other so that they can evaluate and, if necessary, discard or revise ideas and take full responsibility for arriving at mathematical conjectures and/or conclusions. |
| | Questioning | The teacher poses a variety of levels of questions that elicit, engage, and challenge students' thinking, and the students give a variety of types of responses. |

matics beliefs and the underlying philosophy of the reform-oriented curriculum; and (3) teachers' opinions about the new curriculum and its implementation. Each of the three themes was further divided into several sub-themes. For example, the data under the first theme were further divided into three sub-themes: (i) belief about the nature of mathematics; (ii) belief about learning mathematics; and (iii) belief about teaching mathematics. The interview data were triangulated with the teachers' responses in the questionnaire survey. Finally, the teachers' espoused mathematics beliefs were characterized in reference to the theoretical framework (Table 1).

**Table 4** Summary of quantitative and qualitative measures

| Quantitative measures | Qualitative measures |
|---|---|
| Duration of whole lesson and teaching time (in minutes) | Descriptions of social and intellectual climate |
| Number of mathematical tasks that have each task feature | Description of the role of teacher |
| Number of mathematical tasks that involve each kind of cognitive demand | Description of the role of students |
| Number of different kinds of teaching strategies | Descriptions of the teacher-student interactions |
| Percentage and duration of teaching time devoted to different teaching strategies | Descriptions of the student-student interactions |
| Number of times that teacher requires students to give full explanation or justification | |
| Number of times that teacher requires students to memorize mathematical fact, rule, formula or procedures | |
| Number of times that teacher requires students to imitate the procedures suggested by the teacher or the textbook | |
| Number of times that teacher encourage student-student interaction | |
| Number and percentage of different levels of teacher questions | |
| Number and percentage of different types of student responses | |

Roulet (1998) indicates that although mathematics teachers may not describe their personal views of the nature of mathematics in terms of the Platonist, instrumentalist, or social constructivist (problem-solving) positions, such categories may be employed in the analysis of teachers' beliefs about the subject. The same can be said about their views of learning and teaching mathematics. Besides, although individual teacher may simultaneously hold more than one kinds of view of the nature of mathematics, or view of learning mathematics, or view of teaching mathematics (Thompson 1992), his or her views could be approximately characterized as one particular kind according to their dominant inclination. For instance, Chongqing teacher Anna (pseudonym) expressed the following views of mathematics:

> I like mathematics, particularly its rigorous thinking and reasoning … It is good that in mathematics, the result is fixed and does not change as human will changes. One is one, two is two. I like it because it is objective and fair-minded…. Mathematical knowledge is related to each other … Mathematics contains a lot of rules and formulas, stresses use of them for computations and problem solving….

Anna admitted that rules and formulas are essential parts of mathematics, which seems similar to Ernest's instrumentalist view. On the other hand, her statements clearly emphasized the fixedness, certainty and objectivity of mathematical knowledge. She also recognized the connection among mathematical knowledge. Thus, overall, her espoused beliefs about the nature of mathematics inclined towards Ernest's Platonist view.

### *Drawing Conclusion on Teachers' Mathematics Beliefs*

In Chen's (2010) study, it was found that on one hand, as indicated by the questionnaire survey in Hong Kong and Chongqing, the majority of the subjects held informal beliefs about mathematics and mathematics instruction, which seemed to show a relatively high level of consistency between teachers' mathematics beliefs and the underlying philosophy of the reform-oriented curriculum; on the other hand, as indicated by the case studies (classroom observation and interview), most case teachers' mathematics beliefs in both contexts were close to the traditional views, i.e. the Platonist view of the nature of mathematics, the reception view of learning mathematics, and the teacher-centered view of teaching mathematics, which revealed a low level of consistency. Furthermore, it was revealed that the low level of consistency was because almost all factors failed to contribute to the realization of the conditions for conceptual change. Particularly, the role of culture was brought to attention because of its relevance in accounting for some similarities and differences identified between the two places. Based on these findings, Chen (2010) concluded that in the context of curriculum reform, teachers' mathematics beliefs in both Hong Kong and Chongqing are difficult to change in a reform-oriented way unless the conditions for change can be met, and the role of culture in effecting conceptual change cannot and should not be overlooked.

It is noteworthy that in Chen's (2010) study, the disparities between the findings about teachers' mathematics beliefs as elicited in multiple ways were explicitly reported, which enabled a comprehensive and profound understanding of the teachers' mathematics beliefs, and a valid judgment about the level of consistency between their beliefs and the underlying philosophy of the new curriculum.

## Concluding Remarks

In previous sections, we discussed two major methodological issues in studying teachers' beliefs, and also used Chen's (2010) study for illustration purpose. At the end of this chapter, we would like to highlight two suggestions. Firstly, belief researchers should make rational decisions about data analysis methods for their studies. Quantitative, qualitative or both methods can be utilized as long as they serve the research purpose and help address the research questions. Secondly, when drawing conclusion on teachers' beliefs, the disparities between findings about teachers' beliefs as elicited in multiple ways, if any, should be recognized and reported so that a comprehensive understanding of teachers' beliefs can be reached. It is hoped that future belief or affect researchers can benefit from our work, particularly in terms of research methodology. It is believed that with methodological issues taken into full account, the quality of belief or affect research can be enhanced substantially.

## References

Artzt, A. F., & Armour-Thomas, E. (2002). *Becoming a reflective mathematics teacher: A guide for observations and self-assessment*. Mahwah: Lawrence Erlbaum Associates.

Battista, M. T. (1994). Teacher beliefs and the reform movement in mathematics education. *Phi Delta Kappan, 75*(6), 462–463, 466–470.

Beswick, K. (2003). Accounting for the contextual nature of teachers' beliefs in considering their relationship to practice. In L. Bragg, C. Campbell, G. Herbert, & J. Mousley (Eds.), *Mathematics education research: Innovation, networking, opportunity (Proceedings of the 26th annual conference of the mathematics education research group of Australasia* (Vol. 1, pp. 152–159). Melbourne: Deakin University.

Beswick, K. (2005). The beliefs/practice connection in broadly defined contexts. *Mathematics Education Research Journal, 17*(2), 39–68.

Beswick, K. (2007). Teachers' beliefs that matter in secondary mathematics classroom. *Educational Studies in Mathematics, 65*(1), 95–120.

Charmaz, K. (2000). Grounded theory: Objectivist and constructivist methods. In N. Denzin & Y. Lincoln (Eds.), *Handbook of qualitative research* (Vol. 2, pp. 509–535). Thousand Oaks: Sage.

Chen, Q. (2010). *Teachers' beliefs and mathematics curriculum reform: A comparative study of Hong Kong and Chongqing*. Unpublished doctoral dissertation, The University of Hong Kong, Hong Kong.

Collier, C. P. (1969). *The formal-informal dimensions of attitudes toward mathematics and mathematics instruction of prospective elementary teachers*. Wisconsin: University of Wisconsin.

Collier, C. P. (1972). Prospective elementary teachers' intensity and ambivalence of beliefs about mathematics and mathematics instruction. *Journal for Research in Mathematics Education, 3*(3), 155–163.

Cross, D. I. (2009). Alignment, cohesion, and change: Examining mathematics teachers' belief structures and their influence on instructional practices. *Journal of Mathematics Teacher Education, 12*(5), 325–346.

Dobson, R. L., & Dobson, J. E. (1983). *The congruency of teachers' beliefs and practices*. Oklahoma State University.

Ernest, P. (1989). The impact of beliefs on the teaching of mathematics. In P. Ernest (Ed.), *Mathematics teaching: The state of the art* (pp. 249–254). London: Falmer Press.

Green, T. F. (1971). *The activities of teaching*. London: McGraw Hill.

Henningsen, M., & Stein, M. K. (1997). Mathematical tasks and student cognition: Classroom-based factors that support and inhibit high-level mathematical thinking and reasoning. *Journal for Research in Mathematics Education, 28*(5), 524–549.

Hiebert, J., Carpenter, T. P., Fennema, E., Fuson, K. C., Wearne, D., Murray, H., et al. (1997). *Making sense: Teaching and learning mathematics with understanding*. Portsmouth: Heinemann.

Kuhs, T. M., & Ball, D. L. (1986). *Approaches to teaching mathematics: Mapping the domains of knowledge, skills and disposition*. East Lansing: Center on Teacher Education, Michigan State University.

Leder, G. C., & Forgasz, H. J. (2002). Measuring mathematical beliefs and their impact on the learning of mathematics: A new approach. In G. C. Leder, E. Pehkonen, & G. Torner (Eds.), *Beliefs: A hidden variable in mathematics education?* (pp. 95–113). Dordrecht/Boston/London: Kluwer Academic Publishers.

Leder, G. C., Pehkonen, E., & Torner, G. (2002). *Beliefs: A hidden variable in mathematics education*. Dordrecht/Boston/London: Kluwer Academic Publishers.

Lerman, S. (2001). A review of research perspectives on mathematics teacher education. In F.-L. Lin & T. J. Cooney (Eds.), *Making sense of mathematics teacher education* (pp. 33–52). Dordrecht/Boston/London: Kluwer Academic Publishers.

Martino, P., & Zan, R. (2011). Attitude towards mathematics: A bridge between beliefs and emotions. *ZDM, 43*(4), 471–482.

McLeod, D. B. (1992). Research on affect in mathematics education: Areconceptualization. In D. A. Grouws (Ed.), *Handbook of research on mathematics learning and teaching* (pp. 575–596). New York: Macmillan.

Munby, H. (1982). The place of teachers' beliefs in research on teacher thinking and decision making, and an alternative methodology. *Instructional Science, 11*, 201–225.

Pajares, M. F. (1992). Teachers' beliefs and educational research: Cleaning up a messy construct. *Review of Educational Research, 62*(3), 307–332.

Raymond, A. M. (1997). Inconsistency between a beginning elementary school teacher's mathematics beliefs and teaching practice. *Journal for Research in Mathematics Education, 28*(5), 550–576.

Rokeach, M. (1968). *Beliefs, attitudes, and values: A theory of organization and change*. San Francisco: Jossey-Bass.

Roulet, R. G. (1998). *Exemplary mathematics teachers: Subject conceptions and instructional practices*. Unpublished doctoral dissertation, University of Toronto.

Seaman, C. E., Szydlik, J. E., Szydlik, S. D., & Beam, J. E. (2005). A comparison of preservice elementary teachers' beliefs about mathematics and teaching mathematics: 1968 and 1998. *School Science and Mathematics, 105*(4), 197–210.

Skott, J. (2013). Understanding the role of the teacher in emerging classroom practices: Searching for patterns of participation. *ZDM, 45*(4), 547–559.

Speer, N. (2005). Issues of methods and theory in the study of mathematics teachers' professed and attributed beliefs. *Educational Studies in Mathematics, 58*(3), 361–391.

Stein, M. K., & Smith, M. S. (1998). Mathematical tasks as a framework for reflection: From research to practice. *Mathematics Teaching in the Middle School, 3*(4), 268–275.

Stein, M. K., Grover, B. W., & Henningsen, M. (1996). Building student capacity for mathematical thinking and reasoning: An analysis of mathematical tasks used in reform classrooms. *American Educational Research Journal, 33*(2), 455–488.

Strauss, A., & Corbin, J. (1990). *Basics of qualitative research: Grounded theory procedures & techniques*. Newbury Park: Sage.

Strauss, A., & Corbin, J. (1998). *Basics of qualitative research*. Thousand Oaks/London/New Delhi: Sage.

Thompson, A. G. (1982). *Teachers' conceptions of mathematics and mathematics teaching: Three case studies*. Unpublished doctoral dissertation, University of Georgia.

Thompson, A. G. (1992). Teachers' beliefs and conceptions: A synthesis of the research. In D. A. Grouws (Ed.), *Handbook of research on mathematics teaching and learning* (pp. 127–146). New York: Macmillan.

Wilkins, J. L. M. (2008). The relationship among elementary teachers' content knowledge, attitudes, beliefs, and practices. *Journal of Mathematics Teacher Education, 11*(2), 139–164.

Wilson, S. M., & Goldenberg, M. P. (1998). Some conceptions are difficult to change: One middle school mathematics teacher' struggle. *Journal of Mathematics Teacher Education, 1*, 269–293.

Xu, Y. (2003). Kecheng gaige yu shuxue jiaoshi de guannian genxin [Curriculum reform and conception renewal of mathematics teachers]. *Journal of Zhejiang Normal University (Natural Science), 26*(1), 93–96.

Zan, R., Brown, L., Evans, J., & Hannula, M. S. (2006). Affect in mathematics education: An introduction. [Article]. *Educational Studies in Mathematics, 63*(2), 113–121.

# PCK and the Awareness of Affective Aspects Reflected in Teachers' Views About Learning Opportunities – A Conflict?

**Sebastian Kuntze and Anika Dreher**

**Abstract** When teachers design learning opportunities, reflect on instructional situations or when they act and react in the classroom, they are likely to draw on their professional knowledge, including their epistemological beliefs and instruction-related views. Among these are also views related to motivational and affective aspects of learning and instruction. However, the awareness of affective and motivational aspects should be equilibrated with other PCK components. Consequently this chapter aims to explore how the awareness of affective aspects is related to other PCK, and in particular, what emphasis teachers give to aspects of motivation and affect as criteria for evaluating learning and instruction in relation to other relevant aspects for instructional quality and how important they see affective characteristics of representations in tasks. We report results from three empirical studies and discuss their qualitative and quantitative methodologies.

**Keywords** Pedagogical content knowledge • Instruction-related views • Awareness of affective aspects • In-service teachers • Pre-service teachers

## Introduction

Does a 'good' mathematics classroom in the eyes of teachers mean that students should above all have fun with mathematics? Affective variables have certainly shown to play a role in learning. Being aware of affective aspects of learning opportunities is hence a requirement for teachers. However, an overemphasis of affective aspects in the teachers' perceptions might lead to a lack of connectedness with other components of pedagogical content knowledge (PCK). As there is still a substantial need of empirical research in this area, we report results from three studies on views of mathematics teachers about learning opportunities, which focus both on the teachers' awareness of affective aspects and other PCK components.

S. Kuntze (✉) • A. Dreher
Institute for Mathematics and Informatics,
Ludwigsburg University of Education, Ludwigsburg, Germany
e-mail: kuntze@ph-ludwigsburg.de; dreher@ph-ludwigsburg.de



B. Pepin, B. Roesken-Winter (eds.), *From beliefs to dynamic affect systems in mathematics education*, Advances in Mathematics Education,
DOI 10.1007/978-3-319-06808-4_15

In the first study, more than 40 German teachers were asked about the criteria of instructional quality they would use if they had to evaluate the classroom of a peer teacher. Criteria connected to affective aspects were most frequent, suggesting that teachers gave a high importance to motivation and affect in comparison with other criteria for effective classrooms.

As both dealing with representations and evaluating tasks are key aspects of the everyday professional practice of mathematics teachers, the second study uses a complementary approach. It makes a link to more specific PCK and focuses on teachers' views about specific learning opportunities in tasks, which were related to pictorial representations. The results suggest that a number of teachers acknowledged more the motivational character of the pictorial representations than their helpfulness for conceptual learning.

A third deepening study explored whether the awareness of affective aspects might even be in conflict with other PCK components. We found cases which suggest such a conflict, whereas other cases show a more equilibrated professional knowledge.

The three studies use different methodologies, drawing both on qualitative and quantitative methods. In the chapter we will also explain the methodological approaches in relation to our research interests.

## Theoretical Background

Mathematical competency and its development are influenced by affective variables of the learners and characteristics of learning environments which may raise or support motivation (e.g. Helmke and Weinert 1997).

On the theoretical level, the awareness of affect and motivation as impact factors on learning has led to a multi-criteria perspective of instructional goals (e.g. Pekrun and Zirngibl 2004) with a simultaneous focus both on learning goals and goals of supporting motivation.

Moreover, even the construct of mathematical competency has been defined to contain motivational aspects together with abilities (Weinert 2001, p. 27f), which highlights the importance of such affective variables.

Two well-known theories which aim to describe these affective variables are the following:

- The self-determination theory by Deci and Ryan (1985, 1993) emphasizes the spectrum between intrinsic and extrinsic motivation. According to this theory, human beings seek positive experiences of acting autonomously; learning opportunities should hence support the autonomy of learners in order to foster learning and the development of positive affective dispositions of learners.
- According to the theory of Schiefele, Krapp and Prenzel (e.g. Prenzel 1988; Krapp 1992) motivation is a relationship between individuals and objects, which is influenced by situational circumstances and non-situation-specific interest. Accordingly, learning, competency development and affective dispositions of learners are supported by positive conditions in the situational context and by positive intra-personal affective characteristics of the learners.

Well-established affective variables such as interest (e.g. Helmke and Weinert 1997), self-efficacy (e.g. Bandura 1977) or achievement motivation (Heckhausen 1989) can be modeled in both of the approaches introduced above.

Motivation and affect clearly do not only impact on learning processes of students, but also the teachers' motivation may play a role (e.g. Baumert and Kunter 2006). In contrast, the teachers' awareness of the role of motivation and affect for mathematical learning rather belongs to PCK and instruction-related views of teachers. In this chapter, we will focus on such PCK and views related to affective characteristics of learning opportunities in the mathematics classroom, as ways of developing PCK in professional development activities may be informed by needs detected in empirical research. Fostering supportive affective dispositions of learners is an important goal of the mathematics classroom, which also requires the teachers' attention.

Consequently, teachers should be aware of such affective aspects and have integrated them in their PCK, including their views and knowledge related to specific contents or classroom situations. However, even though affective variables are considered as meaningful for learning in the mathematics classroom, empirical research about PCK in this area and its connectedness with other PCK – in particular content- and situation-specific PCK – is still relatively scarce. This study hence aims at responding to this research need, using a multi-layer model of professional knowledge introduced in the following.

## *Teachers' Professional Knowledge and the Awareness of Affect*

The teachers' awareness of affective aspects of the mathematics classroom is connected to various components of professional teacher knowledge including instruction-related views. The model presented in Kuntze (2012, see Fig. 1) can

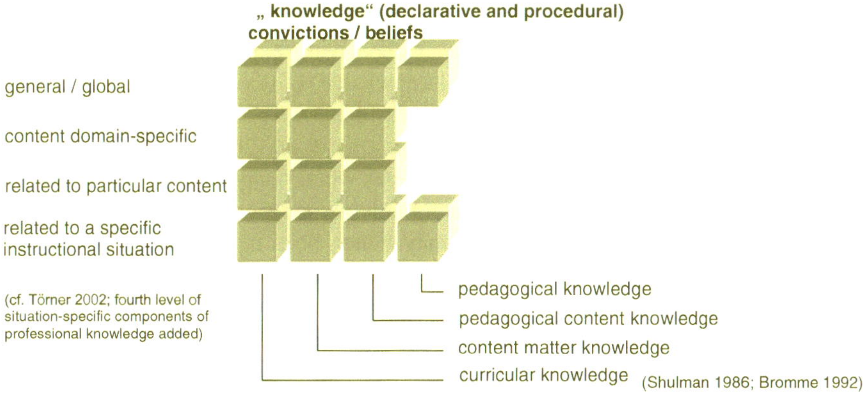


**Fig. 1** Model for components of professional knowledge (Kuntze 2012)

facilitate a structured overview of different professional knowledge components. This model of professional teacher knowledge combines the spectrum between knowledge and prescriptive views, such as instruction-related convictions or epistemological beliefs (Pajares 1992) with the domains of professional knowledge by Shulman (1986, cf. Ball et al. 2008, for the possibility of refinement into further domains). The advantage of this model is that there is a further distinction of globality (cf. Törner 2002) versus situatedness on the horizontal layers: PCK or views relevant for affective aspects of learning opportunities may namely be relatively global on the one hand or specific for particular contents or even for instructional situations on the other hand (Kuntze 2012, cf. Lerman 1990). More detailed explanations of this model including examples of components of professional knowledge are given in Kuntze (2012).

With respect to professional knowledge related to affective characteristics of learning mathematics, we give some brief examples:

- A teacher's view that the content area of probability is more motivating for students than other content areas in school mathematics (c.f. Martignon and Wassner 2005) belongs to PCK on the content-domain-specific level, and rather to convictions and beliefs.
- A teacher's view that a particular task will not be interesting for 6th-graders belongs to the sphere of views on a content-specific level, in the vertical column of PCK (see Fig. 1).
- A teacher's knowledge of empirical survey results that 8th-graders find group work mostly very motivating, can be classified as pedagogical knowledge on a relatively global level.

Teachers' knowledge and views related to affective aspects of learning can hence be associated with various cells in the model shown in Fig. 1. As there is a special relevance for the design of rich learning opportunities, we will in the following focus on the area of teachers' PCK.

Different components of professional knowledge may be connected in the way that they support each other, but they may also be isolated from each other or even in conflict with each other. Consequently, when teachers make use of their professional knowledge, e.g. when they answer classroom-related questions, they may draw only on a part of their professional knowledge (cf. Kersting et al. 2012). This part of their professional knowledge is then dominant over other elements of professional knowledge teachers may possess. Kersting et al. (2012) explain such phenomena by the inert knowledge approach, but when components of professional knowledge are activated, this may not automatically mean that all other, potentially conflicting knowledge is inert. There may hence be a modulation process: If a teacher draws on a specific professional knowledge element, this can indicate that she or he is aware of this professional knowledge element. In this sense, we define teachers' *awareness* of certain elements of professional knowledge as a part of professional knowledge which influences the readiness and ability of teachers to use this professional knowledge element in instruction-related contexts. For instance, if a teacher shows awareness of affective aspects of learning, she or he uses her or his knowledge related to affective aspects of learning. Of course, possessing corresponding professional knowledge is thus a prerequisite for showing awareness related to this professional knowledge element, and the way *how* teachers show awareness affords observing also the elaboratedness of their corresponding professional knowledge.

In this study, we aim to investigate such professional knowledge including the awareness of professional knowledge components on different levels of situatedness in order to get complementary insights. Whereas the first study reported in the following ("study A") concentrates on rather global views about the importance of motivational characteristics among other criteria of instructional quality, studies B and C have a more situated, namely a content-specific and a situation-specific scope. For instance, using the language of Törner's (2002) "levels of globality" of epistemological and instruction-related beliefs, views about tasks (cf. Biza et al. 2007) reflect content-specific beliefs, close to mental representations of classroom situations. As content domain, we chose representations of fractions.

For the reasons given above, these specific views are important components of professional knowledge which complement global orientations in the teachers' PCK. Consequently we will in the following introduce the theoretical background of these more specific approaches to PCK and pedagogical content views.

## *Professional Knowledge and Beliefs: Why Teachers' Evaluations of Tasks and Classroom Situations Give Insight*

When selecting or creating problems for the classroom, teachers have to judge on the learning opportunities that are linked to these tasks, on their motivational potential, on their level of complexity, facilitating or inhibiting factors for the student's understanding and many other aspects linked to specific tasks. These views about problems are situated: Teachers may have in mind a specific group of learners, a specific classroom situation – and of course, tasks are connected to specific contents.

According to the model of professional knowledge presented in the previous section, task-related views of teachers can be described as basically content-specific convictions in the domain of pedagogical content knowledge (c.f. Kuntze 2011).

An important aspect of tasks beyond their level of complexity (e.g. Hosenfeld 2008) and their learning potential (e.g. Kuntze 2011) is their use of representations. Dealing with multiple representations is crucial for the development of mathematical competency (e.g. Duval 2006) and hence can be considered as a big idea in mathematics and mathematics instruction (Kuntze et al. 2011). This big idea reflects strategies used and required in many mathematical domains related to the use of different ways of representing mathematical facts or concepts as well as with changing representations and linking them. Using multiple representations is also an idea emphasized in many national standards (e.g. NCTM 2000; KMK 2003). Especially pictorial representations can add cognitively activating insights to purely formal or algorithmic representations (Duval 2006) and can be conducive to mathematical problem solving, as long as they don't serve a purely decorative function (Elia and Philippou 2004). For the content domain of fractions for example, it can be considered as established in the scientific community that using pictorial representations and linking representations play an important role for building up conceptual knowledge.

Pictorial representations can on a surface level also be used to liven up "dry" contents, hence target on raising or supporting motivational dispositions of the learners. Even beyond a purely decorative function, pictorial representations may connect to the experience world of learners or be designed to catch their interest. However in this case, pictorial representations may still be non-useful for solving the task despite their motivation potential. Prior studies about teachers' views related to pictorial representations (Ball 1993) have suggested that teachers attributed a predominant significance to the motivational potential associated with pictorial representations and appeared to neglect the role of pictorial representations for conceptual learning. However, as pictorial representations are situated e.g. in tasks, there is a need of more specific research about such views, in order to find out to what extent teachers focus only on an illustration effect relevant for motivating students or whether they see the potential of offering mathematical insight and fostering understanding (c.f. Elia and Philippou 2004) – a need which also extends to views and PCK about the use of representations in classroom situations.

Consequently, in the context of using pictorial representations with fractions, study B (presented in section "Study B: teachers' content-specific views about the role of pictorial representations for the learning potential of tasks") examines whether teachers are able to equilibrate motivational and affective aspects of pictorial representations on the one hand with other PCK aspects related to the use of representations such as their potential for conceptual learning on the other hand.

In order to explore whether and how the awareness of affective characteristics can be in conflict with other PCK relevant for considering learning opportunities in classroom situations, study C (section "Study C: qualitative in-depth analysis of teachers' views related to the use of representations in specific classroom situations") deepens this research with a qualitative and interpretive analysis of cases.

## Research Interest

As laid out in the previous section, the awareness of affective aspects of learning and learning opportunities is an important component of PCK, which should be integrated in other PCK, such as individual (global) criteria of instructional quality and knowledge necessary for evaluating the potential for mathematical learning contained in tasks. As corresponding empirical research is relatively scarce, this study focuses on the importance teachers give to affective aspects of learning and learning opportunities. The following research interest is therefore in the focus:

What emphasis do teachers give to aspects of motivation and affect as criteria for evaluating learning and instruction in relation to other relevant aspects for instructional quality?

In particular, this research interest is reflected in the following three more specific research questions:

1. What criteria for instructional quality do in-service teachers have and what importance do they give to affective and motivational criteria?
2. What role do affective aspects play in teachers' views on the role of pictorial representations in tasks? Among teachers' task-specific views, do such affective aspects prevail over other relevant PCK components?
3. Can PCK related to affective aspects be in conflict with other PCK about the use of representations in classroom situations? Does this PCK have to be in conflict with other PCK?

These research questions were explored in three studies: Whereas study A aims at giving an overview of teachers' individual criteria of instructional quality, studies B and C concentrate on a specific area of PCK where affective aspects might even be in conflict with other PCK, namely views about the use of pictorial representations in tasks and in a learning situation.

## Study A: Teachers' Global Views About Criteria for Instructional Quality

Study A aims to explore criteria of instructional quality on a relatively global level. In the analysis presented here, particular interest was devoted to the status teachers give to affective and motivational criteria in comparison to other criteria. As empirical background of criteria of instructional quality we refer to the work of Clausen et al. (2003), who integrated criteria from the perspective of students, teachers and external observers in a relatively comprehensive approach. However, informed by these findings, the present study was designed according to a bottom-up methodology, which afforded addressing the first research question presented in the previous section.

### *Design and Methods*

In order not to drive the attention of teachers to specific aspects or criteria, we chose a relatively open question format in a corresponding questionnaire. Figure 2 shows the design of the questionnaire unit, in which the teachers were asked to note down their individual criteria for good mathematics instruction, which are accessible by external observation. The teachers were also asked to rate the importance given to each of these criteria on a three point Likert scale. In order to

*How would you judge on the quality of mathematics instruction?*
Imagine that a mathematics teacher colleague would ask you to attend and watch one of her/his lessons and to judge on the quality of instruction.
Independently from what feedback you would give to your colleague, you would have to make an honest judgment on the instructional quality of the observed mathematics lesson for yourself.
In order to judge on the quality of mathematics instruction, you have probably a couple of criteria. Please note in the space below such criteria, which are important to you and which you can observe in the lesson. [...] For the criteria you note, please indicate how important they are for you respectively.
*Please limit the time you spend for this page of the questionnaire to about 10 minutes at most.*

| Criteria for the quality of mathematics instruction<br>*(for additional remarks you may use the space on the following page)* | *Importance* |
|---|---|
| | ☐ very important<br>☐ important<br>☐ less important |

The questionnaire provided five more boxes like the one given above in the form of lines of a table.

**Fig. 2** Design of the questionnaire unit for the teachers' individual criteria of instructional quality (cf. Kuntze and Rudolph-Albert 2009, p. 90)

limit the answers to rather non-artificial criteria, which are likely to be present in the every-day life of the teachers, the teachers were told not to spend more than 10 minutes on this part of the questionnaire. The questionnaire unit contained a default space for six criteria (cf. Fig. 2).

In an interpretive analysis, the teachers' answers to the question presented in Fig. 2 were coded following a bottom-up approach. The analysis was done according to an interpretative paradigm similar to "grounded theory" (Straub et al. 1997). Two raters analyzed the teachers' responses in a consensus procedure against the background of possible context information in the questionnaires, so as to establish primary codes very close to the teachers' individual responses. The two researchers subsequently condensed the criteria given by the teachers according to their semantic content: A second step of this interpretative work consisted in grouping the codes according to their semantic proximity or contiguousness, unless there were no codes/responses left allowing to be grouped into a greater common semantic domain. The original responses of the teachers were reviewed again and again during the grouping process of the primary codes as a control of the generation of greater semantic domains against the original data.

This methodology was chosen in order to be able to spot the teachers' criteria for "good" mathematics instruction in an individualized way – and so as to be able to condense core criterion areas in the subsequent analysis.

The sample of this study consisted of 42 German upper secondary in-service mathematics teachers (14 female, 28 male; 19 teachers aged 35 years or less, 8 aged from 36 up to 45 years, 13 aged from 46 up to 55 years, 2 aged more than 55 years) from eleven German academic-track secondary schools. The teachers had been teaching mathematics for on average 11.3 years (SD=9.8 years). The teachers' schools were located in South German small-town environments.

The teachers were participants of an in-service teacher professional development project. However, study A only focused on data from questionnaires the teachers were asked to complete prior to the teacher professional development activity, so that we can assume that their responses were not influenced by any of the interventions. As the participation in the teacher professional development activity was voluntary, a possible process of self-selection of the sample might have occurred, even though the recruitment of the participants was rather top-down, i.e. via the school directors.

In the questionnaire unit, the teachers were asked to give criteria of instructional quality they would use when having to find a personal judgment on a lesson by a colleague. As this situation was clearly fictional (the teachers were not given any kind of material from any lesson), we expected the answers to be on a non-situation-specific level and to reflect the individual criteria the teachers were most aware of.

We hence assert – on the base of the design of the questionnaire and the subsequent bottom-up analysis – that the results give a picture of rather general criteria teachers have in mind when they imagine to have to judge on observed classroom lessons.

## *Results*

The analysis yielded that 299 criteria could be connected to 13 semantic domains (cf. Kuntze and Rudolph-Albert 2009); less than 10 % of all the primary codes were collected as belonging to "other" criteria of instructional quality.

Figure 3 shows the frequencies of teachers' criteria in the corresponding semantic domains with the importance they gave to their criteria. The diagram shows relative frequencies of teachers. If a teacher noted two or more criteria belonging to the same semantic domain, the highest importance ticked was used for the data displayed in the diagram.

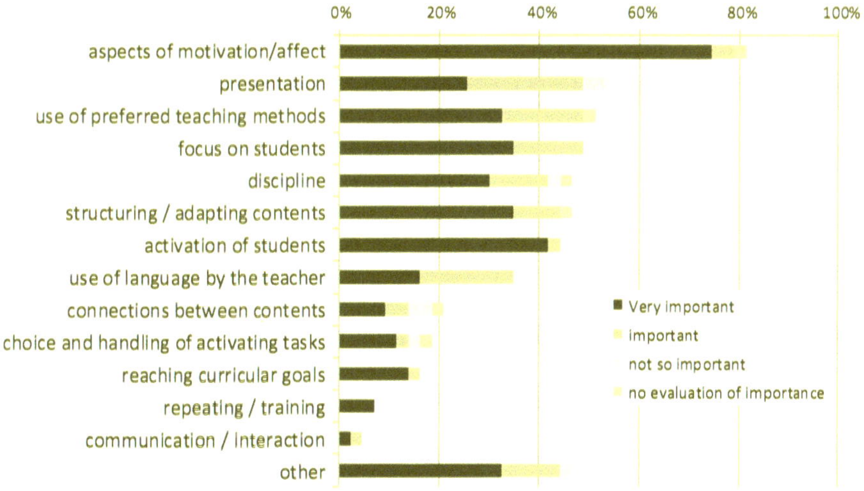


**Fig. 3** Criteria for instructional quality (percentages of teachers making comments assigned to the displayed semantic domains)

As can be seen in Fig. 3, criteria in the semantic domain of motivation/affect appeared with the highest frequency. Moreover, the importance assigned to these criteria by the teachers was mostly high; the corresponding criteria were ticked with "very important" much more often than those for the other semantic domains.

### *Discussion*

The results in Fig. 3 suggest that the teachers gave a high importance to criteria of instructional quality in the area of affect and motivation compared to the other areas of criteria mentioned by the teachers. Affective and motivational aspects of learning environments can have a highly general character and hence can be expected to be among the criteria as spotted by the questionnaire. Against this background it is also interesting that close to 20 % of the teachers did not mention any criterion relevant for the area of affect and motivation in any of the criteria they wrote down.

However, considering the huge variety of possible criteria of instructional quality, motivation and affect appear to have a high priority for what teachers see as "good" mathematics instruction. The percentage of criteria classified as "very important" by the teachers was the highest among all other semantic domains. This means that the large majority of the teachers showed an awareness of affect and motivation when asked what criteria they would use when observing a mathematics classroom and that they saw these criteria as very important. In comparison, only about 35 % of the teachers saw 'structuring/adapting contents' as an important criterion for instructional quality and less than 15 % mentioned a criterion related to the semantic domain 'choice and handling of activating tasks', for instance. This raises the question whether affect and motivation catches the teachers' attention to the disadvantage of other important criteria of instructional quality.

The findings call for follow-up research which takes a closer look at affective aspects together with other PCK related to the content quality of learning opportunities. Such content-specific professional knowledge can be spotted in more situated research designs, e.g. by focusing on task-specific views of teachers – an approach which was chosen for study B.

## Study B: Teachers' Content-Specific Views About the Role of Pictorial Representations for the Learning Potential of Tasks

Professional knowledge related to tasks as well as corresponding epistemological and instruction-related beliefs are likely to play an important role when teachers decide how to design learning opportunities for the mathematics classroom: Tasks perceived as more motivating by teachers may be assigned as having a higher learning potential. When designing learning material, pictorial representations may be designed so as to appeal to the students' interest. However, representations in tasks can also be used to support conceptual learning – a priority which rather requires PCK beyond the focus on affective aspects, when teachers analyse the task's learning potential. Content-specific professional knowledge thus merits attention.

Responding to the research need resulting from study A, study B concentrates on views about pictorial representations in tasks and their structure, as described in the second research question in section "Research interest".

## *Methods*

In this study, a questionnaire was administered to 145 German pre-service teachers (115 female, 27 male, 3 without data) before the beginning of a university course. The pre-service teachers had a mean age of 21.42 years (SD = 3.68 years) and had been studying on average for 1.99 semesters (SD = 1.19). 79 pre-service teachers were preparing to teach in primary schools, 36 in secondary schools for lower-attaining students, 6 in technical-track secondary schools and 21 in schools for students with special needs (3 without data). As the teachers were in a relatively early phase of their university studies, potential career differences related to school types would rather be linked to selection effects when choosing a career type. However, no specific patterns connected with career types could be identified. As the further development of professional knowledge may be influenced by views of the pre-service teachers, exploring these views is of great interest from the point of view of the second research question.

Corresponding to the research question, the participants were asked to evaluate pictorial representations within four tasks by means of multiple-choice items (see Fig. 4). The pre-service teachers could express their approval or disagreement concerning these items on a four-point Likert scale. They were told that the problems were designed for an exercise about fractions in school year six.

All four problems include a pictorial representation which is not necessary for solving the task, i.e. it is possible to give these tasks to students with another or even

Finn-Luca writes $\frac{2}{4}+\frac{2}{4}=\frac{4}{8}$ in his notebook and in addition he draws the following picture:

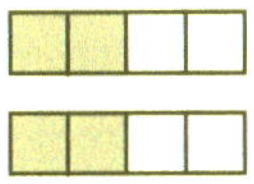

What would you tell him?

| | |
|---|---|
| It is difficult for the pupils to handle the illustration in this problem. |
| In my opinion the role of the illustration is reasonable in order to develop mathematical competences. |
| I would rather use this problem with another illustration. |
| By this illustration pupils can learn a lot here. |
| The illustration confuses pupils rather than it contributes to comprehension. |
| The illustration has a motivating effect. |

**Fig. 4** Sample task for the second type of problems (task 4)

without any pictorial representation. In two problems (tasks 1 and 3) the pictures were selected in a way that on the one hand they were expected to have a motivational impact on students, but on the other hand, as illustrations, they are rather confusing than helpful for solving the problem (c.f. Elia and Philippou 2004). In contrast, in the other two tasks (tasks 2 and 4) the pictorial representations were created to give an extra insight for solving the problem without an emphasis on a particularly motivating character. An example for the first type of task is: "Make up a situation of a word problem which is suitable for the calculation "3 ÷ ¾" together with a picture of a well-known cartoon character slicing three quarters of a cake for three other characters in the background ("task 3"). Obviously, this illustration is not helpful for making up a situation in which one has to divide 3 by ¾. A sample for the second kind of task is shown in Fig. 4. Here the pictorial representation gives an additional insight in Finn-Luca's misconception and is therefore useful for answering the question and also offers an opportunity for reflecting on the rule for adding fractions and the role of the unit. On the other hand the cartoon characters and the cake have a decorative function and are more likely to have a motivational impact on students than the diagram representing fractions.

Based on these differences in the roles that the pictorial representations play for the tasks, we were interested in their evaluations by pre-service teachers. By corresponding multiple-choice items, several aspects of the role of pictorial representations were addressed (cf. Fig. 4). While the third item expresses a rather general attitude towards the pictorial representation, the others explicitly relate to specific aspects which might shape this attitude, namely the learning potential of the pictorial representations, their level of difficulty, their potential for confusion, and – last but not least – their motivational potential.

The design of the study affords an overview on the connectedness of the teachers' awareness of affective criteria with the other criteria, as well as of the status of affective criteria when analyzing the learning potential associated with the representations used in the tasks by comparisons between the views of the different tasks.

## *Results*

The research interest of this study is to explore views of the pre-service teachers related to the role of pictorial representations in tasks and the evaluation of their affective character, in particular. For gaining an overview of the structure of these task-specific views, we used a factor analysis for each problem in order to visualize interdependencies of the different aspects which are expressed by the items. The resulting component diagrams are shown in Fig. 5.

The data in Fig. 5 reflects that the items 1–5 coincide in a common factor, whereas the item related to the motivational potential apparently mainly loads on another component. This means that the overall evaluation of the pictorial representations in the tasks by the participants is closely related to their rating of the learning potential, the level of

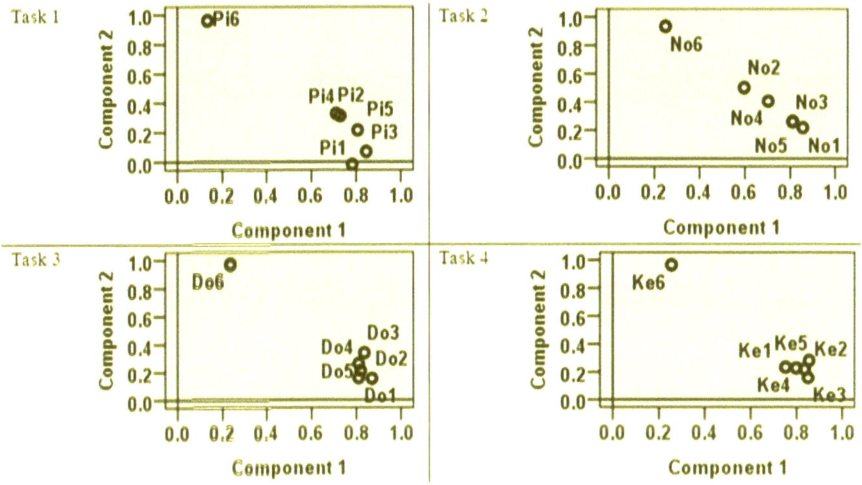


**Fig. 5** Component diagrams (Item 6 is about the motivational potential, respectively)

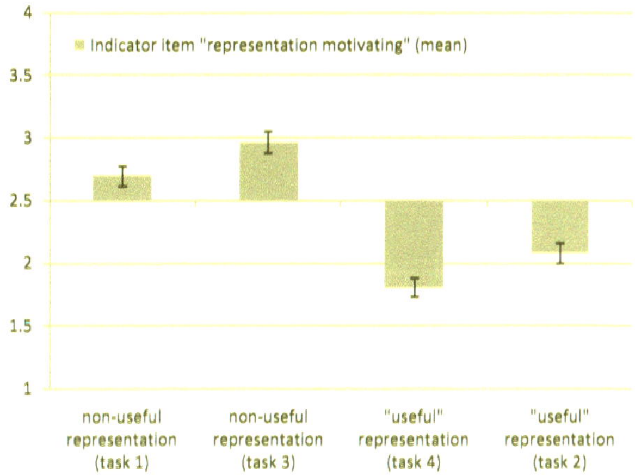


**Fig. 6** Task-related views about the motivation potential of the representations

difficulty and the potential for confusion, whereas it is apparently less interdependent with their views on the motivational potential.

Figure 6 displays the mean values of the motivation potential of the non-useful representations in comparison with the "useful" representations which give additional insight into the problem. The data show that the non-useful representations were seen as more motivating, as expected according to the design of the representations accompanying the tasks.

On the base of the results shown in Fig. 5, items 1–5 were combined to form a scale for each problem, as they loaded on a common factor. The corresponding reliability values (Cronbach's $\alpha$) of these scales range from 0.85 to 0.91. The means of these scales are listed in the upper part of Table 1.

**Table 1** Means of scales, standard deviations, and examples of individual cases

| Positive evaluation of the pictorial representation | Task 1 | Task 2 | Task 3 | Task 4 |
|---|---|---|---|---|
| Mean | 2.33 | 2.15 | 2.41 | 2.86 |
| SD | 0.78 | 0.87 | 0.93 | 0.91 |
| Participant A | 1.6 | 2.6 | 1.2 | 3.4 |
| Participant B | 3.0 | 1.0 | 3.8 | 1.2 |
| Participant C | 1.4 | 1.2 | 1.4 | 3.6 |
| Participant D | 3.6 | 3.5 | 3.8 | 3.8 |

The value 1 means strong disagreement, 4 means strong approval

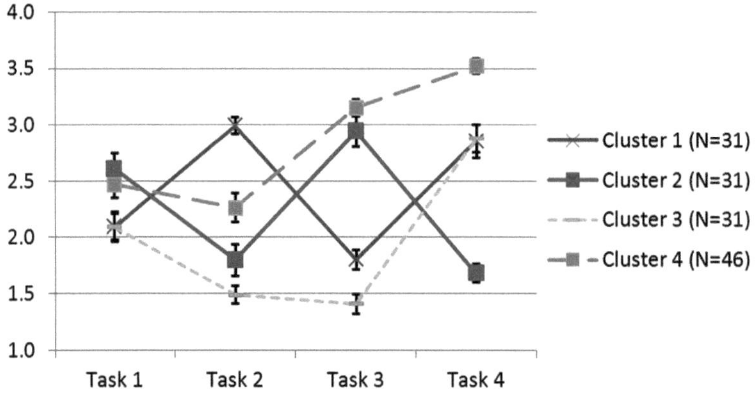


**Fig. 7** Means and standard errors for clusters based on the four scales

The relatively high standard deviation values indicate that the expressed views are spread out over the whole range of the given spectrum. Looking at individual cases reinforces this impression: The lower part of Table 1 shows that completely opposite evaluations can be found. Hence, we were interested in different answering patterns which occurred and chose to explore them by means of a cluster analysis based on the four scales for the views about the four tasks (Ward Method). This yields four clusters showing distinct answering patterns which are presented in Fig. 7.

In Fig. 7 it can be noticed that the lines corresponding to cluster 1 and 2 are almost laterally reversed, which means that these groups of pre-service teachers express opposing views on the pictorial representations in the given tasks: Pre-service teachers of cluster 1 show positive views about the pictorial representations in the problems of the second type ('providing an additional mathematical insight') and negative views about the motivating, illustration-type, but unhelpful representations. Cluster 2 shows a reverse pattern. Pre-service teachers in cluster 3 had positive views only related to task 4, whereas the teachers in cluster 4 had medium or positive views related to all the tasks. The four (extreme) individual cases in Table 1 can be looked at as representative for these four clusters.

Giving an outlook, a possible explanation of these findings may be seen in results from a further part of the questionnaire, in which the pre-service teachers were asked to pick the three reasons for using pictorial representations (in general), that they considered as the most important ones from a selection of given reasons. Considering these more general reasons for using pictorial representations, which were rated as being most important for using pictorial representations by the teachers in clusters 1 and 2, the analysis yields the following frequencies: 56.2 % of the pre-service teacher belonging to cluster 2 ("illustration") included one of the reasons "They can motivate pupils and make mathematics more fun", "They can make it easier to maintain the pupils' attention and interest", or "They can liven up rather dry subject matter like fractions" in their top 2 reasons. In contrast, this is true for only 34.3 % of the participants in the first cluster. The following finding complements this: For 62.5 % of the teachers in cluster 1 ("additional insight"), at least one of the reasons "They can facilitate translation processes between real world situations and mathematical contents" or "They can be conducive to revealing misconceptions" belongs to the two most important reasons, whereas this is the case for just 37.5 % of the pre-service teachers in cluster 2.

Further results from this additional questionnaire part may help to explain clusters 3 and 4: The pre-service teachers in cluster 3 assigned much higher significance to "revealing misconceptions" than to "motivation" when using pictorial representations. This may be seen in line with the participants' aversion to the given pictorial representations in tasks 1–3, except for the fourth one, which assesses misconceptions. The participants belonging to cluster 4 – the cluster with medium or positive evaluations of all given pictorial representations – mentioned "livening up" on average as one of the most important reasons for using pictorial representations, a criterion that was mentioned with a lower frequency in the other clusters.

The clusters' average evaluation of the motivation potential of the representations is shown in Fig. 8. Cluster 1 has a pattern which differs from the other profiles, assigning a relatively higher motivational potential to the representation in task 2, one of the tasks with a "useful" representation.

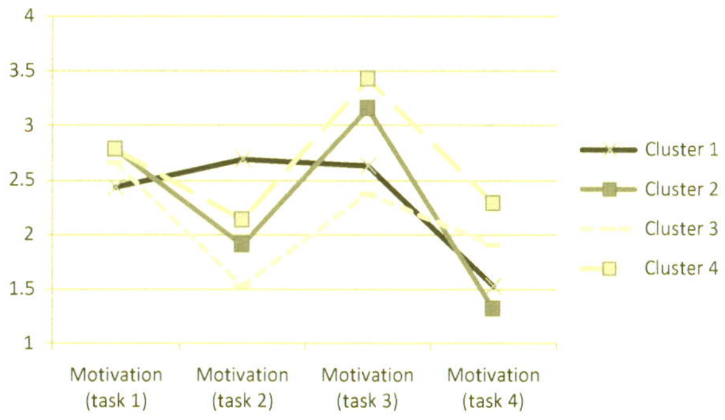


**Fig. 8** Evaluation related to motivation for the clusters

## *Discussion*

The results suggest that the pre-service teachers' views about the pictorial representations in the tasks belonging to the content area of fractions had mainly two dimensions within the aspects implemented in the questionnaire: On the one hand, the learning potential, the danger of confusion, the level of complexity formed a common factor with good reliability values, on the other hand, the motivational potential appeared to open up a separate component. This finding may be interpreted as evidence that – already for pre-service teachers – not only the motivational potential counts, but other aspects play an additional role. Consequently, the results appear to add to the studies cited in Ball (1993) on views about representations – probably as a result of the situatedness of the representations in tasks. However, as relatively heterogeneous aspects such as difficulty and learning potential coincided in one factor, these pre-service teachers still may not have developed a multi-aspect perception when analyzing and evaluating pictorial representations.

Indeed, as there was one cluster of pre-service teachers who rated the motivating but confusing representations such as in task 3 with the cartoon character as better (i.e. to have a higher perceived learning potential and to be less confusing), at least one subgroup of the pre-service teachers showed a need of professional development in order to develop their strategies of analyzing pictorial representations in tasks.

Rating the pictorial representations in problems 2 and 4 better than the others, the participants belonging to cluster 1 had apparently realized which pictorial representations may be more helpful than the others. These teachers might also have a more equilibrated view related to motivational aspects, as they evaluated the motivation potential of the representations differently from their counterparts in the other clusters (Fig. 8). In contrast, the answering pattern of cluster 2 suggests that these pre-service teachers did not notice how confusing the decorative illustrations in tasks 1 and 3 are and that they preferred them over the ones giving an extra insight for solving the problem. The pre-service teachers in cluster 3 might have had a different perspective on the pictorial representations in the problems: The focus on "revealing misconceptions" may have dominated for these teachers, leading to a task-specific evaluation with a positive view on the pictorial representation in task 4 (see Fig. 1). However, the views of the pre-service teachers in cluster 4 could be interpreted as focusing mainly on affective aspects and as somewhat superficial from the point of view of other PCK: This may be reflected in the view that pictorial representations serve, in general, mainly the purpose of "livening up" dry text-like information.

Summing up, even though the findings should be interpreted with care and be considered against the background of the cultural environment of the sample, the evidence suggests four types of answering patterns, namely seeing pictorial representations

- especially as learning opportunities that can provide additional insight and foster understanding, so that quality aspects facilitating learning from the pictorial representations play a key role,
- in the first place as illustrations having to generate motivation, so that surface characteristics are central, such as the affective connotation of the context the representations point at,

- predominantly as possibilities of revealing misconceptions of learners, or
- mainly as elements livening up the presentation of content with possibly little relevance for conceptual learning, so that the quality of the pictorial representations does hardly make a difference.

As dealing with representations and using multiple representations are crucial for building up mathematical competency, teachers' views about representations may play a key role for the ways they will focus on representations in their classroom. Teachers who lack awareness of the role of using multiple representations might not fully exploit the learning potential of representations when designing learning opportunities and dealing with tasks in the classroom.

For instance, teachers who see the main purpose of pictorial representations in illustration and in generating motivation will probably not critically evaluate tasks according to how the pictorial representation is used in the task and how students are encouraged to gain additional insight into mathematical concepts, how they are challenged to enter into reasoning and argumentation processes, or how representations may be an obstacle, an element increasing the level of complexity or a help to learners. These teachers appear to be less aware of such aspects of PCK, probably as a consequence of their awareness of affective aspects, which in this case results in a somewhat superficial way of analyzing learning opportunities contained in representations.

This scope points at even more situated PCK which is necessary for analyzing classroom situations – a domain of professional knowledge which was in the foreground in study C.

## Study C: Qualitative In-Depth Analysis of Teachers' Views Related to the Use of Representations in Specific Classroom Situations

Study C aims at exploring whether PCK related to affective aspects can be in conflict with or even suppress other PCK about the use of representations in specific classroom situations. Complementing the results of study B, even more situated views are addressed and accessed through an open question format and a qualitative analysis. This study thus seeks to answer the third research question as introduced in section "Research interest".

### *Methods*

Corresponding to the research interest of this deepening study, the focus lies on a qualitative analysis of cases. In the context of a larger study exploring teachers' professional knowledge and views about dealing with multiple representations in the mathematics classroom, the participants were given the transcript of a fictitious

P: The fractions $\frac{5}{6}$ and $\frac{7}{8}$ are of equal size! There is always one piece of the whole missing.

T: Hmm, let's have a look at this on the number line: [draws]

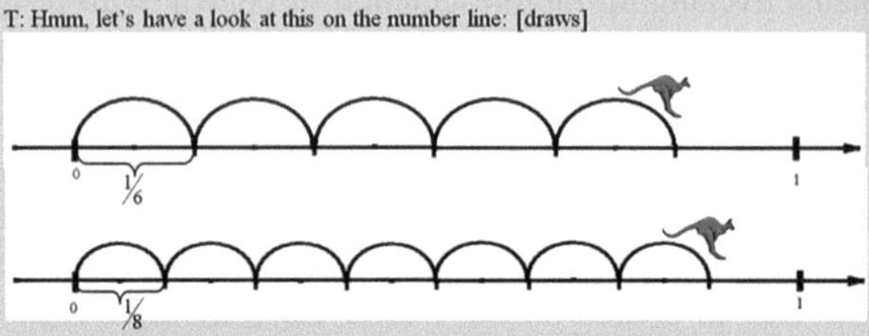


Now you can see that the kangaroo that jumps five times $\frac{1}{6}$ does not get as far as the kangaroo that jumps seven times $\frac{1}{8}$, can't you?

**How much does this response help the pupil? Please evaluate the use of representations in this situation and give reasons for your answer.**

**Fig. 9** Questionnaire item about evaluating a teacher's use of representations in a specific classroom situation

classroom situation (shown in Fig. 9) and were asked to evaluate the teacher's use of representations. In this classroom situation a student reveals a misconception by claiming that the fractions 5/6 and 7/8 were of equal size, since "there is always one piece of the whole missing". As a reaction the teacher draws two number lines with kangaroos jumping five times 1/6, respectively seven times 1/8 starting at zero and argues that the first kangaroo does not get as far as the second one. However, the teacher does not emphasize the fact that the pieces missing of the whole are not of the same size with this representation. Thus, the reaction does not pick up the student's argumentation and one could even say that the representation of fractions on the number line does probably not match the student's mental representation when speaking of missing pieces of the whole. Focusing exclusively on affective aspects of the teacher's representation, a benefit of the jumping kangaroos compared to plain number lines could be that they draw the student's attention and make them engage with the representation more easily and with more interest. Nevertheless, this classroom situation is mainly about dealing with a student's misconception and not so much about motivating learners to engage with a task, since the student is already involved in a specific problem. Against this background we have explored the participant's evaluations of the teacher's reaction with respect to the degree to which their awareness of affective and motivational aspects do not only complement, but suppress other criteria for evaluating the use of representations, the latter belonging to other PCK.

We have analyzed three distinct cases: Anne, an experienced teacher, who teaches mathematics at a German academic track secondary school since 30 years and Linda and Sandra, who are both German pre-service teachers in the first year of their university studies. These cases were selected such that they can be seen as rather extreme examples from the perspective of the studies reported above, i.e. they illustrate different degrees of emphasis on affective and motivational aspects compared to criteria drawn from other PCK components.

## *Results*

We start with the analysis of Anne's answer which is shown in Fig. 10. "Very good" is her rating of the use of representations in this situation and she explains that the teacher's representation is very illustrative and convincing. She seems to consider only the teacher's drawing itself, almost somewhat isolated from the given situation since her reference to the student ("will easily convince the student") does hardly take into account the student's specific problem. The second part of her answer suggests that it may not even be the teacher's drawing that Anne is mainly focusing, but merely the kangaroo and its presumed motivating effect on the students: "the kangaroo itself is motivating for sixth-graders, so that they will engage with the task with pleasure". Hence, Anne's evaluation of the use of representations in this situation can be seen as an example which shows how a teacher's focus on affective aspects of a representation might blind him or her to other important criteria.

Linda's comment (see Fig. 11) takes into account different aspects of the teacher's pictorial representation: Like Anne, she appreciates the affective benefits of the kangaroo ("the kangaroo makes it appealing"), but she also refers to the student's misconception, when analyzing the drawing ("The representation clearly shows that the remaining pieces of the whole are not of equal size"). Moreover, those positive aspects do not keep her from noticing a disadvantage of the teacher's representation:

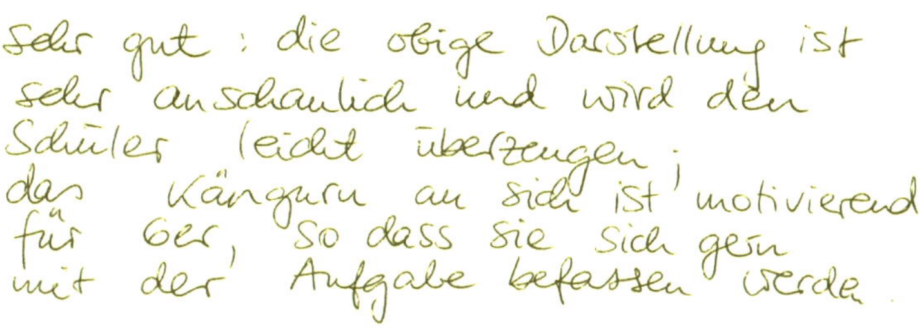


Translation: "Very good: the representation above is very illustrative and will easily convince the student; the kangaroo itself is motivating for sixth-graders, so that they will engage with the task with pleasure."

**Fig. 10** Anne's answer

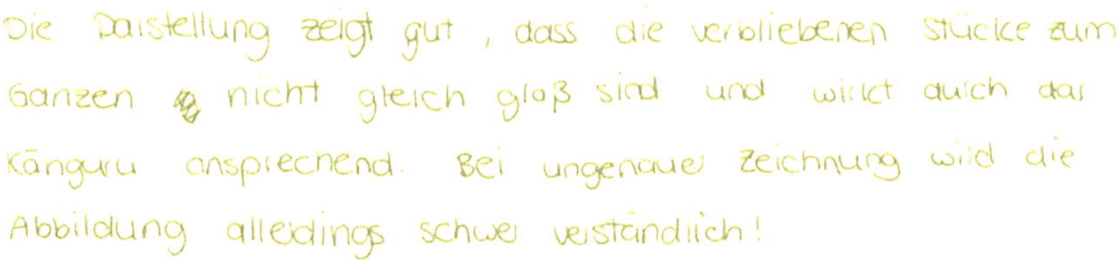


Translation: "The representation clearly shows that the remaining pieces of the whole are not of equal size and the kangaroo makes it appealing. However, imprecise drawing will make the illustration difficult to understand."

**Fig. 11** Linda's answer

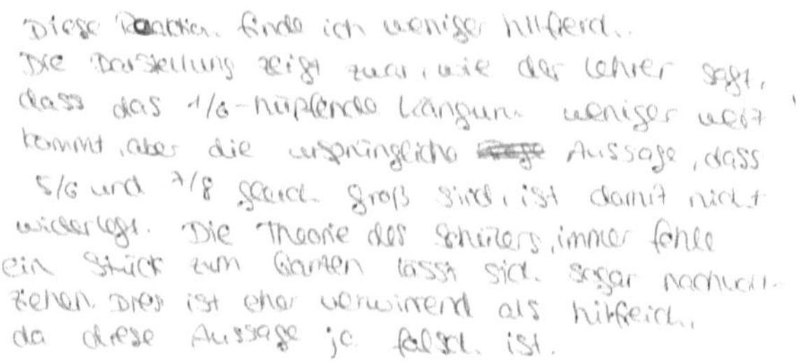
Diese Reaktion finde ich weniger hilfreich.
Die Darstellung zeigt zwar, wie der Lehrer sagt,
dass das 1/6-hüpfende Känguru weniger weit
kommt, aber die ursprüngliche Aussage, dass
5/6 und 7/8 gleich groß sind, ist damit nicht
widerlegt. Die Theorie des Schülers, immer fehle
ein Stück zum Ganzen lässt sich sogar nachvoll-
ziehen. Dies ist eher verwirrend als hilfreich,
da diese Aussage ja falsch ist.

Translation: "I find this reaction less helpful. Although the representation shows, as the teacher says, that the 1/6-jumping kangaroo does not get as far, the original claim that 5/6 and 7/8 are of equal size is not disproved. The theory of the student that there is always one piece of the whole missing is even comprehensible. As this claim is false, this is rather confusing than helpful."

**Fig. 12** Sandra's answer

"However, imprecise drawing will make the illustration difficult to understand". As described above, there are more points of criticism to be made, but nevertheless Linda's answer shows a more balanced consideration of affective and other aspects of the use of representations in this classroom situation.

Sandra (see Fig. 12) clearly focuses on the question which was asked: "How much does this response help the student?". She evaluates the teacher's reaction against the background of how well it tackles the student's misconception. Therefore she points out that neither the pictorial representation nor the teacher's explanations emphasise the different sizes of the missing pieces ("the original claim that 5/6 and 7/8 are of equal size is not disproved. The theory of the student that there is always one piece of the whole missing is even comprehensible"). The possible affective benefits of a kangaroo jumping on a number line don't seem to be relevant to her when evaluating how much the teacher's use of representations helps the student, but she rather might indicate a possibly negative affective aspect, namely that the response might be "rather confusing", since it does not deal with the student's misconception properly.

## Discussion

The results of this deepening case-based study show how the degree to which teachers focus on affective and motivational aspects when evaluating the use of representations in a specific classroom situation can be very distinct. Whereas Anne concentrates in her evaluation almost exclusively on affective characteristics of the teacher's representation, Linda takes into account both motivational aspects and criteria drawn from other aspects of PCK and Sandra focuses mainly on the suitability of the use of representations for dealing with the student's misconception. As the fictitious classroom situation in this study is designed in a way that the teacher's use of representations is not dealing very well with the student's specific problem, it is not essential for an elaborated evaluation to point out the possible affective benefits of the jumping kangaroo. However, letting those motivational benefits turn away one's attention from the fact that the representation is not well suited for tackling the student's misconception might indeed be a problem. Hence, regarding our research question for this study, Anne's case suggests that PCK related to affective aspects can in fact suppress other PCK about the use of representations in specific classroom situations. Linda's case on the other hand can be seen as evidence for the possibility of putting emphasis on motivational benefits of a representation without being blinded for other, possibly negative aspects. And yet one could argue that Linda does not get to the heart of the downside of the teacher's use of representations in her evaluation, whereas Sandra does.

The third research question "Can PCK related to affective aspects be in conflict with other PCK about the use of representations in tasks?" hence can be answered positively, given the cases presented above. The additional question "Does this PCK have to be in conflict with other PCK?" may be answered negatively, if we interpret Linda's answer as an example of a somewhat integrated and equilibrated professional knowledge and admit that Sandra also might have in mind the affective side of the student in the classroom situation when qualifying the teacher's explanation as "rather confusing".

## Conclusions

The three studies presented above have focused on different levels of globality versus situatedness for exploring the relationship between components of professional knowledge related to affect and motivation in comparison with other PCK elements. At the same time, the three studies have used different methodological approaches. Whereas study A used an interpretive bottom-up approach for generating criteria of instructional quality seen by a group of in-service teachers, study B was designed according to a quantitative research paradigm for tapping teachers' views according to criteria defined and set up in advance. However, study B has also a small explorative or bottom-up aspect, namely the grouping of criteria of the tasks' evaluation with the help of factor analyses and the cluster analysis used to discover profiles of task-related views. Study C was again a study using a qualitative approach, this time however with a focus on gathering deepening evidence according to a relatively narrow research question. The prior results of studies A and B allowed a very target-oriented sampling process which afforded identifying interesting cases in a relatively straightforward way. The examples analyzed in study C have a status comparable to a "proof of existence". The lack of generalizability which is often a difficulty in case-based research designs is hence not an issue here, as the existence of corresponding cases can answer the research question. At the same time, the cases also highlight *how* the phenomenon in question can occur. The three studies hence complement each other giving insight into the relationship of affect-related PCK and other PCK components at several relevant levels of professional teacher knowledge.

Beyond methodological considerations, the results of the three studies provide insight into teachers' awareness of affective aspects as PCK reflected in teachers' views about learning opportunities.

Study A shows that most teachers acknowledged affective aspects as important criteria of instructional quality, whereas criteria belonging to other aspects of the mathematics classroom were mentioned at lower relative frequencies. This multi-faceted picture of individual criteria of instructional quality corresponds to a relatively non-situated component of professional knowledge, as the data gathering format was not linked to specific contents or specific classroom situations.

Study B focused on views related to tasks in the content area of fractions and hence addressed a content-specific level of PCK. However, the design still afforded the possibility of multi-faceted evaluations. Somewhat astonishingly, the results revealed a certain (statistical) structure, showing that the motivational aspect of the representations used in the tasks loaded on a different factor than the other criteria for all tasks. Moreover, the results suggest that there might have been a conflict between affect-related views about the representations used in the tasks and views related to other PCK components, at least for a sub-group of the pre-service teachers.

In order to take a closer look at this potential conflict, study C focused on cases – and concentrated on situation-specific components of professional knowledge. Again it gets apparent that different criteria are used by the teachers for evaluating learning opportunities. These criteria appear to support or to restrict the three teachers' in-depth analyses of the quality of the given learning situation. The findings suggest that an excessive awareness of affective aspects of learning situations may "blind" teachers in situations in which they should also draw on other PCK areas for evaluating learning opportunities.

Indeed, an over-emphasis on affective aspects of learning might indicate deficits in other PCK. In a relatively rough consideration, all three studies could be interpreted as indicating specific professional development needs. The comparatively low relative frequencies for many important areas of instructional quality in study A for instance suggest that teachers should be more aware of criteria such as communication and interaction in the classroom rather than focusing mainly on affective characteristics. The results of study B suggest that knowledge necessary for a deepened evaluation of tasks and the use of representations should be strengthened, and study C gives insight into how motivational aspects might prevail over focusing on the understanding of learners, which points to a need of developing corresponding PCK.

Several questions for further research arise from this need of further development of professional knowledge: For instance, the role of prior knowledge or of classroom experience for views related to affective characteristics of learning situations merits deepened investigation. Moreover, circumstances of school culture such as different school types or perceived goals for instruction may impact the teachers' awareness of affective aspects, so that trans-cultural research designs could afford more insight.

Of course, motivational and affective aspects of learning opportunities are important to be considered by teachers and require specific PCK and awareness. The findings presented in this study suggest that this PCK should be well-integrated into other PCK components – and teachers should be aware that an emphasis on affect requires complementary consideration of other characteristics of learning opportunities.

**Acknowledgements** The project ABCmaths was funded with support from the European Commission (503215-LLP-1-2009-1-DE-COMENIUS-CMP). This publication reflects the views only of the authors, and the Commission cannot be held responsible for any use which may be made of the information contained therein.

## References

Ball, D. (1993). Halves, pieces, and twoths: Constructing representational contexts in teaching fractions. In T. Carpenter, E. Fennema, & T. Romberg (Eds.), *Rational numbers: An integration of research* (pp. 157–196). Hillsdale: Erlbaum.

Ball, D., Thames, L., & Phelps, G. (2008). Content knowledge for teaching: What makes it special? *Journal of Teacher Education, 59*(5), 389–407.

Bandura, A. (1977). Self-efficacy: Toward a unifying theory of behavioural change. *Psychological Review, 84*, 191–215.

Baumert, J., & Kunter, M. (2006). Stichwort: Professionelle Kompetenz von Lehrkräften. *Zeitschrift für Erziehungswissenschaft, 9*(4), 469–520.

Biza, I., Nardi, E., & Zachariades, T. (2007). Using tasks to explore teacher knowledge in situation-specific contexts. *Journal of Mathematics Teacher Education, 10*, 301–309.

Bromme, R. (1992). *Der Lehrer als Experte. Zur Psychologie des professionellen Wissens.* [The teacher as an expert. On the psychology of professional knowledge]. Bern: Hans Huber.

Clausen, M., Reusser, K., & Klieme, E. (2003). Unterrichtsqualität auf der Basis hoch-inferenter Unterrichtsbeurteilungen: ein Vergleich zwischen Deutschland und der deutschsprachigen Schweiz [Using high-inference ratings to assess quality of instruction. A comparison between Germany and the German-speaking part of Switzerland]. *Unterrichtswissenschaft, 31*(2), 122–141.

Deci, E. L., & Ryan, R. M. (1985). *Intrinsic motivation and self-determination in human behavior.* New York: Plenum.

Deci, F., & Ryan, R. (1993). Die Selbstbestimmungstheorie der Motivation und ihre Bedeutung für die Pädagogik. *Zeitschrift für Pädagogik, 39*, 223–238.

Duval, R. (2006). A cognitive analysis of problems of comprehension in a learning of mathematics. *Educational Studies in Mathematics, 61*, 103–131.

Elia, I., & Philippou, G. (2004). The functions of pictures in problem solving. In M. J. Høines & A. B. Fuglestad (Eds.), *Proceedings of the 28th conference of the international group for the psychology of mathematics education* (Vol. 2, pp. 327–334). Bergen: PME.

Heckhausen, H. (1989). *Motivation und Handeln* (2nd ed.). Berlin: Springer.

Helmke, A., & Weinert, F. (1997). Bedingungsfaktoren schulischer Leistungen. In F. Weinert (Hrsg.), *Enzyklopädie der Psychologie. Band 3: Psychologie des Unterrichts und der Schule* (S. 71–176). Göttingen: Hogrefe.

Hosenfeld, I. (2008). Diagnostische Kompetenzen von Mathematiklehrkräften und Leistung. [Presentation on 26.08.2009, 71st AEPF conference, Kiel].

Kersting, N., Givvin, K., Thompson, B., Santagata, R., & Stigler, J. (2012). Measuring usable knowledge: Teachers' analyses of mathematics classroom videos predict teaching quality and student learning. *American Educational Research Journal, 49*(3), 568–589.

Krapp, A. (1992). Das Interessenskonstrukt. In A. Krapp, & M. Prenzel (Hrsg.), *Interesse, Lernen, Leistung. Neuere Ansätze der pädagogisch-psychologischen Interessensforschung* (S. 297–329). Münster: Aschendorff.

Kultusministerkonferenz (KMK). (2003). *Bildungsstandards im Fach Mathematik für den Mittleren Schulabschluss*. http://www.kmk.org/

Kuntze, S. (2011). In-service and prospective teachers' views about modelling tasks in the mathematics classroom – Results of a quantitative empirical study. In G. Kaiser et al. (Eds.), *Trends in teaching and learning of mathematical modelling* (pp. 279–288). Dordrecht: Springer.

Kuntze, S. (2012). Pedagogical content beliefs: Global, content domain-related and situation-specific components. *Educational Studies in Mathematics, 79*(2), 273–292. doi:10.1007/s10649-011-9347-9.

Kuntze, S., & Rudolph-Albert, F. (2009). What is "good" mathematics instruction? – Mathematics teachers' individual criteria for instructional quality and attributions for instructional success. In J. Cai, G. Kaiser, B. Perry, & N.-Y. Wong (Eds.), *Effective mathematics teaching from teachers' perspectives: National and cross-national studies* (pp. 71–92). Rotterdam/Boston/Taipei: Sense Publishers.

Kuntze, S., Lerman, S., Murphy, B., Kurz-Milcke, E., Siller, H.-S., & Winbourne, P. (2011). Development of pre-service teachers' knowledge related to big ideas in mathematics. In B. Ubuz (Ed.), *Proceedings of the 35th conference of the international group for the psychology of mathematics education* (Vol. 3, pp. 105–112). Ankara: PME.

Lerman, S. (1990). Alternative perspectives of the nature of mathematics and their influence on the teaching of mathematics. *British Educational Research Journal, 16*(1), 53–61.

Martignon, L., & Wassner, C. (2005). Schulung frühen stochastischen Denkens von Kindern. *Zeitschrift für Erziehungswissenschaft, 8*(2), 202–222.

National Council of Teachers of Mathematics (NCTM) (Ed.). (2000). *Principles and standards for school mathematics*. Reston: NCTM.

Pajares, F. M. (1992). Teachers' beliefs and educational research: Cleaning up a messy construct. *Review of Educational Research, 62*(3), 307–332.

Pekrun, R., & Zirngibl, A. (2004). Schülermerkmale im Fach Mathematik. In Deutsches PISA-Konsortium (Hrsg.), *PISA 2003. Der Bildungsstand der Jugendlichen in Deutschland – Ergebnisse des zweiten internationalen Vergleichs* (S. 191–210). Münster: Waxmann.

Prenzel, M. (1988). *Die Wirkungsweise von Interesse*. Opladen: Westdeutscher Verlag.

Shulman, L. (1986). Those who understand: Knowledge growth in teaching. *Educational Researcher, 15*(2), 4–14.

Straub, J., Kempf, W., & Werbik, H. (Eds.). (1997). *Psychologie. Eine Einführung Grundlagen, Methoden, Forschungsfelder*. München: Deutscher Taschenbuch Verlag.

Törner, G. (2002). Mathematical beliefs. In G. Leder, E. Pehkonen, & G. Törner (Eds.), *Beliefs: A hidden variable in mathematics education?* (pp. 73–94). Dordrecht: Kluwer.

Weinert, F. (2001). Vergleichende Leistungsmessung in Schulen – eine umstrittene Selbstverständlichkeit. In F. Weinert (Hrsg.), *Leistungsmessungen in Schulen* (S. 17–31). Weinheim: Beltz.

# Preschool Teachers' Knowledge and Self-Efficacy Needed for Teaching Geometry: Are They Related?

**Pessia Tsamir, Dina Tirosh, Esther Levenson, Michal Tabach, and Ruthi Barkai**

**Abstract** This chapter focuses on methodological issues related to investigating preschool teachers' self-efficacy for teaching geometry. The first issue discussed is the specificity, as opposed to the generality, of self-efficacy and the need to design instruments which are sensitive to this aspect of self-efficacy. Specificity may be related to content, in this case geometry and the specific figures under investigation. In other words, self-efficacy for teaching triangles may differ from self-efficacy for teaching pentagons. Self-efficacy may also be related to the specific action being performed, such as designing tasks for promoting knowledge versus designing tasks for evaluating knowledge. The chapter also investigates the relationship between preschool teachers' knowledge and self-efficacy for identifying geometrical figures, presenting a method for studying this relationship but also raising questions related to this method.

**Keywords** Preschool teachers • Teachers' self-efficacy • Teachers' knowledge • Specificity • Geometry

## Introduction

Research has shown that promoting young children's mathematics knowledge is important and that the preschool teacher has a significant role in supporting the development of this knowledge (e.g., Ginsburg et al. 2008). Towards the aim of promoting early childhood mathematics education, several position papers have called for advancing preschool teachers' knowledge for teaching mathematics. For example, a joint position paper published in the United States by the National Association for the Education of Young Children (NAEYC) and the National Council for Teachers of Mathematics (NCTM) called for ongoing professional development

P. Tsamir • D. Tirosh • E. Levenson (✉) • M. Tabach • R. Barkai
School of Education, Department of Mathematics, Science,
and Technology Education, University of Tel Aviv, Israel, Tel Aviv, Israel
e-mail: pessia@post.tau.ac.il; dina@post.tau.ac.il; levensone@gmail.com;
levenso@post.tau.ac.il

 
B. Pepin, B. Roesken-Winter (eds.), *From beliefs to dynamic affect systems in mathematics education*, Advances in Mathematics Education,
DOI 10.1007/978-3-319-06808-4_16

that would "move beyond the one-time workshop to deeper exploration of key mathematical topics as they connect with young children's thinking and with classroom practices" (NAEYC and NCTM 2002, p. 6). Teachers' knowledge is one of several factors affecting teachers' actions in the classroom. Studies have also shown that teachers with a high self-efficacy are more enthusiastic and more committed to teaching (Allinder 1994; Coladarci 1992); thus, it is also important to investigate and promote preschool teachers' self-efficacy related to the teaching of mathematics.

For the past several years, our research team has investigated preschool teachers' knowledge and self-efficacy for teaching number and geometry concepts. During our investigation, several issues related to the research methods have arisen. One of these issues relates to the specificity of self-efficacy. Research has shown that self-efficacy is content specific. If so, how specific do the content areas have to be? Is it enough to differentiate between preschool teachers' self-efficacy for teaching number concepts and their self-efficacy to teach geometrical concepts or might there be a difference within the domain of number and geometry, for example, between teaching triangles and pentagons? The same question may be asked related to the specificity of the actions being performed. Teaching mathematics in preschool involves the coordination of several activities on the part of the teacher, among them designing mathematical tasks for the children, holding discussions related to some mathematical situation, and answering children's mathematical questions. Is it enough to differentiate between preschool teachers' self-efficacy for designing tasks and their self-efficacy for answering children's mathematical questions? Within the activity of designing tasks for children, might there be a difference between teachers' self-efficacy for designing tasks aimed at promoting children's knowledge versus designing tasks aimed at evaluating children's knowledge? An additional concern is the relationship between teachers' knowledge and self-efficacy. Theoretically, there are four combinations which may occur: high knowledge together with high self-efficacy, high knowledge together with low self-efficacy, low knowledge together with high self-efficacy, and low knowledge together with low self-efficacy. In reality, do all of these combinations exist? Are knowledge and self-efficacy for teaching mathematics in preschool related? Finally, we ask, how might results of such research impact on professional development programs for preschool teachers. These questions will be discussed in this chapter.

## Teacher Self-Efficacy, Mathematics Self-Efficacy, and Self-Efficacy for Teaching Mathematics

This paper discusses the study of preschool teachers' self-efficacy for teaching mathematics as well as the relationship between self-efficacy and knowledge for teaching mathematics. In a sense, it draws on a combination of studies: studies related to mathematics self-efficacy and studies related to teachers' self-efficacy, as well as studies related to teachers' knowledge for teaching mathematics. This section reviews studies related to self-efficacy. In the next section, when presenting the framework of program, we refer to studies of teachers' knowledge.

According to Bandura's (1986) social cognitive theory, there is a relationship between psychodynamic and behavioristic influences, as well as personal beliefs and self-perception, when explaining human behavior. Bandura defined self-efficacy as "people's judgments of their capabilities to organize and execute a course of action required to attain designated types of performances" (1986, p. 391). It is different than self-concept, which is more related to judgments about one's attributes, rather than what a person can do. It is also important to stress that self-efficacy cannot be measured by an all-purpose measure (Bandura 2006). Self-efficacy beliefs are not only domain specific (e.g., mathematics, history, science) and content specific (e.g., within the domain of mathematics there is numeracy, patterns, geometry, etc.), but action specific (e.g., is the activity implemented in class, outside, individually, in a group) (Pajares 1996; Zimmerman 2000).

Hackett and Betz (1989) defined mathematics self-efficacy as, "a situational or problem-specific assessment of an individual's confidence in her or his ability to successfully perform or accomplish a particular [mathematics] task or problem" (p. 262). With regard to mathematics self-efficacy, research has shown that regardless of mathematical ability, students with a higher self-efficacy tend to expend more effort on difficult mathematics tasks than students with lower self-efficacy (Collins 1982) and that students' self-efficacy beliefs are positively related to mathematics performance (Bandura 1986; Pajares 1996). Even among 6-year old children, mathematics self-efficacy and behavior were found to be positively related (Davis-Kean et al. 2008). Despite Bandura's (1986) claim that self-efficacy cannot be globally measured and that it is action-specific, and despite Hackett and Betz's (1989) assertion that mathematics self-efficacy is problem-specific, some studies which investigated mathematics self-efficacy included general items such as "I'm doing well in mathematics at school" (Merenluoto 2004, p. 299). Other studies were more specific. For example, Pajares and Miller (1994) used a questionnaire which differentiated between domains of mathematics, cognitive demands, and problem contexts. Pajares and Graham (1999) used an even more problem-specific questionnaire where students were shown specific mathematics questions and were then asked to assess their ability to solve them. Likewise, the survey of the Programme for International Student Assessment (PISA) which took place in 2003 assessed students' mathematics self-efficacy by asking them to what degree they felt confident solving each of eight specific mathematics problems such as calculating how much cheaper a television would be after a 30 % discount (Schulz 2005). In short, different studies included questionnaire items with varying degrees of specificity, regarding both domain and problem specificity.

When relating theories of self-efficacy to teachers, Dellinger et al. (2008) differentiated between teacher efficacy and teacher self-efficacy. The first, teacher efficacy, "assesses teachers' beliefs in their ability to affect student performance (outcome)" (p. 752). These beliefs, however, may be confounded by a teacher's sense of control. Many factors affect students' performance, some not within the teacher's control and not necessarily dependent on the teacher's ability to teach. This study does not focus on teacher efficacy but rather on teacher self-efficacy. Teacher self-efficacy may be conceptualized as "what the individual teacher can accomplish given the limitations caused by external factors" (Skaalvik and Skaalvik 2007, p. 612) or as "a teacher's individual beliefs in their capabilities to perform specific teaching tasks at a specified level of quality in a specified situation" (Dellinger et al. 2008, p. 752).

Several studies have investigated teachers' self-efficacy. When constructing items for questionnaires, some of those studies noted that teachers were consulted with regard to identifying situations and tasks encountered in teachers' daily school activities and which were important to them. For example, in a study which took place in Italy, two of the items included were "I can make my students respect rules and codes of conduct" and "I am capable of engaging even the most reluctant and difficult students in my class activities" (Caprara et al. 2006, p. 481). In Norway, a study which investigated teacher self-efficacy and burnout, took into consideration the national curriculum which stresses differential instruction (Skaalvik and Skaalvik 2007). Thus, test items addressed the teacher's belief in his or her ability to "provide good guidance and instruction to all students regardless of their level of ability" and "provide realistic challenge for all students even in mixed ability classes" (Skaalvik and Skaalvik 2007, p. 624). The above mentioned studies investigated teachers' self-efficacy without regard for a specific content domain. We now turn to studies related to self-efficacy and teaching mathematics.

In order to discuss teachers' self-efficacy for teaching mathematics, we differentiate between teachers' mathematics self-efficacy, i.e., self-efficacy related to solving mathematics problems, and teachers' self-efficacy for teaching mathematics. This differentiation was also pointed out by Bates et al. (2011) who investigated the relationship between early childhood (pre-K to third grade) preservice teachers' mathematics self-efficacy and their mathematics teaching self-efficacy. The instruments used in the study conducted by Bates et al. (2011) were the Mathematics Self-Efficacy Scale developed by Betz and Hackett (1993) and the Mathematics Teaching Efficacy Belief Instrument, developed by Enochs et al. (2000). In general, results of the study showed that teachers who reported higher mathematics self-efficacy were more confident in their ability to teach mathematics than teachers with a lower mathematics self-efficacy. Results also showed that teachers who had a higher mathematics self-efficacy performed better on a basic mathematics skills test than participants with a lower mathematics self-efficacy. However, participants with a high mathematics teaching self-efficacy did not necessarily perform well on the mathematics skills test. In other words, some teachers who scored low on the skills test still felt confident to teach mathematics. While the authors pointed out that these results could be due to the inexperience of the preservice teachers, we raise additional questions. For example, how well did the items on the skill test match the items on the teacher self-efficacy questionnaire. The skills test measured participants' ability to solve problems involving integers, fractions, algebra, and geometry. The mathematics teaching efficacy questionnaire included general statements such as "I will continually find better ways to teach mathematics" (Enochs et al. 2000). It could be that in situations where the items on the two questionnaires are more closely related, a correlation would be found. It could also be that early childhood teachers may know that they cannot solve algebra problems but feel confident in their ability to teach the mathematics necessary for young children. These issues are taken into consideration in the next section which presents the framework of our program.

## Framework and Study Background

Our professional development program for preschool teachers is guided by the Cognitive Affective Mathematics Teacher Education (CAMTE) framework (e.g. Tirosh et al. 2011; Tsamir et al. 2014a). This framework takes into consideration the interrelationship between knowledge and beliefs which can affect teachers' proficiency (Schoenfeld and Kilpatrick 2008).

The framework is presented in Table 1. In Cells 1–4, and in Cells 5–8, we address teachers' knowledge and self-efficacy respectively. The same framework guides our research study.

In framing the mathematical knowledge preschool teachers need for teaching, we draw on Shulman (1986) who identified subject-matter knowledge (SMK) and pedagogical content knowledge (PCK) as two major components of teachers' knowledge necessary for teaching. In our previous work with teachers, we found it useful to differentiate between two components of teachers' SMK: being able to produce solutions, strategies and explanations and being able to evaluate given solutions, strategies and explanations (Tabach et al. 2010). Thus our framework takes into consideration both of these aspects of SMK.

Regarding PCK, we draw on the works of Ball and her colleagues (2008) who refined Shulman's theory and differentiated between two aspects of PCK: knowledge of content and students (KCS) and knowledge of content and teaching (KCT). KCS is "knowledge that combines knowing about students and knowing about mathematics" whereas KCT "combines knowing about teaching and knowing about mathematics" (Ball et al. 2008, p. 401). Under this last category, we focus on the design, evaluation, and implementation of mathematical tasks. In Israel, there is a mandatory mathematics preschool curriculum (INMPC 2008), but few curricular materials are available. Teachers often find themselves designing their own tasks to implement in their classes and so it is especially important for them to appreciate the design process and its implication for creating learning opportunities. For each aspect of knowledge in the framework, there is a corresponding aspect of self-efficacy. Thus, the CAMTE framework takes into consideration teachers' mathematics

**Table 1** The cognitive affective mathematics teacher education framework

| | Subject-matter | | Pedagogical-content | |
|---|---|---|---|---|
| | Solving | Evaluating | Students | Tasks |
| Knowledge | Cell 1: Producing solutions | Cell 2: Evaluating solutions | Cell 3: Knowledge of students' conceptions | Cell 4: Designing and evaluating tasks |
| Self-efficacy | Cell 5: Mathematics self-efficacy related to producing solutions | Cell 6: Mathematics self-efficacy related to evaluating solutions | Cell 7: Pedagogical-mathematics self-efficacy related to children's conceptions | Cell 8: Pedagogical-mathematics self-efficacy related to designing and evaluating tasks |

self-efficacy (Table 1, Cells 5 and 6) as well as their pedagogical-mathematics self-efficacy, i.e. their self-efficacy related to the pedagogy of teaching mathematics (Table 1, Cells 7 and 8). What we term pedagogical-mathematics self-efficacy corresponds in a way to what was referred to in the previous section as self-efficacy for teaching mathematics (Enochs et al. 2000). However, in accordance with Bandura (1986) we relate to a more action-specific self-efficacy, i.e., self-efficacy is related to specific, as opposed to general actions being performed. This will be illustrated in the following section. In the following section we also show how the framework was used to design tools to investigate knowledge and self-efficacy for teaching geometry concepts.

Over the years we have gathered data from several groups of preschool teachers who have participated in our professional development programs. The teachers were all practicing teachers at the time they participated in the program, teaching children ages 3–6 years old in municipal preschools, sometimes in mixed-aged groups and sometimes in separate-aged groups. In Israel, children begin first grade at age 6, so we consider kindergarten to be the last year before elementary school. All teachers had a B.Ed., specializing in early childhood education, obtained after completing a 4-year course of study in a teacher-education college. Early childhood programs in these colleges focus mainly on psychology, sociology, and general education, with less attention paid to teaching content such as mathematics.

Two questionnaires were used in this study, one focusing on teaching two-dimensional (2-D) shapes and one on three-dimensional (3-D) solids. The mathematical content of each questionnaire and the subsequent items built for each questionnaire were based on the mandatory Israel National Mathematics Preschool Curriculum (INMPC 2008) which provides guidelines and standards for teaching mathematics to children ages 3–6 years old and on our previous research with young children (e.g., Tsamir et al. 2008). In the next sections we describe in more detail different items of the specific questionnaires, how the data was analyzed, and related results. The section "Preschool teachers' pedagogical-mathematical self-efficacy: content specificity" focuses on the question of self-efficacy being content and action specific. The section "Relating self-efficacy to knowledge" focuses on the relationship between self-efficacy and knowledge.

## Preschool Teachers' Pedagogical-Mathematical Self-Efficacy: Content Specificity

In order to investigate the question of content and action specificity, we focus on Cell 8 (pedagogical-mathematics self-efficacy related to designing and evaluating tasks) of the CAMTE framework, describing related items and results from the 3-D questionnaire.

## *Tools and Data Analysis: Teachers' Self-Efficacy for Designing Tasks*

Teachers design tasks for many purposes. In this study, we differentiated between tasks used to promote children's knowledge and tasks used to evaluate children's knowledge. A four-point Likert scale was used to rate participants' agreements with self-efficacy statements: (1) I do not agree that I am capable; (2) I somewhat agree that I am capable; (3) I agree that I am capable; (4) I strongly agree that I am capable. The statements which teachers were asked to rate their agreement with were:

1. I am capable of designing tasks to *promote* children's knowledge of *cones*;
2. I am capable of designing tasks to *evaluate* children's knowledge of *cones*;
3. I am capable of designing tasks to *promote* children's knowledge of *cylinders*;
4. I am capable of designing tasks to *evaluate* children's knowledge of *cylinders*.

Altogether, we collected questionnaires from 62 practicing preschool teachers. The data collected from the above four questions led to four very specific self-efficacy scores, referring to specific figures as well as to designing tasks for specific purposes. We then calculated the mean self-efficacy score for questions (1) and (2) and then questions (3) and (4), resulting in more general self-efficacy scores for designing tasks for cones and cylinders. In other words, content specificity (i.e., separation of figures) was preserved but action specificity (i.e., separating designing tasks for promoting knowledge from designing tasks for evaluating knowledge) was generalized. We then calculated mean scores for questions (1) and (3) and then questions (2) and (4), resulting in more general self-efficacy scores for promoting children's knowledge of 3-D figures and evaluating children's knowledge of 3-D figures. In other words, we kept the activity very specific and generalized the content.

## *Results: Specificity of Self-Efficacy for Designing 3-D Geometry Tasks*

Taking into consideration that the self-efficacy scale ran from 1 (lowest) to 4 (very high), in general, teachers did not have a very high self-efficacy when it came to designing tasks related to three-dimensional figures (Table 2). This was true for both cones and cylinders as well as for designing tasks for promoting knowledge and designing tasks for evaluating knowledge. In general, it also seemed that teachers' self-efficacy related to designing cylinder tasks was greater than teachers' self-efficacy for designing cone tasks and that self-efficacy related to designing tasks for promoting knowledge was greater than self-efficacy related to designing evaluation tasks.

**Table 2** Self-efficacy for designing different types of tasks per figure

| Designing tasks for... | Promoting knowledge | | Evaluating knowledge | |
|---|---|---|---|---|
| Figures | M | SD | M | SD |
| Cones (N=61) | 2.33 | .87 | 2.18 | .85 |
| Cylinders (N=60) | 2.52 | .79 | 2.45 | .81 |

**Table 3** Comparing self-efficacy: different figures and different activities

| Cones versus cylinders | Mean difference | t-value | df | p-value |
|---|---|---|---|---|
| Designing tasks for *promoting* knowledge: cones versus cylinders | –.17 | –1.80 | 59 | .077 |
| Designing tasks for *evaluating* knowledge: cones versus cylinders | –.22 | –2.43 | 59 | .018 |
| Designing tasks in general: cylinders versus cones | –.17 | –1.92 | 60 | .060 |
| Promoting knowledge versus evaluating knowledge | | | | |
| Cones: designing tasks for *promoting* knowledge versus tasks for *evaluating* knowledge | .15 | 2.87 | 60 | .006 |
| Cylinders: designing tasks for *promoting* knowledge versus tasks for *evaluating* knowledge | .07 | 2.05 | 59 | .045 |
| 3-D figures: designing tasks for *promoting* knowledge versus tasks for *evaluating* knowledge | .12 | 2.95 | 61 | .004 |

In order to analyze if the general results outlined above were significant, paired-samples t-tests were carried out. Results are presented in Table 3. Differentiating between cones and cylinders, we see that teachers had a significantly lower self-efficacy for designing tasks to evaluate children's knowledge of cones than for designing tasks to evaluate children's knowledge of cylinders. However, when it came to designing tasks to promote knowledge or just designing tasks in general, the specific figure, whether it was a cylinder or cone which was the object being discussed, did not seem significant. Focusing on the types of tasks being designed, significant differences were consistently found between teachers' self-efficacy for designing tasks to promote knowledge and their self-efficacy for designing tasks to evaluate knowledge, regardless of the figures being targeted. In other words, for this group of preschool teachers, task-specificity seems to be more of an issue than the specific 3-D figure at stake. Furthermore, it seems that teachers have a higher self-efficacy when it comes to designing tasks for promoting knowledge than they do for designing tasks to evaluate knowledge. That being said, although some of the results were significant, they were relatively small. Thus, teacher educators should consider promoting teachers' self-efficacy for designing both types of tasks.

## Relating Self-Efficacy to Knowledge

### *Tools and Data Analysis: Identifying Two and Three Dimensional Figures*

In order to investigate the possible relationships between knowledge and self-efficacy, we focus on items related to Cells 1 and 4 of the CAMTE framework (knowledge and self-efficacy for solving problems) from the 2-D and 3-D questionnaires. Within the context of two-dimensional shapes, we chose to focus on identifying triangles, pentagons, and circles. Within the context of three-dimensional figures, we focused on cones and cylinders. Identifying these two and three-dimensional figures is mentioned specifically in the preschool mathematics curriculum as a competency expected of kindergarten children (INMPC 2008). Both of these questionnaires consisted of two parts. The first part of the 2-D questionnaire began with the following self-efficacy related questions: If I am shown a triangle, I will be able to identify it as a triangle. If I am shown a figure which is not a triangle, I will be able to identify it as not being a triangle. This was repeated for pentagons and circles. Likewise, the 3-D questionnaire inquired about teachers' ability to identify cones and cylinders as well as their ability to identify nonexample of cones and nonexamples of cylinders. As previously described, a four-point Likert scale was used for these questions, 1 meaning the teacher was not in agreement that she was able to identify the figure and 4 meaning that she was in complete agreement that she was able to identify the figure.

When analyzing the data from these items, a mean self-efficacy score was created for each figure from the two self-efficacy questions related to identifying examples and nonexamples of that figure. A more general mean self-efficacy score was then calculated reflecting self-efficacy for identifying two-dimensional figures and for identifying three-dimensional figures.

After the first part of the questionnaire was collected, the second part was handed out. The second part of each questionnaire consisted of a series of examples and nonexamples of different figures. Each figure was accompanied by a question: Is this a triangle (or pentagon or cylinder) Yes/No? Figures 1, 2, and 3 present the figures used when investigating triangles, pentagons, and circles. Figures 4 and 5 present the figures used when investigating cones and cylinders. In choosing the figures, both mathematical and psycho-didactical dimensions were considered. That is, we not only considered whether the figure is an example or a nonexample, but whether or not it would intuitively be recognized as an example or a nonexample (Tsamir et al. 2008). When considering triangles, for example, the equilateral triangle may be considered a prototypical triangle and thus intuitively recognized as a triangle, accepted immediately without the feeling that justification is required (Hershkowitz 1990; Tsamir et al. 2008). The narrow and long scalene triangle may be considered a non-intuitive example because of its "skinniness" (Tsamir et al. 2008). The nonexamples were chosen so that for each figure one critical attribute

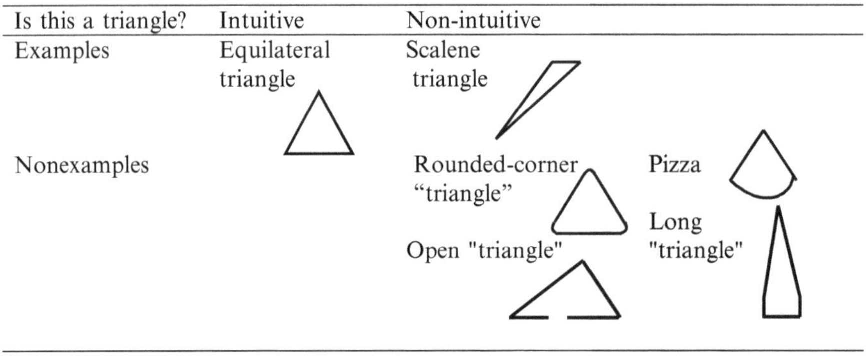


**Fig. 1** Is this a triangle?

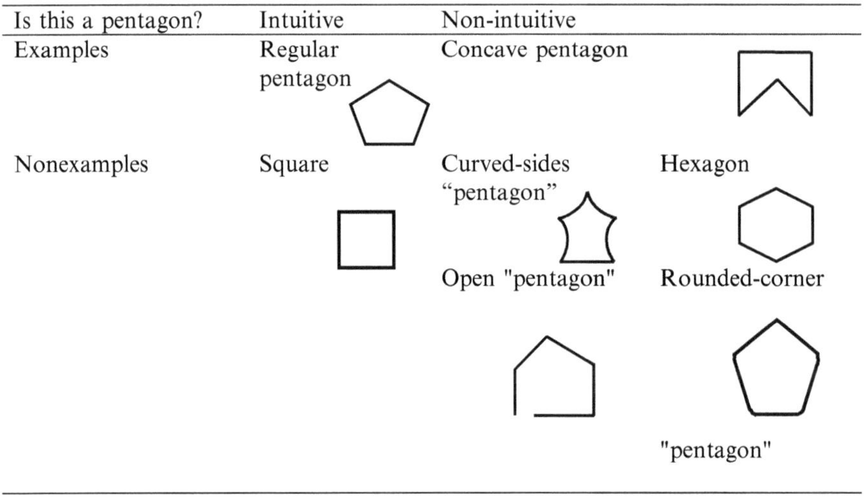


**Fig. 2** Is this a pentagon?

is contradicted. Thus, one figure is open, another has five sides; one has a curved side and another has rounded corners. Whereas a circle may be considered an intuitive nonexample of a triangle, the pizza-like "triangle" may be considered a non-intuitive nonexample because of visual similarity to a prototypical triangle. Similarly, the regular pentagon was thought to be easily recognized by children who had been introduced to pentagons whereas, the concave pentagon is more difficult to identify.

As few studies have investigated young children's knowledge of solids, our differentiation between intuitive and nonintuitive solids is based on our experience and studies regarding how children identify them (Tirosh and Tsamir 2008) and related studies with two-dimensional shapes. For example, studies have shown that

**Fig. 3** Is this a circle?

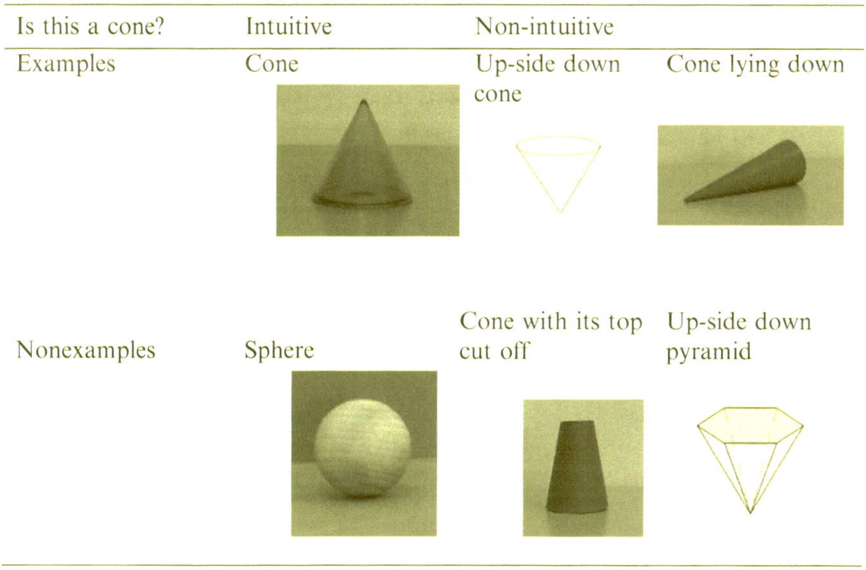


**Fig. 4** Is this a cone?

a circle may be considered as an intuitive nonexample of a triangle (Tsamir et al. 2008) and that for many children, being able to give a name to one shape guarantees that it will not be some other shape. Likewise, because most children can name a ball, we classified the ball as an intuitive nonexample of a cone.

When analyzing data from these items, a mean score was configured for identifying each of the different figures. For example, when investigating identification of a cone, six figures were presented. Thus, a participant who correctly identified (either as an example or as a nonexample) three out of the six figures, received a score of 50 %. As with the self-efficacy scores, a general mean knowledge score was configured separately for the two and three-dimensional figures, reflecting teachers' knowledge for identifying two-dimensional figures and their knowledge for identifying three-dimensional figures.

| Is this a cylinder? | Intuitive | Non-intuitive | |
|---|---|---|---|
| Examples | Cylinder  | "Coin-like" cylinder  | Cylinder lying  down |
| Nonexamples | Sphere  | Cone with its top cut off  | Cylinder cut on a slant  |

**Fig. 5** Is this a cylinder?

**Table 4** Mean knowledge scores and self-efficacy scores per 2-D and 3-D figure

| | Correct identification | | Self-efficacy | |
|---|---|---|---|---|
| | M | SD | M | SD |
| Triangle (N = 19) | .95 | .14 | 3.82 | .38 |
| Pentagon (N = 18) | .88 | .13 | 3.47 | .58 |
| Circle (N = 17) | .98 | .07 | 3.34 | .58 |
| General 2-D (N = 19) | .94 | .08 | 3.55 | .43 |
| Cone (N = 63) | .93 | .14 | 3.04 | .73 |
| Cylinder (N = 62) | .87 | .13 | 3.22 | .58 |
| General 3-D (N = 63) | .90 | .11 | 3.10 | .65 |

## *Results: Relating Knowledge and Self-Efficacy for Identifying Two and Three Dimensional Figures*

We begin by presenting overall results of participants' self-efficacy and knowledge for identifying the various specific figures. Recall that self-efficacy was rated on a scale of 1–4, 1 being very low and 4 being very high. Results (Table 4) indicated that in general, participants were able to identify two and three dimensional figures and had a high self-efficacy regarding their ability to do so.

In order to investigate the question of whether preschool teachers' knowledge for identifying some figure is related to their self-efficacy for identifying that figure, Pearson correlations were carried out for each figure. For example, we compared teachers' knowledge of identifying triangles with their self-efficacy for identifying triangles. For the most part, knowledge and self-efficacy were not found to be

**Table 5** Levels of knowledge versus self-efficacy – identifying cones

| Self-efficacy knowledge | Low | High | Total |
|---|---|---|---|
| Low | 12 | 5 | 17 |
| High | 24 | 16 | 40 |
| Total | 36 | 21 | 57 |

**Table 6** Levels of knowledge versus self-efficacy – identifying cylinders

| Self-efficacy knowledge | Low | High | Total |
|---|---|---|---|
| Low | 25 | 8 | 33 |
| High | 12 | 11 | 23 |
| Total | 37 | 19 | 56 |

related. There were two exceptions. Results indicated a significant positive correlation between teachers' knowledge for identifying cylinders and their self-efficacy for identifying cylinders ($r = .30$, $p = .03$) and for identifying, in general, 3-D figures and their self-efficacy for identifying 3-D figures ($r = .30$, $p = .02$).

Being that significant results were only found related to 3-D figures, we decided to further analyze the distribution of results for the cones and cylinders. Specifically, we were interested in the possibility that teachers who were knowledgeable had a low self-efficacy and/or teachers who were less knowledgeable, nevertheless had a high self-efficacy.

Tables 5 and 6 describe the distribution of low and high knowledge scores for identifying cones and cylinders, respectively, versus low and high self-efficacy, where low and high was determined by the mean score for each variable. We acknowledge that the mean knowledge scores for both the cones and cylinders were above 85 % and that it might seem harsh to claim that a score of less than 85 % is low. However, taking into consideration that all participants were already practicing teachers, and that the means were indeed high, we feel that a score below the mean may be considered in this case, to be low. In general, we see that for cones and cylinders, all four possible combinations of high and low knowledge and self-efficacy exist. We also note that for both figures, few teachers exhibited a low level of knowledge with a high level of self-efficacy, meaning that there were few teachers who could not identify the figures but thought that they could do so. Finally, we note that the phenomenon of being able to correctly identify figures but yet not being aware of this knowledge, was more prevalent for cones than for cylinders.

## Discussion

There were two main issues investigated in this chapter: the specificity of self-efficacy and the relationship between knowledge and self-efficacy. When organizing this chapter, the dilemma arose regarding which section should be presented first, the section focusing on specificity or the section focusing on the relationship between knowledge and self-efficacy. On the one hand, we felt that before discussing teachers' self-efficacy for designing geometry tasks, we should first investigate their geometric knowledge. After all, being able to identify cylinders is a prerequisite for being able to design tasks to promote children's knowledge of cylinders. And yet, as we began to design an instrument to investigate different elements of teachers' self-efficacy for teaching geometry, we found ourselves grappling with the question of specificity. That is, yes, investigating knowledge is of prime importance and the relationship between knowledge and self-efficacy is a relevant question. But before this can be investigated, we have to address the issue of how specific the self-efficacy instrument need be. And so, as the focus of this chapter is on methodological questions, we decided to present the sections in the order of which we grappled with the questions.

Two aspects of specificity were discussed in this paper. The first related to the specific figure, cones versus cylinders, and the second related to the specific activity, designing tasks for promoting knowledge versus tasks for evaluating knowledge. Building very specific questions was carried out in accordance with Bandura's (1986) theory that self-efficacy cannot be measured globally. As mentioned previously, Hackett and Betz (1989) asserted that mathematics self-efficacy is problem specific.

The issue of specificity in self-efficacy also arose in other studies we conducted with preschool teachers. For example, when studying teachers' self-efficacy for teaching number concepts, we differentiated between teachers' self-efficacy for teaching verbal counting versus their self-efficacy related to object counting (Tsamir et al. 2014b). While both types of counting are related, they involve different skills. Verbal counting includes being able to say the number words in the proper order and knowing the principles and patterns in the number system as coded in one's natural language. Object counting refers to counting objects for the purpose of saying how many. Gelman and Gallistel (1978) outlined five principles of counting objects: the one-to-one correspondence principle, the stable-order principle, the cardinal principle, the abstraction principle, and the order-irrelevance principle. Recognizing the complexity of counting, one item addressed teachers' self-efficacy to promote children's skill in verbally counting up to 30 while a different item addressed teachers' self-efficacy to promote counting eight objects. Notice also, that in those two questions, the specific number to which children should count to and the specific number of objects to be counted is also related. In other words, the question addressed very specific counting skills and not general ones. In addition to differentiating between self-efficacy related to verbal and object counting, we also related to the specific skills involved with each type of counting. For example, saying which number comes before and after a given number, are two separate important skills related to verbal counting. Thus, one item investigated teachers' self-efficacy related to promoting children's knowledge of which number *follows* each of the numbers from 0 to 9 while a separate item was directed at teachers' self-efficacy for promoting the skill of saying which number comes *before* each of the numbers 1–10. Other number skills promoted during preschool are composing and decomposing numbers and recognizing number symbols. Once again, a separate item on the questionnaire addressed teachers' self-efficacy for teaching each of these skills. Preliminary results indicated that teachers' self-efficacy varied with the items. For example, teachers had a higher self-efficacy for building tasks that assess children's knowledge of enumerating eight items than they did for building tasks that assess children's knowledge related to the counting sequence up till 30.

In the current study, we separated not only between two and three-dimensional figures but investigated self-efficacy related to specific figures, cones versus cylinders. The issue of specificity arose, even in this case, when designing items to investigate teachers' self-efficacy related to identifying each of the figures. One question was directed at teachers' self-efficacy for identifying examples of, for instance, a cone, while a separate question addressed teachers' self-efficacy for identifying nonexamples of a cone. No significant differences were found between teachers' self-efficacy for identifying examples and nonexamples of cones or of cylinders and thus we configured a more general self-efficacy score for identifying each shape. However, at the start, specificity of self-efficacy was taken into consideration. The question which arises from these results is how many items ought to be used in order to insure specificity, taking into consideration, perhaps, that the more items there are on a test, the more general the test might be considered. Hackett and Betz (1989) related to this issue by dividing mathematics self-efficacy into three sub-scales, each containing between 16 and 18 items, in order to measure three sub-constructs of mathematics self-efficacy. This is an open question which needs further investigation.

In general, for this group of preschool teachers, the type of activity (in this case, designing tasks for promoting knowledge versus designing tasks for evaluating knowledge) seemed to have more of an effect on self-efficacy than the specific figure being discussed. Of course, we only differentiated between cones and cylinders. A next step would be to investigate additional 3-D figures. It could also be that the figure is less important when the action being taken is designing tasks, but for other actions, such as responding to children's questions, the specific figure may be very relevant. In the case when a difference was noted between cones and cylinders, designing tasks for promoting knowledge, teachers' had a higher self-efficacy with regard to cylinders. We take note of this as we consider the second issue of this study, the relationship between knowledge and self-efficacy.

The relationship between knowledge and self-efficacy was investigated with regard to identifying 2-D and 3-D figures. No correlations were found within the group of 2-D figures. Within the group of 3-D figures, a significant correlation was found between knowledge and self-efficacy for identifying cylinders. Once again, significant results were found with regard to cylinders but not for cones. In general, it seems that teachers were more aware of their knowledge of cylinders than of cones. Perhaps it was this awareness which affected their higher self-efficacy with regard to designing tasks for promoting knowledge of cylinders than for cones. This is in line with Bates et al. (2011) who found that teachers with a higher mathematics self-efficacy were more confident in their ability to teach mathematics than teachers with a lower mathematics self-efficacy. However, in that study, questionnaire items related to general teaching abilities. In our study, we attempted to pinpoint the different activities a teacher must perform. A next step for us might be to investigate the relationship between mathematics self-efficacy (Cell 5 of the CAMTE framework, Table 1) and teachers' pedagogical-mathematics self-efficacy related to designing tasks (Cell 8 of the framework).

While some significant results were noted, for the most part, knowledge and self-efficacy were not significantly correlated. Non-significant results could mean very plainly that no correlation exists. However, as noted in the background, previous studies found mathematics self-efficacy positively related to performance (Hackett and Betz 1989; Bates et al. 2011). This raises methodological questions. Insignificant correlations may be due to insufficient variance among the variables. While it could very well be that teachers have no difficulties identifying various examples and nonexamples of figures, it could also be that a questionnaire, with more examples and nonexamples of the different figures, would differentiate more clearly between levels of knowledge among teachers. When investigating self-efficacy, nearly all teachers rated their self-efficacy for identifying figures as high (3) or very high (4). It could be that a finer scale is necessary and that the results of this study were limited by a ceiling effect.

Another methodological issue which needs to be investigated is the order of the presentation/administration of the self-efficacy and performance questions. In accordance with previous studies which investigated self-efficacy and performance (e.g. Hackett and Betz 1989) our questionnaire began with self-efficacy questions and then proceeded to performance questions. On the one hand, this makes sense. If I see that I can successfully complete a given activity then I will believe in my ability to complete the same activity again. Thus, if we placed the performance questions first, it could affect how teachers answered the self-efficacy questions. But does that mean that the self-efficacy questions were not influenced by other factors? According to Bandura (1986) one of the sources for self-efficacy beliefs are performance attainments; success raises self-efficacy while failure lowers it. In other words, it is possible that the teachers' past experiences with geometric activities, affected how they responded to the self-efficacy questions. For example, teachers were asked to estimate their ability to identify nonexamples of circles. What nonexamples came to their mind when answering this self-efficacy question? Perhaps they recalled a time when they incorrectly identified an ellipse as a circle. This would then affect their self-efficacy to identify nonexamples of circles. Finally, we also question the assumption that a person's knowledge is unshakeable. Knowledge, or performance on tasks, might be influenced by several factors other than the individual's ability to perform the activity. Might it be that answering the self-efficacy questions affected teachers' performance on the tasks that followed? This needs further investigation.

To summarize, this chapter focused on methodological issues related to investigating teachers' knowledge, self-efficacy, and the relationship between them. We showed how one can design questionnaires that allow for different levels of specificity, both in content and in actions. With relation to content, we began with very specific items to investigate both knowledge and self-efficacy and gradually generalized the investigation. For example, we presented for identification very specific examples and nonexamples of different figures and from the specific items configured a general knowledge score for cones and cylinders. A similar process was carried out with self-efficacy items. Likewise, we differentiated between specific actions, for example, between designing tasks for promoting knowledge and designing tasks for evaluating knowledge. In all cases, there is also the issue of scaling self-efficacy. The scale we used ran from 1 to 4. Perhaps a wider scale would have been more sensitive to differences in self-efficacy. All of these issues influence the results of such an investigation, of knowledge, self-efficacy, and the relationship between them. Our task, as mathematics educators is to design questionnaires that are both specific enough and yet general enough to investigate these issues.

How can the results of this study inform professional development programs for preschool teachers? Teachers in this study were able to identify most of the examples and nonexamples presented to them. This presents a possible starting point from which teachers can begin to explore additional aspects of geometric figures such as definitions, critical and non-critical attributes, and an expanded example space of these figures. In general, teachers' self-efficacy with regard to 3-D figures was lower than their self-efficacy for identifying 2-D figures. This might indicate that during professional development more attention should be paid to promoting teachers' self-efficacy for identifying 3-D figures. Within the group of 3-D figures, a correlation was found between teachers' knowledge and self-efficacy for identifying cylinders but not between their knowledge and self-efficacy for identifying cones or two-dimensional figures. Some teachers were knowledgeable of cones, yet their self-efficacy for identifying cones was low. Studies have shown that mathematics self-efficacy predicts children's choices of the types of problems they prefer to engage (Bandura and Schunk 1981). Likewise, teachers with a low self-efficacy related to cones, may avoid planning activities that involve this figure. Professional development may benefit these teachers by not only increasing their self-efficacy but increasing their self-awareness. This would also benefit those few teachers who had a low self-efficacy but nevertheless thought they were knowledgeable. Wheatley (2002) claimed that teachers' efficacy doubts may cause a feeling of disequilibrium which in turn may foster teacher learning. Results of this study also indicated that teachers had a higher self-efficacy when it comes to designing tasks for promoting knowledge than they did for designing tasks to evaluate knowledge, regardless of the specific figure being addressed. This might indicate that preschool teachers have less experience with designing tasks to evaluate children's knowledge. This issue could be raised and explored during professional development. Teachers can be encouraged, within the supporting environment of professional development programs, to design such tasks, implement them with children in their classes, and discuss together the results. In conclusion, while this paper raised several methodological questions regarding the study of preschool teachers' self-efficacy for teaching geometry, it also led to results which may be used to inform professional development aimed at promoting preschool teachers' knowledge and self-efficacy for teaching geometry.

**Acknowledgements** This research was supported by THE ISRAEL SCIENCE FOUNDATION (grant No. 654/10). We would also like to thank Dr. Sigal Levy, from The Academic College of Tel Aviv Yaffo, for her assistance and advice regarding statistical analysis.

## References

Allinder, R. M. (1994). The relationship between efficacy and the instructional practices of special education teachers and consultants. *Teacher Education and Special Education, 17*, 86–95.

Ball, D., Thames, M., & Phelps, G. (2008). Content knowledge for teaching. *Journal of Teacher Education, 59*(5), 389–407.

Bandura, A. (1986). *Social foundations of thought and action: A social cognitive theory*. Englewood Cliffs: Prentice Hall.

Bandura, A. (2006). Guide for constructing self-efficacy scales. In F. Pajares & T. Urdan (Eds.), *Self-efficacy beliefs of adolescents* (pp. 1–43). Greenwich: Information Age.

Bandura, A., & Schunk, G. H. (1981). Cultivating competence, self-efficacy, and intrinsic interest through proximal self-motivation. *Journal of Personality and Social Psychology, 41*(3), 586–598.

Bates, A. B., Latham, N., & Kim, J. (2011). Linking preservice teachers' mathematics self-efficacy and mathematics teaching efficacy to their mathematical performance. *School Science and Mathematics, 111*(7), 325–333.

Betz, N. E., & Hackett, G. (1993). *Mathematics self-efficacy scale*. Palo Alto: Mind Garden Press.

Caprara, G. V., Barbaranelli, C., Steca, P., & Malone, P. (2006). Teachers' self-efficacy beliefs as determinants of job satisfaction and students' academic achievement: A study at the school level. *Journal of School Psychology, 22*, 473–490.

Coladarci, T. (1992). Teachers' sense of efficacy and commitment to teaching. *Journal of Experimental Education, 60*, 323–337.

Collins, J. (1982). *Self-efficacy and ability in achievement behavior*. Paper presented at the Meeting of the American Educational Research Association, New York.

Davis-Kean, P. E., Huesmann, L. R., Jager, J., Collins, W. A., Bates, J. E., & Lansford, J. (2008). Changes in the relation of beliefs and behaviors during middle childhood. *Child Development, 79*, 1257–1269.

Dellinger, A., Bobbett, J., Livier, D., & Ellett, C. (2008). Measuring teachers' self-efficacy beliefs: Development and use of the TEBS-self. *Teaching and Teacher Education, 24*, 751–766.

Enochs, L. G., Smith, P. L., & Huinker, D. (2000). Establishing factorial validity of the mathematics teaching efficacy beliefs instrument. *School Science and Mathematics, 100*, 194–202.

Gelman, R., & Gallistel, C. (1978). *The child's understanding of number*. Cambridge: Harvard University Press.

Ginsburg, H. P., Lee, J. S., & Boyd, J. S. (2008). Mathematics education for young children: What it is and how to promote it. *Social Policy Report, XXII*(I), 1–22.

Hackett, G., & Betz, N. (1989). An exploration of the mathematics self-efficacy/mathematics performance correspondence. *Journal for Research in Mathematics Education, 20*, 261–273.

Hershkowitz, R. (1990). Psychological aspects of learning geometry. In P. Nesher & J. Kilpatrick (Eds.), *Mathematics and cognition* (pp. 70–95). Cambridge: Cambridge University Press.

Israel National Mathematics Preschool Curriculum (INMPC). (2008). Retrieved April 7, 2009, from http://meyda.education.gov.il/files/Tochniyot_Limudim/KdamYesodi/Math1.pdf

Merenluoto, K. (2004). The cognitive – Motivational profiles of students dealing with decimal numbers and fractions. In M. J. Høines & A. B. Fuglestad (Eds.), *Proceedings of the 28th conference of the international group for the psychology of mathematics education* (Vol. 3, pp. 297–304).

National Association for the Education of Young Children & National Council of Teachers of Mathematics (NAEYC & NCTM). (2002). *Position statement. Early childhood mathematics: Promoting good beginnings*. Available: www.naeyc.org/resources/position_statements/psmath.htm

Pajares, F. (1996). Self efficacy believes in academic settings. *Review of Educational Research, 66*, 543–578.

Pajares, F., & Graham, L. (1999). Self-efficacy, motivation constructs, and mathematics performance of entering middle school students. *Contemporary Educational Psychology, 24*, 124–139.

Pajares, F., & Miller, M. D. (1994). Role of self-efficacy and self-concept beliefs in mathematical problem solving: A path analysis. *Journal of Educational Psychology, 86*, 193–203.

Schoenfeld, A. H., & Kilpatrick, J. (2008). Toward a theory of proficiency in teaching mathematics. In D. Tirosh & T. Wood (Eds.), *The international handbook of mathematics teacher education: Tools and processes in mathematics teacher education* (Vol. 2, pp. 321–354). Rotterdam: Sense Publishers.

Schulz, W. (2005, April 11–15). *Mathematics self-efficacy and student expectations*. Results from PISA 2003. Paper prepared for the annual meetings of the American Educational Research Association in Montreal.

Shulman, L. S. (1986). Those who understand: Knowledge growth in teaching. *Educational Researcher, 15*(2), 4–14.

Skaalvik, E., & Skaalvik, S. (2007). Dimension of teacher self-efficacy and relations with strain factors, perceived collective teacher efficacy, and teacher burnout. *Journal of Educational Psychology, 99*(3), 611–625.

Tabach, M., Levenson, E., Barkai, R., Tirosh, D., Tsamir, P., & Dreyfus, T. (2010). Secondary school teachers' awareness of numerical examples as proof. *Research in Mathematics Education, 12*(2), 117–131.

Tirosh, D., & Tsamir, P. (2008). *Starting right: Mathematics in preschool*. Unpublished research report. In Hebrew.

Tirosh, D., Tsamir, P., Levenson, E., & Tabach, M. (2011). From preschool teachers' professional development to children's knowledge: Comparing sets. *Journal of Mathematics Teacher Education, 14*, 113–131.

Tsamir, P., Tirosh, D., & Levenson, E. (2008). Intuitive nonexamples: The case of triangles. *Educational Studies in Mathematics, 69*, 81–95.

Tsamir, P., Tirosh, D., Levenson, E., Tabach, M., & Barkai, R. (2014a). Employing the CAMTE framework: Focusing on preschool teachers' knowledge and self-efficacy related to students' conceptions. In C. Benz, B. Brandt, U. Kortenkamp, G. Krummheuer, S. Ladel, & R. Vogel (Eds.), *Early mathematics learning – Selected papers from the POEM 2012 conference* (pp. 291–306). New York: Springer.

Tsamir, P., Tirosh, D., Levenson, E., Tabach, M., & Barkai, R. (2014b). Developing preschool teachers' knowledge of students' number conceptions. *Journal of Mathematics Teacher Education, 17*(1), 61–83.

Wheatley, K. (2002). The potential benefits of teacher efficacy doubts for educational reform. *Teaching and Teacher Education, 18*(1), 5–22.

Zimmerman, B. J. (2000). Attainment of self-regulation: A social cognitive perspective. In M. Boekaerts, P. R. Pintrich, & M. Zeidner (Eds.), *Handbook of self-regulation* (pp. 13–39). San Diego: Academic.

# A Specific Language Towards a New Conceptual Framework for Networking Methodologies in the Field of Affect

**Chiara Andrà**

**Abstract** This chapter focuses on ways of networking different methodologies used in empirical research in the field of affect. It also aims to develop a specific language and a conceptual framework for the networking of these methodologies. Two situations are presented to exemplify the extent to which two methods can be integrated without losing their characteristic features. The first example regards a case in which two methodologies, with strong commonalities between them, can be synthesized. The result of the synthesis is a new methodology for the examination of the interplay between cognitive-related and emotional/motivational-related variables impacting the outcome of an assessment test. The second example, conversely, presents two methodologies that have much less overlap, yet the commonality of language and shared experiences still allows us to compare them, although at a lesser extent of integration.

**Keywords** Networking theories • Multiple methodologies • Qualitative and quantitative methods

## Introduction

### *Motivation*

The starting point of this chapter is related to the availability of a huge amount of data about undergraduate students at the University of Torino, since the academic year 2001/02. My first concern was about the students' difficulties with respect to the transition from secondary school to university (see Andrà et al. 2013), an issue which is the focus of much research around the world (see e.g. Rylands and Coady 2009), and within the MAthematical-VIew community (Gòmez-Chacòn et al. 2012; Griese et al. 2012). Data gathered in this study regard both cognitive-related

C. Andrà (✉)
Department of Mathematics, University of Torino, Torino, Italy

Polytechnic of Milan, Milan, Italy
e-mail: chiara.andra@gmail.com



B. Pepin, B. Roesken-Winter (eds.), *From beliefs to dynamic affect systems in mathematics education*, Advances in Mathematics Education,
DOI 10.1007/978-3-319-06808-4_17

and emotional/motivational-related variables, and different methods for analysing the phenomenon need to be applied and coordinated in order to look at the phenomenon as a whole. The focus of this chapter is on ways of combining different methodologies.

## *Different Purposes in Employing Multiple Sources of Evidence*

In mathematics education in general, the use of multiple sources of evidence has been discussed, for example by Schoenfeld (2007), and it is common in the field of affect in particular (see e.g. Chen and Leung, this book). The use of different methods can serve various purposes: for example, Di Martino and Zan (2011) *contrast* two methodologies, questionnaires versus narrative accounts, with respect to their usefulness in analyses of the emotional flow in classroom experiences, while Chen and Leung (this book) talk about the *integration* of multiple methodologies. In its theoretical aspect, affect is by its very nature a *combining* of cognition, motivation and emotion: Hannula (2011) identifies researches in the field of affect as those which perceive "emotions, cognition and motivation in a synergic relationship" (p. 35), rather than concentrating only on the cognitive aspects. Theoretical constructs and their methodology are deeply intertwined; hence such a *combining* is central at both theoretical and methodological levels. Whether the purpose is combining, contrasting, comparing or integrating, I feel the need to find a suitable conceptual framework to deal with, and to reflect upon, the practice of using multiple methods. A first step towards a new conceptual framework for networking methodologies in the field of affect is to find a specific language for it. The aim of the present chapter is, thus, to provide a meta-language, a space where different methodologies can dialogue and integrate, to an appropriate extent. I will resort to Prediger et al.'s (2008) and Radford's (2008) contributions in the domain of networking theories.

## *Networking Methodologies*

The term 'networking theories' in mathematics education indicates a research field that contributes to systematizing a collection of case studies using different strategies for combining theoretical approaches that frame empirical research (Prediger et al. 2008). The term 'networking methodologies' is used in this chapter to conceptualize the connecting strategies among multiple methods in empirical research. I will talk about networking methodologies by addressing the dimensions of identity and integration as they have been put forward by Radford (2008), and considering the specificity of methodologies with respect to theories (Radford 2008). As a consequence, the longitudinal study on undergraduate students will serve as an example to illustrate and bring to light some interesting issues.

### *Method and Methodology*

The terms 'method' and 'methodology' can be used as synonyms, but the relationship between method and methodology is like the relationship between the words psyche and psychology, or between derma and dermatology. The Greek word λογοζ (from which '-logy' comes) derives from the verb λέγω, meaning 'to tell, say, speak'. In this chapter I understand methods as the techniques or processes we use to conduct our research. The methodology is the body of knowledge, the discourse about these methods. Given that methodology is theory about the method(s) being employed, we can see networking of methodologies as discourses about combining research methods. Once again, with networking of methodologies being a *discourse*, then the aim of finding a suitable language is central for the conceptual framework.

### *What's Next*

The remainder of the chapter is organized as follows. First, I recall background knowledge on theories and their networking in mathematics education, and I briefly present the theoretical constructs that are used in the field of affect. The methodological framework illustrates some methods and networking strategies at different levels of integration. Two examples, their relative analysis and discussion, follow. The chapter ends with some closing remarks that further elaborate on the language for discourses about connecting methodologies.

## Theoretical Background for Discourses on Networking Methodologies

### *Theories in Mathematics Education, and the Relationship with Methodologies*

Radford (2008) suggests defining a theory as a mode of understanding and acting grounded on a system of basic principles, a methodology, and a set of paradigmatic research questions. According to this view, methodology entails techniques of data collection and data interpretation *as supported by the system of principles*: "data relevance is dictated by the exigency of coherence between the principles and the methodology of a theory" (p. 321). This suggests considering the influence between the principles of a theory and methodology as a two-way affair: following Schoenfeld (2007), who quotes Albert Einstein, "there is no empirical method without speculative concepts and systems; and there is no speculative thinking whose concepts do not reveal, on closer investigation, the empirical material from which they stem" (p. 3), we maintain that each methodology produces meaningful data only within a certain theoretical framework, in an unavoidably selective manner that "helps the researcher to select some data among the data that was produced but also helps the researcher to forget or to leave some other data unattended" (Radford 2008, p. 321).

## *Networking Theories*

Given this close intertwining between principles and methodology, an issue of interest can regard the kind of relationship between different methods when either two theories are combined, or two methodologies are employed, to examine *the same phenomenon*. Radford (2008) suggests creating a conceptual space where theories and their connections can become the objects of discourse: the semiosphere. It is the space where the networking of theories takes place and Prediger et al. (2008) depict various levels of integration. Integration and identity are two fundamental themes (Radford 2008): the constitution of the semiosphere's meta-language resides in the dialectical tension between them. Identity is construed as being relational and refers to the awareness of oneself – as a product of the process of identification and differentiation from significant others. Integration (from the Latin *integer*, meaning 'whole' or 'entire') refers to combining parts so that they work together or form a whole. The main focus in Radford's work is on the system of principles, which plays a prominent role. Agreeing with this standpoint, I would like to dig deeper into details on the applicability of his considerations with respect to the methods: hence, a discussion about the boundaries among methodologies, when networked, will be the focus of the following paragraphs. Radford (2008) defines the *boundaries of a theory* as "the edge that a theory cannot cross without a substantial loss of its own identity" (p. 323), the "limit of what a theory can legitimately predicate about its object of discourse" (p. 324). He proposes also an example on networking of methodologies that turns out to be a hierarchical (re)organization of the systems of principles the two methodologies respectively refer to. He also suggests that, while systems of principles are characterized by the hierarchical structure of their own key concepts, methodologies pivot around operability and coherence. Hence, it is necessary to define the boundaries of a methodology not only as the limits beyond which a theory conflicts with its own principles, but also as the edge beyond which a method loses its own usefulness and operability. Taking Radford's (2008) and Prediger et al.'s (2008) work on networking theories as a foundation stone, I focus on identity and integration of methodologies, showing different modes of networking.

## *Theoretical Constructs in the Field of Affect*

According to Hannula (2011), affect is the interplay of cognition, emotion and motivation. *Cognition* is anything to which it is possible to assign a truth value (either subjective or not), and consequently beliefs, memories, scripts and concepts belong also to this domain (see also Furinghetti and Pehkonen 2002 for a discussion about a distinction between knowledge and beliefs). *Emotions* are various – fear, joy, pride, sadness and so on; they cannot be counted as one, and they variously intertwine with cognition in the learning processes. Roth and Radford (2011) see emotions as "a holistic expression of the subject's current state with respect to the object/motive and the subject's sense of likelihood of success in realizing the object/motives it has subscribed to." *Motivation* reflects personal preferences and helps in explaining choices (Hannula 2011).

Radford (this book) further invites us to see emotions as linked to the meaning of life rather than as merely biological experiences or as irrational forces: emotions are part of our worldview that we come to share through our participation in cultural and social activities. Emotions, furthermore, become related to the students' motives in mathematical classroom activities. This has a counterpart in the following methodological framework.

## Methodological Framework

In this section I describe briefly the methods of data collection and interpretation I have used in the examples that serve the purpose of illustrating situations when multiple methods are employed. I will deliberately use simple language, so as to be understood also by a non-expert reader. For those interested in deeper details, references to more technical work are provided. Simplifications have been made, as they mirror the reality of assessment rather than the advanced status of psychometrics.

### *Multiple-Choice Assessment Tests*

*Multiple-choice assessment tests* measure a subject's ability, for example mathematical ability, by posing a question and providing some possible answers (among which usually only one is correct). The Rasch model (see Baker 1992) assumes that a student with a certain level of ability has a high probability of answering correctly the questions that have a level of difficulty lower than his ability level. The Rasch model provides goodness-of-fit indices, and the response patterns which deviate too much from the model are eliminated from the data set: this process is called *calibration*. In a test administered periodically, year after year, the set of questions to be posed becomes more and more calibrated, to the point that the probability of finding a misfitting item is very low: this is the case of the data under examination in my study. At this point, let me suggest shifting our perspective from what fits the Rasch model to what does not (which is seen as non-explained variance for example). Even if the assessment test is assumed to measure a unique ability according to the Rasch model, misfit indices (that compare the most probable pattern of responses with each actual pattern by summing up the square differences) can be seen as indicators that more than one ability/dimension intervene in the process of answering the test (see Baker 1992). In other words, the same statistical model for the analysis of the students' answers during an assessment test takes into account a certain degree of integration between the cognitive variables being measured and other (even non-cognitive) ones. Such 'other than the measured ability' variables are taken as noise in the model, but the very point in my view here is that it is by considering integration between cognition and affect that the Rasch model becomes more adaptive in describing the students' answers.

## Likert-Scale Tests

*Likert-scale tests* consist of "a series of statements typically rated from 'strongly agree' to 'strongly disagree' and five divisions are very common. The divisions are converted to numerical scores" (Leder and Forgasz 2002). Multiple-choice assessment tests and Likert-scale ones differ in the dimension they intend to measure: the former focus on cognition, while the latter are concerned with emotions and motivation (and sometimes beliefs). As regards integration, not only do they both take into account the possibility that other dimensions may intervene in the process of answering the test (even if both them assume unidimensionality), but also they share two important features: (1) they identify what is measured by the test, that is, a score, with the dimension (the latent variable) they intend to measure (e.g. mathematical ability *is* the score earned in the test); and (2) they assume monotonicity and unidimensionality for the estimator of the latent variable they measure. Table 1 reports a brief description of the variables used in the present study.

Variables psy01–04 relate to Savickas et al.'s (2009) Career Adaptability inventory. Students were asked to rate from 1 (very low) to 6 (very high) the perceived ability level. Variables psy05–09 come from Zimmerman and Kitsantas's (2005) Perceived Responsibility Scale. The students have to grade from 1 (student) to 7 (teacher) the corresponding attributed responsibility. A score of 4 indicates that both, teacher and student, are equally responsible. Variables psy10–14 relate to socio-cultural issues concerning affect, representing a rather psychosocial dimension, while psy16–20 represent a psychological one. Both groups of items have been developed with the research group in the psychology of job counselling at the University of Padua. Soresi and Nota (2007) is the reference for psy10–20. The students have to grade from 1 to 5 how well each heading describes their way of thinking and behaving. According to Hannula's (2011) terminology, psy01–02 and psy16–17 refer to emotional disposition, a rather trait-like aspect of emotions, while psy03 and psy13–14 refer to the state-like one. Psy04–09 can be linked to perceived competence (see also Di Martino and Zan 2011), cognition in Hannula's framework. Also, psy10 refers to beliefs, hence to cognition in Hannula's view. Motivation rather relates to psy11, psy19 and psy20. Savickas et al. (2009), Zimmerman and Kitsantas (2005) and Soresi and Nota (2007) have shown that these variables are

**Table 1** The list of variables used in the first example

| | |
|---|---|
| psy01 | Attitude to thinking positively about one's own professional future |
| psy02 | Attitude to considering oneself responsible for determining one's own professional future |
| psy03 | Curiosity and desire to explore new opportunities and possibilities |
| psy04 | Self-confidence in fostering one's own professional realization |
| psy05 | Ability to formulate learning goals |
| psy06 | Ability to schedule one's own homework and free time |
| psy07 | Ability to search for advice and help |
| psy08 | Writing ability |
| psy09 | Ability to detect the key ideas |
| psy10 | Efficacy beliefs with respect to scientific subjects |
| psy11 | Expectations with respect to scientific subjects |
| psy12 | Interest with respect to scientific subjects |
| psy13 | Perceptions of social support with respect to scientific subjects |
| psy14 | Perceptions of barriers with respect to scientific subjects |
| psy15 | Educational goals with respect to scientific subjects |
| psy16 | Hope about self-realization in the future |
| psy17 | Tendency to be optimistic |
| psy18 | Negative view of future possibilities |
| psy19 | Temporal frame |
| psy20 | Resiliency |

good predictors of cognitive achievement, hence I would rely on these results to show an example coming from data mining: the decision tree (for further details see Andrà and Magnano 2012).

## *An Example of Integration*

Let us now see how the variables listed in Table 1 jointly contribute to predicting mathematical ability as it is measured by an assessment test, and how the method of decision tree works. To build a classification tree, we need: (1) a population; (2) a set of quantitative or categorical factors $X_i$ (features), whose values are known for all individuals in the population; (3) a categorical response variable Y, which represents the factor that should be predicted; on the given population the response values are known. For simplicity, assume that Y is a binary variable (as it is in my analysis), representing success or failure. The whole population is divided into two groups, the true positives (Y = success) and the true negatives (Y = failure). The first basic concept for understanding the method is the *splitting* operation. A splitting criterion consists in the choice of a specific feature $X_i$ and in a partition of the possible values of $X_i$ into two sets. Then, the population is split into two groups, A and B, according to the individual values of $X_i$. Assume, for instance, that the majority of individuals in A are true positives: then, we say that belonging to A *predicts success*. The second basic concept is the *node impurity* associated with a split. An individual belonging to A but being a true negative becomes a *false positive*; conversely, if the value of $X_i$ predicts failure but the individual is a true positive, we call that individual a *false negative*. The misclassification rate associated to the given split is the total proportion of false positives and false negatives in the population under that particular splitting criterion.

Within Prediger et al.'s (2008) terminology, the decision tree can be seen as an example of *synthesizing* between two methodologies: multiple-choice assessment tests and Likert-scale tests. Prediger et al. define the synthesizing of two theories as a networking strategy that consists in taking two equally stable theories and connecting them in such a way that a new theory evolves. This strategy has strong preconditions: for example, it is necessary that different parts of incompatible theories are not synthesized. Within our methodological realm, multiple-choice and Likert-scale tests can be synthesized according to their strong commonalities, and their synthesis leads to a new methodology. This new methodology helps the researcher to integrate the information separately provided by each method.

## An Example of Comparison

Data mining techniques provide us with percentages that can help us to understand the different weights the variables have in order to 'predict' the test score, but there could be a need for deeper interpretation of these percentages. A qualitative, case-study-wise analysis may help the researcher to better ground his/her interpretation, and provide also new information. The methodological framework in this respect is taken from Di Martino and Zan (2011), who propose a characterization of attitude towards mathematics based on the students' narratives about their own experiences. Three main themes emerge: emotional disposition (like/dislike), perceived competence (can do/can't do) and visions of mathematics. Di Martino and Zan further try to characterize causal links among the three dimensions, as they emerge from the students' narratives. A drawback of the methodology suggested by Di Martino and Zan (2011) is that students are free to mention a certain factor or to ignore it – even if it is crucial for them. In a questionnaire, conversely, each student is invited to reflect upon the same list of arguments. This last comment is related to the internal diversity between tests and narratives, and the list of features that render the two methodologies different is much longer. Integration, by contrast, needs to be discussed. Only if there is an overlapping between the two methodologies, in fact, can we proceed to networking them and using them as multiple sources of evidence with respect to the same phenomenon. I would like to focus on the social nature of both methodologies. Bruner (2004) suggests that in autobiographical stories the forms of self-telling are a product of social, cultural and historical development, not only the expression of the narrator's inner space. Hence, individuals share a common cultural basis for the language and the ways of structuring their own narrative. As well, multiple-choice questionnaires such as Likert-scale ones can be seen as a crystallization of a similar process: ways of telling become socially and culturally shared to the point that they are taken as universally valid. We assign a certain meaning to a word or a statement because we share not only a language, but also a set of analogous experiences connected with such a word/sentence. Radford (this book) provides us with a theoretical basis for this claim, arguing for the dual nature of emotions: both individual and social.

This last comment on the possibility of integration between tests and narratives (as socially and culturally shaped forms of sense-making) points to the need for *making understandable* (Prediger et al. 2008) each methodology to the other, given the different systems of principles: this is, in fact, a precondition for comparing them. *Comparing* is a networking strategy that seems to be suitable in this case: it refers to similarities and differences in perceiving each other's components. *Comparing* is less demanding with respect to *synthesizing* in terms of the commonalities shared by the two methodologies to be networked. It can provide a base for communication between methodologies with strong internal diversities. It can also contribute to a better understanding of each methodology. It can offer a rational basis for choosing proper methodologies.

I would like to end this section by underlining that I am aware that there exist other, significant, methodologies that can be used. Classroom observation, for example. The purpose of this paper, however, is not to state which methodologies are worth using in affect-related studies in mathematics education, but to discuss the issue of identity and integration among methodologies. Of course, this is just a start: my hope is to bring to the fore some relevant issues that need to be taken into account in general, and specifically in the field of affect. I now present the results from the decision tree and from the perspective of teachers' narratives, and in the discussion I dig deeper into networking methodologies.

## The Data Under Consideration

The data come from a longitudinal study on undergraduate students at the Faculty of Sciences (undergraduate courses in mathematics, physics, chemistry, computer sciences, natural sciences, geology) at the University of Torino, which have been monitored since the academic year 2001/2002. For all the students, I have information regarding: the high school period (diploma grades and type); the performance during the non-selective test for the assessment of minimum requisites (TARM) they took when enrolling in the undergraduate course; students' marks and credits from the university examinations; and phone interviews with drop-out students, aimed at knowing their choices after leaving the undergraduate studies and outlining the main reasons for drop-out. In the first example, however, I do not consider the entire sample of 2,500 students enrolled in the academic years from 2001/2002 to 2012/2013, but only one cohort (the 2012/2013 one), and only a subset of variables, namely the information we have from the entrance test. The reason for this choice is to show how a methodology taken from educational data mining can result in the *synthesis* of multiple-choice assessment tests and Likert-scale tests which had been calibrated and validated in previous studies (Andrà and Magnano 2012).

The second example regards a group of 43 prospective secondary school teachers during the academic year 2012/13. They were attending a 1-year master's course that would enable them to teach math at secondary school. During a lesson within the Mathematics Education course, I asked them to describe their relationship with mathematics, focusing on the transition from secondary school to university. I use a narrative lens to analyse their written reports, and then I compare this methodology with Likert-scale tests of the first example, to show how two methodologies can be *made understandable* to each other.

## The First Example: Beginning Undergraduate Studies

Figure 1 shows the estimated decision tree regarding all the variables listed in Table 1, normalized to have mean = 50 and s.d. = 10, plus sex, meant as factors predicting the score of TARM (0 = less than 14, 1 = greater or equal to 14), for 870 students enrolled on undergraduate courses in mathematics, physics, mathematics for business and insurance, geology, computer sciences and materials sciences in the academic year 2012/13 at the University of Torino. The median score for TARM is 14, and it has been taken as a sufficiency threshold: the students with

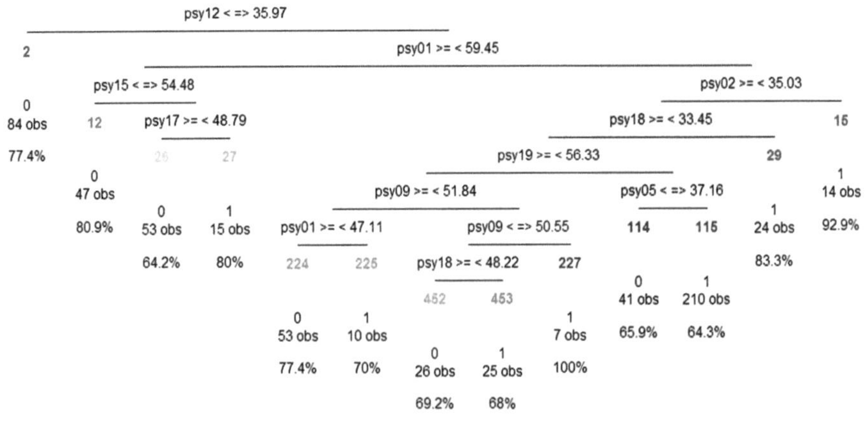


**Fig. 1** The decision tree for the entrance test of the students enrolling in fall 2012

lower scores should take a tutorial course in mathematics. The first split of the tree is determined by psy12 (interest with respect to scientific disciplines): the students who have psy12<36 (that is, quite low, since 36 is smaller than the mean, 50, minus one standard deviation, 10) consistently earned a TARM score lower than 14; psy12 predicts 'failure' for 77.4 % of 84 students. Knowing the degree of interest with respect to scientific disciplines is the most predictive information about whether a student's score would be above or below the median. This can be seen as not surprising, given the kind of students who answered the test. Going on along the branches of the tree, if psy12 is not so low, then the inclination to think positively about professional future (psy01, a career-adaptability variable) determines the second split; and if it is high, psy15 (educational goals) below the median (50) predicts failure for 80.9 % of 47 students, while psy15 above the median and psy17 (optimism) above the median predict failure for 64.2 % of 53 students. On the other branch of the tree, namely where psy01 is not very high, low values of psy02 (below the mean minus 1.5 times the standard deviation) predict success for 93 % of 14 students. Higher values of psy02 (perceived responsibility about one's own future), of psy18 (negative view of future possibilities), of psy19 (time frame), of psy09 (ability to detect the key ideas), and of psy01, predict 'failure' for 77.4 % of 53 students, while lower levels of at least one of these variables predict 'success' for the rest of the sample.

## *What Can We Learn from the First Example? A Discussion of the Results*

The commonalities between assessment tests and Likert-scale tests allow us to integrate them without facing internal contradictions. Furthermore, their joint use allows us to resort to a new, different methodology: the decision tree. This new methodology is still consistent and useful with respect to the systems of principles that inform the two starting ones: it is a synthesis between two well-established methodologies. Of course, the nature of the comments that I made in analysing the example is dual: on the one hand, I have focused my attention to the specific variables under examination, looking at how they jointly contribute to predicting the outcome of the assessment test, and showing that both cognitive-related and emotional/motivational-related variables play important roles. On the other hand, I have come back to the main purpose of this study, a discussion about the use of multiple methods in the field of affect, seen as a combining of cognition, emotion and motivation.

However, the different methods employed within the same research study do not in every situation allow for a synthesis of this extent: in many cases the commonalities between the two methodologies are less, and hence other ways of networking should be used, as is shown in the second example below.

## The Second Example: Prospective Secondary School Math Teachers

In this example I search for possible overlapping between Likert-scale tests (previously analysed) and the written reports by prospective secondary school teachers. I apply the framework suggested by Di Martino and Zan (2011), and I search for similarities in the words that are used in this second example. A similarity of language to describe similar experiences is a condition for networking: hence, if the students in the two examples can be regarded as sharing a common language, this would further support the integration of two different methodologies such as questionnaires and analysis of stories.

At first sight, I have been struck that only 11 prospective teachers among 49 mentioned the intention of becoming a teacher in their written answers about their relationship with math. This fact, in a sense, seems however to confirm what emerges from the first example: career-related emotional/motivational variables (psy01–04) seldom intervene in predicting success at the beginning of the university. Hence, I would like to look at some answers given by these 11 prospective teachers, to find out more.

Irene: *In my class at secondary school there were many students who were very good in math, and this stimulated me to continuously try to go beyond my limits. Since primary school I had an excellent relationship with my math teachers, and hence also with math. Moreover, I enjoyed it and I practised gladly, also because it allowed me to feel helpful with respect to my classmates. When I arrived at university, I felt floored, because my expectations were to do more exercises. Conversely, I stared at blackboards full of theorems and proofs. Then I started to see math in a new light, and I continued to like it. Sometimes I had to bite the bullet and tell myself that I had to finish my path, because to teach I needed the degree. Now I think that I have made the right choice for my future, and I hope that I will continue to think this.*

In the first part of her answer Irene mentions her relationship with her classmates, which had been competitive, and thus motivated her to do well in math. Motivation is related to career-adaptability: "I had to bite the bullet and tell myself that I had to finish my path, because to teach I needed the degree." Irene's words (underlined, by me) sound like 'being unflinching' and 'planning how to reach one's own goals', which are present in the headings concerning psy01–04. Furthermore, Irene says: "I started to see math in a new light", which points to 'considering different ways of doing things', another heading in the psy01–04 list. A strong motivation helped Irene when facing difficulties, and taking the degree: this confirms the results obtained by Savickas and his colleagues, but also tells us that a lack of such a dimension would not necessarily lead to failure, as these variables do not determine a relevant split in the decision tree from the first example (TARM).

Other words in Irene's protocol point to other links with the questionnaires presented in the first example: to "tell myself that I had to finish my path" can be linked with 'being explicit about educational goals I intend to pursue', one of the psy05–09 statements; to "try to go beyond my limits" can be seen as Irene's 'self-confidence with respect to her abilities' (psy10–15); finally, "I think that I have made the right choice for my future" entails 'being sure that one would obtain what one desires' (psy16–20).

Di Martino and Zan (2011) suggest looking at the interplay of emotional disposition (like/dislike), perceived competence (can do/cannot do) and vision of mathematics (math is …): Irene uses the terms "enjoyed" and "gladly" to express her liking, "helpful" for perceived competence (to the point that she felt able to help her classmates), and "exercises" for her vision of mathematics. The causal link among the three dimensions is established by her view of mathematics: she likes math, she likes exercises, she succeeds in doing her math exercises, and she feels a sense of helpfulness with respect to her classmates, which fosters her enjoyment. However, at university Irene faced contrasting views of mathematics ("exercises" versus "theorems"), and these words in her protocol point to the issue of (difficult) transition from secondary school to university (see also Andrà et al. 2012). The last sentence in Irene's protocol is related to psy16 and psy17, a positive emotional disposition ("hope") as regards her choice. The same is present in Sara's protocol:

Sara: *During the last year at secondary school I fought with my math teacher; she didn't encourage me and she told me that it's better to be a comic actor rather than a teacher, because math in her opinion is sterile, void, dead. I was willing to teach and I loved math and physics, and I had chosen math for my university studies also because it had represented a challenge for me. I am happy with my choice; I would never change it. I would like to transmit to my students the love for this subject and the joy of learning, of knowing.*

We can notice that both Irene's and Sara's answers begin with a focus on the relationships with other persons, influencing their relationship with math: for Irene her classmates, for Sara the teacher. I think that Di Martino and Zan (2011) would have seen an external causal attribution in both cases: the relationship with other people is interpreted by the prospective teachers as influencing their relationship with mathematics. In both Irene's and Sara's cases, a contrast also emerges, and a challenge. Irene had to compete with her classmates, while Sara had to differ from her teacher, who had a negative view of math ("sterile, void, dead"). Sara's vision of mathematics (Di Martino and Zan 2011) emerges: mathematics is a "challenge" for her, as well as to become a teacher is a challenge for her. Moreover, I see an emergence of psy01–04 dimensions: 'counting on oneself' and 'defending one's own point of view'. Positive emotional disposition ("loved") and perceived competence ("challenge") relate with her being "willing to teach", a motivation that sustained her path towards the degree. I see also an emergence of resiliency (psy20) in these words. Sara is "happy" with her choice and she wants to spread her "joy" of learning among her future students, words that express emotions associated with 'being sure that in the future I will succeed in doing interesting things' (psy16–20).

Finally, Francesco's protocol shows a case of negative emotional disposition:

Francesco: *My relationship with math during secondary school was very difficult, and I have begun to fall in love with it in the last two years of secondary school, thanks also to higher commitment I dedicate to studying, commitment that has brought better results and personal satisfaction. This, jointly with the desire I already had to dedicate myself to teaching, led me choosing the math degree.*

I see a time frame (psy16–20): at the beginning, Francesco didn't like math and he felt a low perceived competence ("difficult"), then he fell in love with math, and the causal link established by him is related to his higher "commitment", which in fact brought "better results" (perceived competence) and "satisfaction" (emotional disposition). No vision of mathematics is made explicit. Francesco's desire to "dedicate myself to teaching" points to psy16–20 dimensions, while other dimensions are not mentioned in this protocol.

## *What Can We Learn from the Second Example? A Discussion of the Results*

Francesco's words tell us that also for "successful" students (i.e. the ones that take a math degree) there could have been difficulties with math, and Sara highlights that teachers do not always encourage such students to pursue their desires to teach math as their professional choice. This kind of information, of course, would have not been retrieved by resorting only to questionnaires such as the ones presented in the first example. Moreover, different causal links, of different nature, are established in the students' narratives, and this can help in enriching the analysis of the decision tree. For example, the awareness about their professional choices (mainly psy02) in Irene's and Sara's narratives helped these prospective teachers in overcoming difficulties and struggles.

A deep analysis on the different nature of causal links is beyond the purposes of this contribution, but I would like to stress (on methodological grounds) that it is possible to compare quantitative (questionnaires) and qualitative (analyses of narrative accounts) studies regarding such a complex phenomenon as the transition from secondary school to university. This comparison of methodologies allowed us to focus mostly on the similarities of languages, which point to similarities of lived experiences. Comparison allows dialogue between two methods, as well as helping to shape their identity in order to better understand their features and their uses.

## Final Remarks on Networking Methodologies

This chapter addresses the interplay of identity and integration of methodologies. We have seen an example of methods that can be synthesized (Prediger et al. 2008): assessment tests and Likert-scale tests. The decision tree is seen as a *synthesis* of the two. We have also seen an example of methodologies that can be *compared* (Prediger et al. 2008), but that are too different to be integrated: Likert-scale tests and narratives. The reader may have perceived an initial difficulty in dealing with quantitative data from Likert-scale tests, and qualitative data from the students' narratives. The difficulty has been overcome by resorting to networking strategies that allowed the two kinds of methodologies to communicate. Hence, the focus of attention has been shifted from boundaries, peripheries and centres, to neighbour interactions and similarities amongst methodologies. This sort of consideration applies to the practice of combining methods and provides the researcher with a language for dealing with and reflecting upon this practice. In the examples presented in this chapter we have seen that, when two methodologies share ways of measuring (e.g. a score), of interpreting results and dealing with unexpected outcomes (e.g. unidimensionality versus the presence of other dimensions), a synthesis between them can lead to the creation of a new methodology that can help us better understand the interplay of cognition and emotion. Conversely, when the methodologies do not share such fundamental assumptions on how to measure and interpret, we can search for a commonality of experiences they refer back to, commonality that is conveyed by using consonant terms and language. In this case, thus, the product of the coordination of different methods is no longer a synthesis, but a comparison. Comparison fosters a better understanding of the characteristic features of the methodologies employed, their limits and their potentials. This is particularly important in the field of affect, which according to Hannula (2011) considers the interplay of cognition, emotion and motivation as the focus of its inquiries, and hence needs suitable methodologies (not always synthesizable) to deal with such a combination.

In the phenomenology of the researcher, a proper methodology is not a priori given, but is progressively found as the shortlist of alternatives grows and differentiates, and this process of finding and applying a method to understand a phenomenon is social and cultural in nature. With Radford (2008), in fact, I see a theory as a *social practice*, framed in forms of self-reflectiveness, a way of producing understanding when common sense is no longer of help. Following Radford (2008), I maintain that, as a 'theory of all', that is, a super-theory, cannot exist, also there is not a 'methodology for all', namely a methodology which is always appropriate for collecting data and interpreting them. In my study I have proposed some examples that can open up the possibility to extract more general considerations. And with Radford (2008), who focused on networking of theories, I would say that the networking of methodologies contributes to the development of research in mathematics education: integration can give rise to new multi-methodological frameworks, capable of revealing and dealing with the complexity of teaching and learning, whilst identity can enhance our understanding of methodologies in our field, their similarities and differences. The development of suitable methods for understanding the objects of its inquiry is a substantial part of the efforts made in the field of affect, as is testified also by Di Martino and Zan (this book) with respect to attitudes.

**Acknowledgements** The research presented in this chapter is part of a research project on undergraduate students' career at the Department of Mathematics at the University of Torino, coordinated by Prof. Guido Magnano. I further wish to thank Francesca Morselli, Lionel LaCroix, Giorgio Santi, Dina Tirosh and Peter Liljedahl for their insightful comments about the chapter, which helped improve its quality. I would also like to thank the anonymous reviewers and the editors of the book for their precious feedback.

## References

Andrà, C., & Magnano, G. (2012). Understanding undergraduate mathematics students' career: Insights on an innovative quantitative methodology. *Nordic Studies in Mathematics Education, 17*(3–4), 75–90.

Andrà, C., Magnano, G., & Morselli, F. (2012). Dropout undergraduate students in mathematics: an exploratory study. In *Current state of research on mathematical beliefs XVII. Proceedings of the MAVI-17 conference, September 17–20, 2011* (pp. 13–22). Ruhr-Universitat Bochum, Germany.

Andrà, C., Magnano, G., & Morselli, F. (2013). Undergraduate mathematics students' career as a decision tree. In *Proceedings of the 18th mathematical views international conference, 12–17 September 2012,* University of Helsinki, Helsinki, Finland.

Baker, F. (1992). *Item response theory: Parameter estimation techniques.* New York: Marcel Dekker.

Bruner, J. (2004). Life as narrative. *Social Research, 71*(3), 691–710.

Di Martino, P., & Zan, R. (2011). Attitude towards mathematics: A bridge between beliefs and emotions. *ZDM – The International Journal on Mathematics Education, 43*, 471–483.

Furinghetti, F., & Pehkonen, E. (2002). Rethinking characterisations of beliefs. In G. Leder, E. Pehkonen, & G. Törner (Eds.), *Beliefs: A hidden variable in mathematics education?* (pp. 39–57). Dordrecht: Kluwer.

Gòmez-Chacòn, I., Garcìa Madruga, J. A., Rodrìguez, R., Vila J. O., & Elosùa, R. (2012). Mathematical beliefs and cognitive reflection: Do they predict academic achievement? In B. Roesken & M. Casper (Eds.), *Current state of research on mathematical beliefs XVII. Proceedings of the MAVI-17 Conference, September 17–20, 2011* (pp. 64–73). Ruhr-Universität Bochum, Germany.

Griese, B., Glasmachers, E., Harterich, J., Kallweit, M., & Roesken, B. (2012). Engineering students and their learning of mathematics. In B. Roesken & M. Casper (Eds.), *Current state of research on mathematical beliefs XVII. Proceedings of the MAVI-17 conference, September 17–20, 2011* (pp. 85–96). Ruhr-Universität Bochum, Germany.

Hannula, M. S. (2011, February 9–13). The structure and dynamics of affect in mathematical thinking and learning. In M. Pytlak, E. Swoboda, & T. Rowland (Eds.), *Proceedings of the CERME7, Reszow, Poland* (pp. 34–60). University of Reszow, Poland: CERME.

Leder, G. C., & Forgasz, H. (2002). Measuring mathematical beliefs and their impact on the learning of mathematics: A new approach. In G. C. Leder, E. Pehkonen, & G. Törner (Eds.), *Beliefs: A hidden variable in mathematics education?* (pp. 95–113). Dordrecht: Kluwer.

Prediger, S., Bikner-Ahsbahs, A., & Arzarello, F. (2008). How can networking strategies for connecting theoretical approaches help to develop theories in mathematics education? *ZDM – The International Journal on Mathematics Education, 40*(2), 165–178.

Radford, L. (2008). Connecting theories in mathematics education: Challenges and possibilities. *ZDM – The International Journal on Mathematics Education, 40*(2), 317–327.

Roth, W.-M., & Radford, L. (2011). *A cultural historical perspective on teaching and learning.* Rotterdam: Sense.

Rylands, L., & Coady, C. (2009). Performance of students with weak mathematics in first-year mathematics and science. *International Journal of Mathematical Education in Science and Technology, 40*(6), 741–753.

Savickas, M., Nota, L., Rossier, J., Dauwalder, J. P., Duarte, M. E., Guichard, J., Soresi, S., van Esbroeck, R., & van Vianen, A. E. M. (2009). Life designing: A paradigm for career construction in the 21st century. *Journal of Vocational Behavior, 75*(3), 239–250.

Schoenfeld, A. H. (2007). Method. In J. Frank & K. Lester (Eds.), *Second handbook of research on mathematics teaching and learning* (Vol. 1, pp. 69–107). Charlotte: Information Age.

Soresi, S., & Nota, L. (2007). Le capacità di studio [Abilities to study]. *Giornate di studio AIP Sezione di Psicologia dello Sviluppo e dell'Educazione, Verona, settembre 2007.*

Zimmerman, B. J., & Kitsantas, A. (2005). Students' perceived responsibility and completion of homework: The role of self-regulatory beliefs and processes. *Contemporary Educational Psychology, 30*, 397–417.

# Studying the Factorial Structure of Ghanaian Twelfth-Grade Students' Views on Mathematics

**Emmanuel Adu-tutu Bofah and Markku S. Hannula**

**Abstract** Researchers often import and adopt surveys from one cultural setting to another in order to collect comparative data or to simplify the laborious process of instrument development. Even when the instrument has been proven to have high reliability in the original setting, the reliability may prove to be much weaker in the new setting, especially when Western instruments are imported into non-Western countries. In this chapter, we discuss the problems of importing an instrument from one culture to another and associated methodological challenges. More importantly, we present a detailed account of using structural equation modeling (SEM) and MPlus software to validate a survey instrument imported to Ghana. The students' Views of Mathematics (VOM) instrument is based on earlier Western research and was further developed in Finland, where it had been validated to have high reliability. First, we used confirmatory factor analysis to test whether the seven factors identified in Finland were identifiable in Ghana. As the original factor structure was found not to fit the Ghanaian data, we continued with an explorative approach to identify the Ghanaian factor structure, resulting in a four-factor structure. For cross-validation purposes, the sample was randomly split into two, one-half of the sample assigned as the calibration sample and the other half as the validation sample. Measurement invariance was established at the configural, metric and structural levels between the calibration and validation sample. We further discuss the measurement artifacts and cultural differences as possible causes for the observed differences in the factor structures between the Ghanaian and the Finnish sample.

**Keywords** Cross-cultural affects • Views of mathematics • Affect and students' beliefs structures • Factor analysis • Measurement invariance • Multigroup analysis • Construct reliability and validity • Structural equation modeling • Survey instrument

---

E.A. Bofah (✉) • M.S. Hannula
Faculty of Behavioral Science, Department of Teacher Education, University of Helsinki, Helsinki, Finland
e-mail: emmanuel.bofah@hotmail.com; markku.hannula@helsinki.fi

 
B. Pepin, B. Roesken-Winter (eds.), *From beliefs to dynamic affect systems in mathematics education*, Advances in Mathematics Education,
DOI 10.1007/978-3-319-06808-4_18

## Introduction

The important role of students' beliefs on learning mathematics is widely acknowledged. We know that students' affective dispositions influence their learning of mathematics, for the better or for the worse (for an overview, see Hannula 2012). The declining numbers of students that are studying science, technology, engineering and mathematics at university level in the Organization for Economic Cooperation and Development (OECD) countries (Ainley et al. 2008) can at least be partially attributed to students' negative views towards mathematics.

Surveys are an important method for studying students' mathematics-related beliefs. Examples of surveys include The Trends in International Mathematics and Science Study (TIMSS) and The Programme for International Student Assessment (PISA). The first widely used instrument for studying mathematics-related affect was the Mathematics Anxiety Rating Scale (Richardson and Suinn 1972). In the long run, the Mathematics Attitude Scales (Fennema and Sherman 1976) has been much more influential in the field with a total of nine scales, namely: scales for students' anxiety, confidence, success, and effectance motivation, students' perceptions of mathematics as a male domain, the perceived usefulness of mathematics, their ratings of their respective teacher's perceptions of themselves and their parents' (mothers' and fathers') interest in mathematics. In addition, the "Indiana Mathematics Belief Scales," which was designed for secondary school and college level students (Kloosterman and Stage 1992), has been influential in the field. The multiscale instruments presented in the aforementioned studies and others inspired by them are useful in exploring the structural properties of mathematics related beliefs (McLeod 1992; Op't Eynde et al. 2006; Roesken et al. 2011).

Although international comparative studies such as TIMSS and PISA have measured student's mathematics related beliefs worldwide, their instruments have been developed in Western countries, usually in North America. Their approach relies on an unwarranted assumption that the structure of affect is cross-culturally invariant (Van de Vijver and Leung 2000). Empirical studies have revealed that the reliabilities of TIMSS and PISA scales vary across countries, being highest in the Western countries and lower in non-Western countries (Metsämuuronen 2012a; Rutkowski and Rutkowski 2010).

The present study reports the implementation and utility in Ghana of one such instrument "Views On Mathematics (VOM)" scale, which was developed in Finland by Pehkonen's research team (Hannula et al. 2005; Roesken et al. 2007, 2011). Finland, a Nordic welfare state and a member of the OECD, has repeatedly scored very high in human development indexes (e.g., Malik 2013). Specifically, Finland is known to have an excellent educational system, which has achieved eminence in the recent TIMSS and PISA results. Ghana is a sub-Sahara African country that has medium level human development index, slightly above the Africa average (Malik 2013). Ghana has not participated in PISA, and it has been performing poorly in TIMSS (Mullis et al. 2012). First, we will discuss the methodological issues relating to validation of an instrument in a new cultural context. Second, we will also provide a detailed account of applying exploratory and confirmatory factor analysis to analyze the structure of beliefs. Third, we will report our empirical findings regarding the belief structure in Ghana for mathematics in twelfth-grade Ghanaian students.

## Theoretical Background

According to McLeod (1992), four structural qualities distinguish students' mathematics belief systems: (a) beliefs about mathematics; (b) beliefs about the self; (c) beliefs about mathematics teaching and (d) beliefs about the social context. The first classification, beliefs about mathematics, includes students' beliefs such as thinking that mathematics is difficult, and the belief about the usefulness of mathematics. The second categorization, beliefs about the self, includes the self-concept, confidence, and causal attributions. These, in turn, include success and failures related to mathematics. The third category, beliefs about teaching, includes beliefs about what is expected of a teacher to help students learn mathematics. In other words, this measures the importance that students attach to mathematics instruction (Op't Eynde et al. 2002). McLeod's fourth category, "beliefs about the social context", includes the cultural issues associated with mathematics education, influence of parents and others outside the school on one's mathematics learning in addition to one's home environment.

Op't Eynde and colleagues (2002, 2006) further developed a framework of students' mathematics-related belief systems. Based on relevant literature reviews, they clustered students' mathematics-related beliefs systems into implicitly or explicitly held subjective conceptions students hold to be true for:

1. "Beliefs about mathematics education: (a) beliefs about mathematics, (b) beliefs about mathematical learning and problem solving, (c) beliefs about mathematics teaching;
2. Beliefs about the self as a mathematician: (a) intrinsic goal orientation beliefs, (b) extrinsic goal orientation, (c) task-value beliefs, (d) control beliefs, (e) self-efficacy beliefs;
3. Beliefs about the mathematics class context: (a) beliefs about the role and the functioning of the teacher, (b) beliefs about the role and the functioning of the students in their class (c) beliefs about socio-mathematical norms in their own class." (Op't Eynde et al. 2006, p. 63)

Studies on Finnish teacher training students (Hannula et al. 2005) and upper secondary school students (Roesken et al. 2007, 2011) have provided data on beliefs and motivation. Roesken and her colleagues argued that it is possible to empirically distinguish between students' cognitive beliefs, motivations, and their emotional relationship with mathematics. They reported five dimensions for students' cognitive beliefs (*ability, success, teacher quality, family encouragement,* and *difficulty*), and separate dimensions for student motivation (*effort*) and emotions (*enjoyment of mathematics*).

Due to the high reported reliability of the scales for a similar age group (Cronbach's alpha ($\alpha$): 0.800–0.910: Roesken et al. 2011), and inclusion of emotional and motivational dimensions, we decided to use the VOM instrument for our study in Ghana for measuring upper secondary students' mathematical beliefs systems. Moreover, we used existing measurement scales, which allowed us to compare and synthesize what is already known. This study is also based on the fact that indigenous research and theorizing, as well as research that integrates different cultural perspectives, are crucial to the establishment of more useful and universal theories (Leung and Zhang 1995; Van de Vijver and Leung 2000). Many researchers have lamented about the Western bias in cross-cultural research (e.g., Van de Vijver and Leung 2000). The bias is reflected in the methods used, and the theoretical orientations adopted. For example, there has been severe criticism of validity and reliability problems associated with the importation of Western instruments into non-Western countries (e.g., Cheung 1996; Van de Vijver and Leung 2000).

The cultural backgrounds of students' in Ghana differ from Finland in many respects (e.g., school types, educational resources, disparity between and within schools, socialization norms, daily experiences). Ghana has had relatively stable economic development, which is reflected in its comparatively high human development in relation to its gross national income per capita (Malik 2013). In their educational structures, these two countries have similarities and differences. Compulsory education in Finland starts from age seven, whereas in Ghana compulsory education starts from age 4. Both countries have 6 years of secondary education. In Finland, all teachers are required to have a master's degree including at least 1 year of pedagogical studies, whereas in Ghana, teaching requires a diploma or a Bachelor's degree. The gross enrolment ratio in senior high schools is 34 % in Ghana whereas that of Finland is above 100 % (UNESCO 2011). The share of girls' enrolment in senior high schools in Ghana is 44 % (Ghana Education Service 2013) and in Finland 57 % (Statistics Finland 2013). These vast differences makes it interesting to investigate how the students' in these two countries view themselves as learners of mathematics.

Studies on mathematics related affects in Ghana have been using various survey instruments. For instance, Eshun (2004) and Nyala (2008) used The Mathematics Attitude Scales (Fennema and Sherman 1976) to measure students Mathematics self-belief. Asante (2012) used the Attitude Towards Mathematics Inventory scale (ATMI), (Tapia and Marsh 2000). Asante (2012) reported the Cronbach's alpha reliability for ATMI to be 0.940, and Nyala (2008) reported 0.630 for Fennema & Sherman Mathematics Attitude Scale. In those studies, on the scale of mathematics self-confidence, Asante (2012) and Eshun (2004) reported significantly higher scores for males at the senior secondary school whereas Nyala (2008) reported no significant different between both sexes at the junior secondary school level. Also on the usefulness of mathematics scale Eshun (2004) reported higher scores for males at the senior secondary school whereas Nyala (2008) reported higher scores for females at the junior secondary school. Similar findings were reported for the mathematics as a male domain and anxiety at both school levels with girls reporting lower scores (Eshun 2004; Nyala 2008).

Multiscale construction and development is usually a multistage multifaceted process. Over the past several decades, scales for measuring students' affective structure have become the norm. Their possible widespread usefulness is because they provide multiple converging pieces of information about the studied constructs and can involve unlimited sample size in addition to robust methods for analyzing the sample to facilitate generalizing the findings. Most of these instruments or constructs are imports from Western research. Translations of such constructs are an inevitable tool to conduct such studies. However, translation does not guarantee that the translated instruments will measure the same as in the original. Differences in linguistics, cultural or both can make translations of the instruments difficult and meaningless. As such, the adaptation of these instruments should be based on theory, construct reliability analysis, exploratory and confirmatory factor analysis (Marsh et al. 2012). As Marsh and his colleagues argued "from a construct validation perspective, theory, measurement, statistical analysis, empirical research, and practice are inexorably intertwined, so that the neglect of one will undermine the others." (ibid. p. 111)

Researchers, policy makers and educators interest in cross-national comparative studies such as the TIMSS and PISA have gained considerable attention recently. However, challenges to TIMSS and PISA studies are that the target populations have unique social conventions-cultures, school systems and cognitive structures and styles (Metsämuuronen 2012a, b). Implementing the instruments developed in one cultural setting into a new cultural setting is problematic regardless of their high reliabilities in the original settings. For example, Metsämuuronen (2012a), and Rutkowski and Rutkowski (2010) reported that some scales (e.g., math self-concept) that had been used in PISA and TIMMS studies showed less reliable scores in East Asia, the Middle East, and Europe, when compared to data from North America, where the scales were originally constructed. Metsämuuronen (2012b) found that in the TIMSS2007, the math attitudes scale were not invariant and manifested "*fragmentation*" in most of the participating countries (in most low achieving countries) due to different cultural values. With empirical examples, Rutkowski and Rutkowski (2010) found that the possible cause to this was too much missing data: a possible sign of respondent misinterpretation.

Other researchers have argued that TIMSS and PISA uses robust psychometric, sampling methods, and translations methods, yet the math motivational construct is still affected by construct bias, method bias and item bias (Van de Vijver and Leung 2011). Rutkowski and Rutkowski (2010) argued that, the country composition of PISA makes it impossible to have motivational constructs that will measure the desired goals for non-OECD countries. On the other hand, for each successive cycle, TIMSS have been dropping or adding new mathematics motivational constructs in response to reported validity, reliability, psychometric properties of the data and feedback from various countries coordinators (Marsh et al. 2012; Rutkowski and Rutkowski 2010).

### *Aims of the study*

The present study objectives are:

- to test for the factorial validity of the VOM for Ghanaian upper secondary students,
- in the event of a model misfit (i.e. the seven-factor structure), propose and statistically test an alternative factorial structure,
- to cross-validate the new factorial structure across a second independent sample from the Ghanaian data,
- to test for factorial and structural invariance across a subsample (gender) from the Ghanaian sample, and
- to affirm the theoretical structure of the VOM construct.

## The Present Investigation: *A Priori* Predictions and Research Questions

The present study examined three research questions that give support for construct validity and reliability. First, we will compare the reliabilities of the scales in the Ghanaian sample to the reliabilities observed in Finland. Second, a more robust approach, confirmatory factor analysis will be used to validate the constructs. If the theoretical model is not supported, exploratory factor analysis will be used to determine the factor structure of the Ghanaian sample. The sample will be split into calibration and validation sample. Third, confirmatory factor analysis will be used to test the equivalence (measurement invariance) of the derived constructs with the validated sample.

*Hypothesis 1* Research have showed that imported constructs regardless of their high reliabilities in the original settings, often shows a very low reliabilities when imported to a different cultural setting (e.g., Cheung 1996; Metsämuuronen 2012a; Rutkowski and Rutkowski 2010; Van de Vijver and Leung 2000). Given that the constructs come from Western research, we hypothesized that reliability estimates (e.g., Cronbach's alpha (α) will be lower than in the reported studies in Finland.

*Hypothesis 2* All things being equal, we could have hypothesized that the students' views of themselves as mathematics learners using VOM could be explained by seven factors *(ability, effort, success, teacher quality, family encouragement, difficulty, and enjoyment)*. However, since these items have not been fully used in other countries apart from Finland, we will leave open the research question as to whether there is support for the seven-factor structure identified in previous studies (Hannula et al. 2005; Roesken et al. 2011). We hypothesized *a priori* that: (a) each item has a non-zero loading on the VOM factor it was designed to measure, and zero loadings on all other factors, (b) the factors are correlated and, (c) the error/uniqueness term

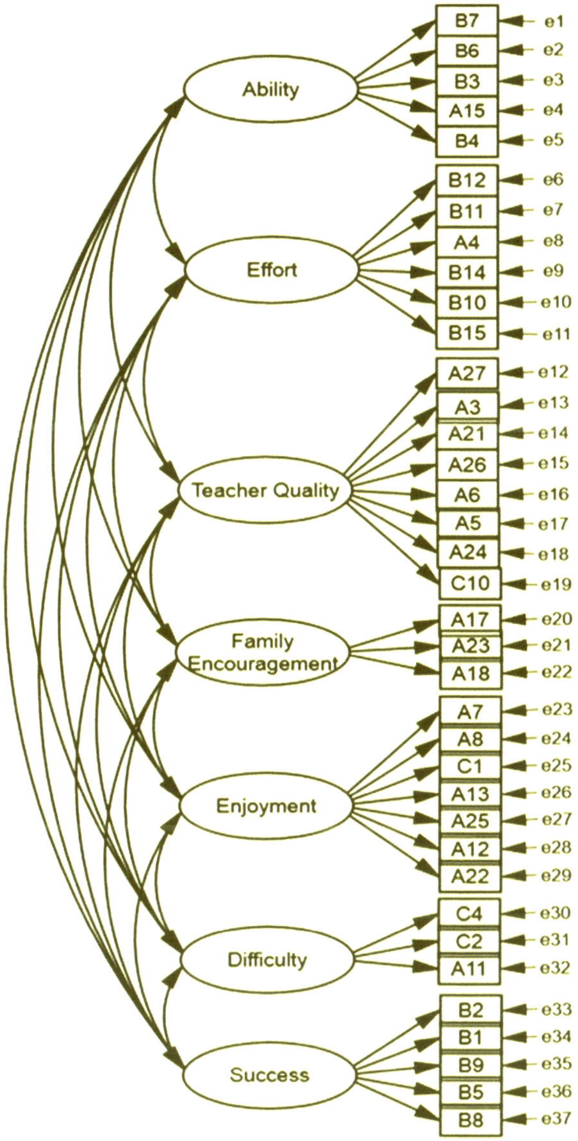


**Fig. 1** The seven-factor model of students' view of themselves as learners of mathematics as identified in Finland. Short one-way arrows denote measurement error terms associated with the observed measures. This diagram was drawn based on the findings in Roesken et al. (2011)

for the item variables are uncorrelated. A schematic representation of this postulated model is presented in Fig. 1.

*Hypothesis 3* In the confirmatory factor analysis (CFA) that follows the EFA, we expect support for the invariance of factor loading, and factor variance-covariance (structural invariance) of the new proposed factor structure for the calibrated and validated independent sample including students' gender. We hypothesized a low to moderate correlation between the constructs.

# Methods

Given the cultural background differences between students of the Finnish and Ghanaian samples, it is possible that some items of the instrument function differently, which may lead to different factorial structures. The factorial validity of VOM factors has been examined using only Principal Component Analyses (PCA) approach. Based on deficiencies associated with exploratory factor analysis (EFA) in general (Marsh et al. 2009, 2011; Sass and Schmitt 2011; Schmitt 2011) and PCA in particular (Marsh et al. 2009; Schmitt 2011), Confirmatory Factor Analysis (CFA) procedures were used to provide a more robust test of factorial validity. Moreover, the factorial structure of the VOM had not been validated using two independent samples of any data.

## *The Need for Factor Analysis*

Factor analysis (FA) has become a highly popular statistical method in the behavioural sciences. In fact, it is especially relevant for test construction and development, as we will see throughout the rest of this chapter. FA is a method generally used to help uncover the relationships between assumed latent variables and manifest variables.

There are two main types of factor analysis: EFA and CFA. EFA is a data-driven approach such that no *a priori* specifications are made concerning the number of common factors and the indicators (i.e. factor loadings). In contrast, CFA is used to test the extent to which *a priori*, theoretical model of factor loadings provides an adequate fit for the actual data. Thus, in EFA the statistical method determines the factors and loadings, whereas CFA detects how well our theoretical model matches reality (the actual data) (Hair et al. 2010). Thus, CFA is a tool that enables one to confirm or reject *a priori* theory. FA bears resemblances to a statistical approach often used in the behavioral and social sciences for data reduction, and has been used in all analyses involving VOM which is called principal component analysis (PCA) (Raykov and Marcoulides 2010). PCA has often been assumed to be a factor analytic method. However, from a technical perspective, PCA is not a member of the FA family (Schmitt 2011; Raykov and Marcoulides 2010). One main difference between PCA and FA is that PCA assumes no measurement of error, whereas FA methods account for the measurement error (Schmitt 2011). Moreover, in FA the common factors are interpretable in addition to reduction of complexity whereas PCA is only for data reduction. Schmitt argued that though evidence suggests that PCA can produce similar results as FA when measurement reliability is high and when factor items are many, estimations of PCA will be less close to CFA than any factor analysis method (Schmitt 2011).

There are some limitations associated with EFA such as (a) not being able to yield a unique factor structure, (b) not defining a testable model, (c) not assessing the extent to which a hypothesized model will fit given data, (d) not being able to suggest model improvements, and, (e) not offering a strong analytic framework for evaluating the equivalence of measurement models across distinct groups (e.g., gender) (Byrne 1991; Brown and Moore 2012). Thus, CFA is a more powerful tool for the testing of factorial validity and construct validation, which necessitated its use in the present study.

It is also important to know that structural equation models (SEMs) based on CFAs may produce very different structural coefficients and model fit statistics than EFAs, as the CFA approach can depict the factor structure differently (Marsh et al. 2011; Sass and Schmitt 2011; Schmitt and Sass 2011). Therefore, specifying the appropriate measurement model (EFA or CFA) has direct implications for replicating factor structures and interpreting structural coefficients (Marsh et al. 2009; Marsh et al. 2011; Sass and Schmitt 2011).

In determining the number of factors, different statistical methods were used, the Minimum Average Partial (MAP) method using the IBM SPSS Statistic 21 (O'Connor 2000) and Parallel Analysis (PA) (Henson and Roberts 2006) procedure in Mplus. The use of the above factor retention methods were used as recommended by Schmitt (2011). The determining of the factors was also guided by the quality of the variables measuring the factors, size of the loadings (>0.300) on the standardized scale, size of indicator communalities, number of variables that load on the factor (min 3), factor homogeneity, and factor determinacy—correlation between the estimated factor score and the factor.

## *Participants*

The sample consisted of 2034 twelfth-grade Ghanaian students (*M*age = 18.49, *Md*nage = 18, *SD*age = 1.25; 58.2 % girls). Nine Senior High Schools were selected from urban and rural schools in Ghana based on their rankings by the Ghana Education Service. The first author gave the questionnaire to the students during their normal class hours in the summer of 2011. Participants' permissions were collated and received by the heads of institutions. The participation in the survey was voluntary and students had the right to withdraw or skip any question that they did not wish to answer. The schools were selected to represent the most representative variety of school types in Ghana, and they included single-sex, coed, private, religious and public schools. Some schools fell under more than one of these categories. The students were enrolled in different mathematics classes; core mathematics (49.3 %) and elective mathematics (50.7 %). They were enrolled in either General Arts (33 %), Business (19.2 %), Science (29.1 %), or Vocational Science (18.7 %) streams.

We cannot claim that the sample is representative of the entire student population of secondary schools in Ghana, but the schools were chosen to represent the most commonly occurring types of high schools in terms of the social intake, disciplines and rates of academic success and failure. Therefore, the results cannot be applied to the students of all schools, though they are representative of students in a range "typical" of the secondary school system in Ghana. There were 63 different student classrooms with an average class size of 32 students.

## *Measures*

We used the VOM instrument (Hannula et al. 2005; Roesken et al. 2011). The instrument consists of 55 items, most of which had originated from a qualitative study on student-teachers' views of mathematics (Pietilä 2002). An additional four items originated from a previous study on Finnish comprehensive schools (Nurmi et al. 2003), and 10 items originated from the self-confidence scale of Fennema-Sherman mathematics attitude scales (Fennema and Sherman 1976), and some novel items developed by the team to measure student perceived success in mathematics. Apart from the 10 Fennema-Sherman items, all the other items were originally in Finnish and had been translated into English.

Items were assessed using a 5-point Likert scale. The statements in the questionnaire were grouped around the following topics: (1) Experiences as a mathematics learner (A1–A29), (2) Image of oneself as a mathematics learner (B1-B15), and (3) View of mathematics and its teaching and learning (C1–C11). The instrument has been successfully implemented in teacher and student settings, whereby reliability and validity of the instrument have been demonstrated. Cronbach's alpha ($\alpha$) reliability in a study of Finnish upper secondary students (Roesken et al. 2011) was between (0.800–0.910) and in a study of student teachers (0.780–0.910) (Hannula et al. 2005). Abbreviated four-item versions of the scales for *success, ability, effort, difficulty* and *enjoyment* were also used in a study of Finnish comprehensive school students (Hannula and Laakso 2011), and again, the reliabilities were found to be good: for eighth-grade students (0.780–0.880) and with exception of *effort* (0.660) reliabilities were also good for fourth-grade students (0.750–0.810).

The studies reported high correlations between core dimensions of the beliefs (up to 0.790; Roesken et al. 2011). We are aware that the high correlation (>0.750) between constructs is a possible sign of multicollinearity (Byrne 2012; Hair et al. 2010). Multicollinearity violations may lead to the wrong interpretation of the findings because it makes it difficult to predict the individual importance of a predictor. Moreover, instances where even a proper solution can be obtained, multicollinearity can lead to inaccurate parameter estimates and a high incidence of Type II errors, particularly when reliability is weak, sample size is small, and explained variance is low (Grewal et al. 2004, p. 526). Although multicollinearity was high, it was not a concern for the authors because reliabilities were high, a high $R^2$, and the large sample size offset the problems caused by the multicollinearity (M. Hannula, personal communication, October 8, 2013). Other literature also supports the argument (see: Grewal et al. 2004; Mason and Perreault 1991). Grewal and colleagues, further argue that the problem of multicollinearity should not be viewed in isolation unless the multicollinearity is severe.

## Analyses

For cross-validation purposes, the whole sample was randomly split into two, with one-half of the sample, ($N=1{,}017$) assigned as the calibration sample and the other half ($N=1{,}017$) as the validation sample. The reason for this split was to ascertain whether a model that has been specified in one sample could be replicated over a second independent sample from the same population. The objective was to find a robust model that was replicable among the sample and avoid the problem of capitalization on a chance outcome that can appear when only one sample is analyzed.

Data were analyzed in three-stages. First, CFA procedures were conducted on the whole sample to investigate whether the established dimensionality (seven-factor structure) and factor-loading pattern fitted the Ghanaian twelfth-graders' sample. This was the confirmatory aspect of the analysis.

Second, the data did not fit the hypothesized model, therefore analyses proceeded in an exploratory mode using both EFA and CFA approaches to identify the course of the misfit, and specify an alternative model for the factor structure. The EFA method was used to examine the number of underlying factors and the CFA-post hoc procedures were used to identify item parameters that contributed to the model misfit. Information from the exploratory analyses (both EFA and CFA-post hoc) was used to propose a final factorial structure based on the calibration sample. CFA was then used again to investigate whether the established dimensionality and factor-loading pattern fitted the independent validation sample. Third, VOM equivalency across the calibration and validation sample was tested in respect of (a) factor form invariance or configural invariance- that freely estimated the item loadings on both samples, (b) factor loading invariance or metric invariance for the calibrated and validated samples, and (c) the common characteristics of individuals by examining factor variances and covariances (FVCV) relationship in both samples (structural invariance). FVCV invariance will also help to ascertain the homogeneity (unidimensionality) of the constructs.

## Goodness of Fit and Reliability Estimates

Evaluation of a model fit was based on multiple criteria that reflected statistical, theoretical, and practical perspectives. A goodness of fit was evaluated by using Chi-Square Difference Testing using the Satorra-Bentler Scaled Chi-Square test statistic ($SBS\Delta\chi^2-MLR\chi^2$), the Root Mean Square Error of Approximation (RMSEA), the comparative fit index (CFI) and the Tucker-Lewis Index (TLI), which are relatively independent of sample size (Chen 2007).

The hypothesized model and the final model were compared for the best fit using, the information Criteria indices such as Akaike (AIC), Bayesian (BIC), and Sample-Size Adjusted (SSBIC) because the models were not nested. The CFI and TLI vary along 0–1 and values ≥0.90 and 0.95 are deemed acceptable and excellent threshold respectively, and RMSEA ≤0.08 and 0.05 for close and reasonable fitting model (Brown 2006). For AIC, BIC, and SSBIC, the model with the smallest value information criterion is preferred. When evaluating the worth of individual parameters, statistical significance values as indicated by the Mplus $z$–values, goodness-of-fit based on the normalized residual values, modifications indices (MIs), and model meaningfulness were also taken into account.

The SEM analyses in the present study were done using Mplus 7.11 (Muthén and Muthén 1998–2012). All analyses were based on the Mplus robust maximum likelihood estimator (MLR), with standard errors and test-of-fit that were robust to non-normality of the observations to control for the non-independence of observation (Muthén and Muthén 1998–2012). In addition, the choice of MLR, rather than categorical variable estimator procedure was based on research studies (Rhemtulla et al. 2012) that indicated how categorical methods make little or no differences when Likert scales of five or more categories are treated as categorical variables or continuous variables. In order to include all of the observed data, missing data patterns were handled with Mplus feature of full information maximum likelihood (FIML).

We analyzed the normality assumptions, by investigating the normality of each variable in terms of its kurtosis and skewness. With guidelines of normality (i.e., skewness: <3; kurtosis: <7) proposed by Curran et al. (1995), there were few non-normality items that supported the use of robust maximum likelihood estimator (MLR).

Cronbach's alpha has been used as the standard measure of reliability for a long time, although it is known to either underestimate or overestimate reliability (Geldhof et al. 2014; Novick and Lewis 1967). Composite reliability (ω) (Geldhof et al. 2014; Raykov 2012) used in conjunction with structural equation modeling (SEM) will be estimated to complement the α estimates of the new VOM scales. Composite reliability (ω) takes into account the computed factor loadings, and produces more precise estimates of reliability than those provided by α (Geldhof et al. 2014; Raykov 2012). It is interpreted in the same way as Cronbach's alpha. Generally, ω values of 0.600–0.700 are acceptable in exploratory research (Hair et al. 2010).

## *Invariance Model Testing*

Measurement invariance is the equivalence of a measured construct in two or more groups, such as people from different cultures (Chen 2008). It assures that the same constructs are being assessed in each group (Sass 2011). Invariance model testing usually begins with a baseline model often called the *configural model* in which all parameters in the model are freely estimated across groups. When the baseline model fits adequately in each group, this indicates that the same number of factors best represents the data for all, and the same variables define each factor across groups. Then one can test if the factor structures are equal by restricting the factor loadings to be equal across groups. The model in which the factor loadings are held equal is usually called the *metric invariance or weak invariance.* When metric invariance holds, we can conclude that the constructs are similarly manifested in each of the groups. Finally, we imposed constraints to factor variance and covariance to test for structural invariance. A non-invariance structural model would suggest a differential structure for the construct being measured across the groups (i.e. the associations among the underlying factors varying across groups). Thus, structural invariance indicates the homogeneity (unidimensionality) of the constructs, which is a necessary condition for both reliability and validity.

## Results

### *Stage 0: Computing Cronbach's Alpha (α) for the Hypothesized Scales*

Our *a priori* model (Fig. 1) posited that the VOM constructs could be explained by seven-factors. The seven factors were the *ability, effort, teacher quality, family encouragement, enjoyment of mathematics, difficulty of mathematics, and success* (Roesken et al. 2011). The first confirmatory approach was to compute the Cronbach's alpha for each factor. The Cronbach's alpha coefficients were calculated as indicators of factor reliability. These alpha coefficients for the Ghanaian sample were within the acceptable standard for *ability* (0.863) and *enjoyment* (0.764), with the rest below the acceptable threshold: *effort* (0.538), *teacher quality* (0.190), *family encouragement* (0.623), *difficulty* (0.565) *and success* (0.661). The Cronbach's alpha values indicated that most scale reliabilities were considerably lower than those of the Finnish sample. The scale for *teacher quality* was unacceptably low, but two of the scales were above the usual considered 0.700 reliability threshold and two of the remaining three were sufficiently reliable for some researchers to consider them acceptable (Hair et al. 2010), or even all three when their content coverage and unidimensionality were sufficient (Schmitt 1996). Since Cronbach's alpha does not index unidimensionality of the constructs together with what have been discussed earlier, there is good reason to apply a more robust approach (CFA-stage 1; EFA-stage 2) to test the whole model, which was stage 1.

### *Stage 1: Test for Factorial Validity; Confirmatory Factor Analyses*

CFA indices for the hypothesized seven-factor model were poor from both statistical ($MLR\chi^2_{(608)} = 1922.993$) and a practical (CFI = 0.843, TLI = 0.828, RMSEA = 0.046) perspective. This model was therefore rejected. We also tested the model fit after removing the scale with the lowest reliability (*teacher quality*), but the model fit was only marginally improved (MLR$\chi^2_{(362)}=1361.447$, CFI = 0.849, TLI = 0.831, RMSEA = 0.052).

A further look at the correlations indicated very high factor correlations between some of the constructs, which indicated multicollinearity: the correlations between the *ability* and *difficulty* factors ($r=0.853$), and *ability* and *enjoyment* ($r=0.847$), *difficulty* with *enjoyment* ($r=0.871$), suggested that the factor structures were not statistically distinguishable, thus they measured the same dimension.

## *Stage 2: Exploratory Factor Analysis*

After rejecting the *a priori* model, the next logical step was to take an exploratory approach to analyze these data in order to identify a better fitting model. A particularly important question was (a) whether the Ghanaian data could be described more reasonably by a model that specified less than, or more than the seven factors, and (b) whether an independent sample from the Ghanaian data exhibits the same pattern of loadings for all factors. The data were reanalyzed using exploratory factor analysis (EFA) to answer these questions. Previous research indicates different dimensions of mathematics-related beliefs correlate (e.g., Roesken et al. 2011), thus Geomin (oblique) rotation was used as the rotation procedure to get a cleaner simple factor structure that is similar to CFA (Schmitt 2011).

All 55 items from the original questionnaire were used in the EFA analysis. The results from the Minimum Average Partial (MAP) method indicated a six-factor solution whereas Parallel Analysis indicated a seven-factor solution**.** EFA analyses for 4–7 factors were run on the data simultaneously to determine if there were plausible models that could explain the relationships among the items. A four-factor solution was included and tested because of the high correlation that was identified early by pre-supposing three factors to be measuring the same dimension. The residual variance of all items, i.e. the proportion of variance in the indicators that has not been explained by the latent variables, were checked in respect of all the proposed factor structures (four, five, six, and seven factors). Items with very high residual variance (>0.800), loadings of less than 0.300 and high cross-loadings were deleted. When the EFA was re-run, no item loaded for the seventh factor and analysis was continued with four, five and six factor models. Again, items with a high residual variance and which loaded less than 0.300 or high cross-loadings were deleted and the EFA was re-run. Whereupon only two items loaded on the sixth factor and thus we removed the items and continued comparing four and five factor solutions. The high factor correlation (unstandardized: $r=0.914$, standardized $r=0.849$) indicated that two of the constructs in the five-factor solution were not statistically distinguishable. The EFA for a four-factor structure was acceptable as the final model.

## *The Four-Factor Structure*

Given both substantive and statistical considerations discussed above, the EFA suggested the four-factor solution as the most optimal to represent the Ghanaian data. The *a priori* hypotheses model included 37 items, whereas only 29[1] items out of 55 items exceeded the threshold for inclusion in the analysis and were included in the present EFA model. The four factors were labeled as *self-confidence, self-concept, family encouragement,* and *teacher quality.*

The *self-concept* factor includes all five *ability* items, five out of seven *enjoyment* items, two out of three *difficulty* items, one *success* item, one *effort* item and two new items making a total of 16 items. The two new items were item A10: *My eagerness to study mathematics is seasonal* and item A19: *Mathematics has been a clear and precise subject to study.* Therefore, this study has demonstrated that the *ability, success,* and *enjoyment* factors loads on the same factor and therefore can be treated empirically as the same construct. Absolute target loadings were high between (0.395 and 0.789) with non-target loadings between (0.003 and 0.159). The *self-confidence* factor include three items from the *success* factor (B9, B2, B1), and one *effort* factor item (B15). Absolute target loadings on the *self-confidence* factor were between 0.435–0.688, and very low non-target loadings of between 0.003 and 0.088. The items of the *teacher quality* factor were all from the Finnish based factor except that, two items from the original solution were not included (A5, C10), because they failed to surpass the threshold value in addition to being cross loading items. Target loadings were between 0.358 and 0.740, non-target loadings between 0.011 and 0.118. All items on the *family encouragement* factor exceeded the threshold for inclusion in the analysis. Target loadings were between 0.496 and 0.596, whereas non-target loadings were between 0.002 and 0.218. All factor loadings were statistically significant ($p$ <0.001). The patterns of the correlations were consistent with a low-moderate (0.141–0.430) correlation in line with the *a priori* hypothesis. These results support the assertion that the VOM structure of the Ghanaian data is different.

The four factors can be identified within McLeod's (1992) structural qualities associated with VOM. The *self-confidence* and *self-concept* factors corresponded to 'beliefs about self', whereas the *teacher quality* factor corresponded to that of 'beliefs about mathematics teaching' and the *family encouragement* factor to 'belief about the social context'. Factor determinacies were 0.958 for self-*concept*, 0.878 for *teacher quality*, 0.862 for *self-confidence* and 0.798 for *family encouragement.*

[1] Four items were deleted due to content overlap detected from the post hoc confirmatory factor analysis in the next section.

## *Post hoc Confirmatory Factor Analysis*

The EFA suggested a four-factor structure for the VOM. We ascertained the extent to which the newly specified model fitted the data over the hypothesized model, by using the CFA approach. For a better model over the hypothesized model,[2] we compared the AIC, BIC, and SSBIC information criteria between two models because they were not nested. This is because the new four-factor model had been structurally revised (i.e., factors had been partially collapsed into a single latent variable—for example, *ability*, *enjoyment* and *difficulty* factors). To improve the VOM constructs, modification indices (MI) were consulted. Six consecutive CFA analysis guided by the MIs, correlations between items and item residual variance, led to items B7, A7 and C1 (i.e. all on the self-*concept* factors) being deleted and the error covariance between items B4 and B3 included in the final model (for detail analysis, see Bofah and Hannula 2014).

There is considerable discussion in the CFA literature regarding the interpretation and need for the inclusion of the error covariance in addition to what the appropriate solution to the problem is (Byrne 2012; Byrne 1993; Marsh et al. 2012). Studies have indicated that including the error covariances in a model improves the model fit whereas excluding them is likely to bias and inflate parameter estimates (Byrne 2012; Marsh et al. 2004, 2012). The inclusion of the error covariance has been justified when these parameters represented non-random measurement errors (Byrne 1991) due to *method effects; and* as such their presence was expected. Brown and Moore (2012, p. 362), and Edelen and Reeve (2007) argued that the possible causes for such covariation are results from the following: common assessment methods (e.g., questionnaires); reverse items, or similarly worded items, items that are presented sequentially, and items with high content overlap. They also listed items prone to differential susceptibility to other influences such as self-report items, demand characteristics, reading difficulties, item translation, acquiescence, and the format of the instrument or social desirability. In our study data set, translation, and content overlap was the case in all but one of the error covariance (i.e., between B3 and B4).

Moreover, there seems to be a good explanation for the error covariance between items B3 ("Mathematics is my weakest subject") and B4 ("Mathematics is difficult for me"). Crosstabulation of responses revealed that most respondents tended to have similar agreement with both items, though there was a significant subset of respondents who disagreed strongly with B3 and yet agreed with B4. We assume that these respondents are the students who find mathematics difficult, and that who also struggle more with other subjects. Another explanation was that items B3 and B4 were presented sequentially in the questionnaire and both measure students' weakness in mathematics that display local dependence of both items. In addition, both items were negatively worded. Including error covariance between these items improves the model fit because it adjusts for this response pattern.

Reviewing the error covariance, we see that not only is it very highly significant ($p < 0.0001$) in the model, it was also very high ($r = 0.444$). We evaluated the

[2] Indices not reported.

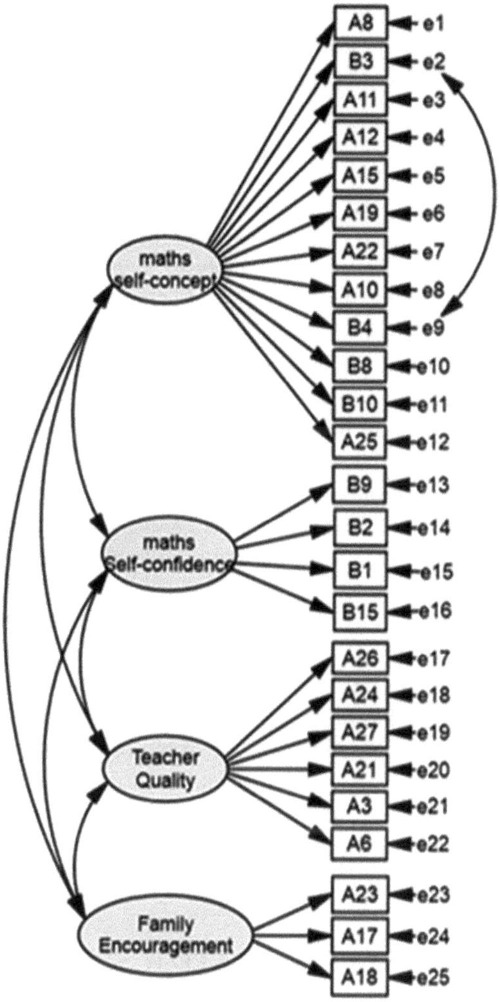


**Fig. 2** Final hypothesized model and Baseline model of the VOM for Ghanaian twelfth-grade students. Short one-way arrows stand for measurement error terms associated with the observed measures

strength associated with the error covariance term linking item B3 and B4 together with the indicated rationale, and considered it to be more realistic to include this parameter in the model, rather than ignoring its presence. We tested models with and without the error covariance B3 and B4 to show that including the error covariance in the final model improved the model. Fit indices for model with error covariance ($MLR\chi^2_{(268)} = 612.874$, CFI = 0.932, TLI = 0.923, RMSEA = 0.036), and model without error covariance ($MLR\chi2_{(269)} = 690.419$, CFI = 0.916, TLI = 0.907, RMSEA = 0.039). All the fit indices together with the $SB\Delta\chi2 - MLR\chi^2$ value of 61.314 and $\Delta df = 1$ indicate that the model with the error covariance provides a significantly better fit to the data than the model without the error covariance (the critical value for $SBS\Delta\chi^2 - MLR\chi^2$ is 3.84; $\alpha = 0.05$, $df = 1$).

Therefore, our final model illustrated in Fig. 2 became the final model that represented the VOM structure for the Ghanaian data. It provided the baseline model for subsequent analyses related to cross-validation and multi-group invariance testing of the Ghanaian data.

## Stage 3: Cross-Validation Analyses

We continued to cross-validate the factor structure using the results obtained from both the EFA and CFA analyses. Our hypothesized new model illustrated in Fig. 2 provided the baseline model. Cross-validation of our hypothesized model was achieved by testing for invariance across separate calibration and validation samples of the data. Measurement invariance modeling starts with testing for configural invariance. There were support[3] for the *configural, and metric models*, which indicated support for the construct validity across the calibrated and validated samples. Of substantial interest were the two specified residual covariances and the extent of their invariance across the calibration and validation samples. We considered it worthwhile for psychometric reasons and to remove any doubt of capitalizing on chance for their inclusion in the model (MacCallum et al. 1992). We postulated a model in which, the factor loadings, factor variances, factor covariances and residual covariances were constrained to be equal call the structural model.

The sample supported the structural invariance model. Consistent with our study hypotheses, all correlations were in the low to modest range ($r = 0.191$–$0.535$) between the dimensions (see Table 1). Support for the structural invariance, indicated the unidimensionality of the constructs..

In addition, gender invariance (not reported) was tested and there was support for *configural*, *metric*, and *structural invariance*, which gives a further support for the validity and reliability of the constructs.

## Reliability of the New VOM Scales

In response to the reliability hypothesizes, we began by evaluating the Cronbach's coefficient alpha (α) reliability and the Composite reliability (ω) used in conjunction with SEM models of the new four factor VOM scale.

Given that the original items and the constructs come from Western research, it would be expected to find lower reliability estimates. Moreover, the brevity of some of the constructs, (e.g., *family encouragement*) coupled with the positive relationship of Cronbach's alpha reliability values to the number of items on a construct, led some of the coefficient alpha (α) estimates to be below the acceptable threshold. Reliability values of some scales reached the desirable standard of 0.700 (self-concept, $\alpha = 0.872$; $\omega = 0.868$), (*teacher quality*: $\alpha = 0.706$; $\omega = 0.716$). However, they also fell below an acceptable value of 0.700. Although the thresholds may decrease to 0.600 (*self-*

[3] Fit indices not reported.

**Table 1** Factor structure relating the VOM items

| Item | Factor loadings | *SE* | Item wording |
|---|---|---|---|
| Mathematics self-concept factor(1) | | | |
| A8 | 0.589 | 0.028 | Doing calculations has been enjoyable. |
| RB3 | 1.124 | 0.027 | Mathematics is my weakest subject. |
| RA11 | 0.862 | 0.027 | Mathematics has been difficult in upper secondary school. |
| A12 | 1.118 | 0.024 | Mathematics is my favorite subject. |
| A15 | 0.851 | 0.025 | I have made it well in mathematics. |
| A19 | 0.882 | 0.029 | Mathematics has been a clear and precise subject to study. |
| RA22 | 1.044 | 0.028 | Mathematics has been the most boring part of my study. |
| RA10 | 0.685 | 0.032 | My eagerness to study mathematics is seasonal. |
| RB4 | 1.061 | 0.027 | Mathematics is difficult for me. |
| B8 | 0.641 | 0.029 | I can handle advanced Mathematics tasks. |
| RB10 | 0.948 | 0.030 | I have a wrong attitude about mathematics |
| A25 | 0.518 | 0.033 | I enjoy pondering over mathematics tasks. |
| Mathematics self-confidence factor (2) | | | |
| B9 | 0.586 | 0.038 | I know that I can do well in mathematics |
| B2 | 0.530 | 0.036 | I can get a good grade in mathematics. |
| B1 | 0.497 | 0.037 | I am certain that I can learn mathematics. |
| B15 | 0.260 | 0.032 | It is important for me to get a good grade in mathematics. |
| Teacher quality factor (3) | | | |
| A26 | 0.961 | 0.031 | The teacher has so far been a positive example. |
| RA24 | 0.929 | 0.032 | The teacher rushes through the teaching of mathematics. |
| RA27 | 1.029 | 0.034 | I would need a better teacher. |
| RA21 | 0.787 | 0.038 | The teacher does not inspire me to study mathematics. |
| A3 | 0.544 | 0.033 | The teacher explains the studied topics. |
| A6 | 0.470 | 0.037 | Teacher explains what the studied topics are needed for. |
| Family encouragement factor (4) | | | |
| A23 | 0.799 | 0.044 | My family encourages me to study mathematics. |
| A17 | 0.863 | 0.045 | The importance of competence in mathematics has been emphasized at my home. |
| A18 | 0.825 | 0.046 | The example of my parent (s) has had a positive influence on my motivation. |
| | | | Items removed during CFA. |
| B6: I am not the kind of person that knows mathematics well. | | | |
| B7: I am not good in Mathematics. | | | |
| A7: Studying mathematics is boring. | | | |
| C1: Mathematics is a mechanical and boring subject. | | | |

(continued)

**Table 1** (continued)

| Item | Factor loadings | *SE* | Item wording | |
|---|---|---|---|---|
| Factor correlations | | | | |
| | 1 | 2 | 3 | 4 |
| 1 | 1.000 | | | |
| 2 | 0.423 (0.026) | 1.000 | | |
| 3 | 0.535(0.024) | 0.232(0.030) | 1.000 | |
| 4 | 0.260(0.033) | 0.248(0.037) | 0.191(0.034) | 1.000 |

NB: All significant at $p$ <0.001. These results are based on the metric invariance model. Factor loadings are unstandardized estimates. Correlations are constrained to be equal across calibration and validation sample. For model identification, all factor variances were fixed at 1. VOM = students' views on mathematics. For the correlations, parenthesis are standard errors (*SE*), items with R in front are reversed coded

*confidence*: $\alpha$=0.690; $\omega$=0.697), (*family encouragement*: $\alpha$=0.619; $\omega$=.621) as reported in exploratory research and adapted constructs such as those being used in our present study (Hair et al. 2010).

Results for $\omega$ and $\alpha$ were roughly the same, with $\alpha$ values slightly underestimating the *teacher quality, self-confidence, family encouragement* constructs and overestimating the *self-concept* construct. The lower reliabilities we obtained may imply substantial error of measurement and/or limited true individual differences, hence may attenuate the validity of interpretations based on manifest scale scores, weaken statistical power, and effect sizes (Raykov 2012; Schmitt 1996). It is thus advisable to base any comparisons on latent-variable models that account for the unreliability and measurement errors as suggested by Marsh and colleagues (2012).

## Discussion

The factorial validity of the Finnish View of Mathematics (VOM) instrument was tested on a sample of Ghanaian twelfth-grade students. The original seven-factor model fitted the data poorly. In addition, further exploratory factor analysis (EFA) revealed that the Ghanaian data can best be explained by a four-factor structure. The alternative factorial structure was validated, further refined and then cross-validated with an independent sample from the Ghanaian data using a confirmatory factor analysis (CFA) approach. Moreover, measurement of invariance was established at the configural, metric and structural parameter levels. In respect of the new four-factor model, two scales from the Finnish model were partly confirmed and one other scale fully confirmed. The reliability values of the new scales were not very much higher than the values of the original scales. These may be due in part to the brevity of scale consisting of only three or four items. However, the overall fit of the four-factor model was significantly better than the fit of the original seven-factor model. We do know and attest to the use of Cronbach's alpha reliability ($\alpha$) as a non-dependable general index of reliability for multidimensional scales, irrespective of whether their component errors are correlated

(Raykov 2012). Cronbach's alpha underestimated the reliabilities when there was no error term and overestimated the reliability when there was an error term within the construct. Moreover, because α is sensitive to the number of items in a scale, it underestimated the *family encouragement* and the *self-confidence* constructs. The reliabilities for the Ghanaian sample were generally acceptable.

The differences between the Finnish and Ghanaian models were interesting. In the Finnish samples, the scales for *ability, enjoyment* and *effort* were discrete, which provided support for separating emotions and motivation from more cognitive beliefs (Roesken et al. 2011). In Ghana, *ability, success,* and *enjoyment* factors were loaded on to the same factor and therefore can be treated empirically as one construct. The Ghanaian *self-concept* scale included three *ability* items, one *effort* item, four items from the *enjoyment* factor, one item from the *difficulty* factor, one item from the *success* scale and two items (item A10 and A19) that had not originally been part of the factorial structure of the Finnish samples. The *effort, enjoyment,* and *difficulty* factors were not confirmed in the Ghanaian data. Similarly, a study by Kaldo and Hannula (2012) failed to confirm the reliability for the scale of *effort* in a sample of Estonian university students. Hence, it is possible that the separate scale of *effort* is a characteristic feature of Finnish students.

The observed differences in the factor structures between the Ghanaian and the Finnish sample are not necessarily indicative of cultural variation, since such differences may be due to one or more measurement artifacts unrelated to the constructs. Indeed, careful reading of the content associated with each deleted item-pair reveals a strong content overlap. We do admit that in scale construction, it is important to look for items that are highly inter-correlated in order to establish a high degree of internal consistency and reliability. Nevertheless, in our study there was a translation issue, whereby originally two Finnish terms that emphasized different aspects of boredom (unpleasant versus tiresome) were both translated as 'boredom', which reduced the necessary variation between the different items. This finding highlights that item translation is critical when implementing an instrument into a new linguistic setting.

Also, this study has confirmed early studies (Edelen and Reeve 2007) that when items are measuring the same construct and are negatively worded, placing both items next to each other in a questionnaire can lead to local dependency—the response of the items are based on the response of each other.

The psychometric approach in Finland had been PCA, which technically is not able to determine and evaluate measurement error or to indicate alternative model specifications. Using Structural Equation Modeling (SEM) helped us to detect measurement error and bias, while also understanding the disparities. Although the identified factor structure does not conform to the original seven-factor structure of the VOM, it does reflect those important aspects of students' belief systems defined by McLeod (1992) and by Op't Eynde and colleagues (2006). In the light of the extremely stringent approach used to test the validity, the new 25-item scale proved to be quite sound from a psychometric perspective and it theoretically supports many of the dimensions suggested in the literature (McLeod 1992; Op't Eynde et al. 2006; Roesken et al. 2011). However, our results also indicate that important cultural differences are not restricted to how strongly students hold different beliefs, but also to what the actual belief constructs are.

We also conclude that, differences between the present study and previous studies on VOM reflect genuine differences in how the Finnish and Ghanaian upper secondary students' views of mathematics are structured. Realistically, given the dramatic cultural differences between these two countries it was not surprising that the mathematical self-concept among Finnish students seems to have an underlying structure, whereas for Ghanaian students the construct is a single entity. However, the findings can be partially attributed to the choice of statistical analyses. For example, in the present study, by the combined use of EFA and CFA we assumed that the best and correct factor structure for the Ghanaian sample was identified. The original approach in Finland applied PCA and less robust approaches to identify the factor structure. We do expect our conclusions to be more reliable than the results from previous studies in which these important modeling considerations were neglected.

Being able to independently validate the factor structure in two independent samples and for both genders, allowed us to conclude that (a) there is strong empirical support for a new four factor structure, (b) the same variables define each factor across all subsamples (c) all the latent variables have the same relationship within the sample and any differences in the covariance between the measured variables are due to the common factors. In addition, the finding that the newly formed four-factor model supports metric invariance across students' gender as well as single-sex and coeducational schools (Bofah and Hannula submitted) increases its value as an assessment instrument. This indicates that the instrument educes responses to questions that are being asked in the same way by different groups within a sample.

To summarize, the current study supports the measurement and structural invariance of VOM, as measured in the Ghanaian sample, across student gender, and suggests that further mean comparisons within the belief constructs can be interpreted as representing the underlying mean differences in the Ghanaian data.

Moreover, the present study has shown that, translation of a construct into a different language is more than just producing a text in another language. Knowing the linguistic and cultural differences can help reduce the problems associated with responses to translated adapted constructs. In addition, this paper has raised three important issues educational researchers face when they adopt and validate a construct cross-culturally. The three issues discussed below have been similarly argue for and discussed in Geisinger (1994) and Lin et al. (2005):

First, adaptation issue: an important question that researcher need to ask is, "Does a given construct need to be adapted?" Reliability estimates have been used as a yardstick to circumvent this question. Often or not, researchers adopt survey construct in a new cultural setting because it has a strong reliabilities in the original setting. Although, this issue is not problematic when no marked differences exist between the original population and the target population. Translation is needed when administering a survey instrument to respondents who speak another language either than the language used in original setting. This is where the cultural differences as well as the linguistics of the original and the target populations need to be taken into account. Moreover, a more difficult issue concerning test adaption is subpopulation within a given nation as well as cultural adaption within a single language. For instance, Ghana has five strong ethnic groups with over 100 linguistic and cultural groups within these five ethnic groups (Bodomo 1996). Ghana has adopted English as their national language; will an adopted Western construct translated into the national Language be adapted for use across the whole nation. This "... can be a difficult question, sometimes with more complex answers" (Geisinger 1994, p. 305).

Second, construct validity: this issue mainly deals with the question: "Does the construct measures what it intended purposes were in the new language or culture settings?" To answer this question, the construct validation and the reliability should be demonstrated. This can help establish if the assess construct have the same meaning in new target population. This is an important issue when the new population differs from the original in terms of cultural development. Researchers are encourage to use a more robust methodology such as SEM to validate constructs in cross-cultural research. This can help reduce method effects such as construct bias, method bias, and item bias associated with cross-cultural research (Lin et al. 2005; Sass 2011; Van de Vijver and Leung 2000). This can also help detect problems such as content overlap, item local dependency, and acquiescence.

Third, interpretation issue: importantly, after adapting and validating the instrument, how to interpret the scores of outcome on the new target population, i.e., "what do scores on the adapted measure mean?" Does the outcome support the literature meaningfully? Were the results driven by "acquiescence or substantive cultural differences?" Are the construct different across cultures due to religion and method effect? Can there be any referent group effect?—(see Marsh 2007). As discussed earlier, cultural and linguistic differences, can lead to different interpretations. Thus, the construct and instrument comparability across the cultures should be examined critically before giving interpretations (Lin et al. 2005).

One potential limitation of the study is that 12 items were removed from their designated factors because of model misfit and dimensionality concerns. In addition, two new items were included in the final model. This indicates that the rotation criterion used and how the factor analysis (i.e., EFA or CFA) is parameterized can significantly alter construct correlations and loadings/cross-loadings (Sass 2011; Sass and Schmitt 2011). For these reasons, Sass and Schmitt (2011, p. 301) urge, "model specification, modification, and verification decisions should be made judiciously and researchers must be cognizant of how the modeling approach influences the statistical and theoretical conclusions" (see also, Jöreskog 1993). Although our purpose was not to refine previous measures, implementing this modification should benefit future research using these scales. Another limitation is that the multilevel nature of the data was not taken into account. Because the data were derived from students in intact classes (students' in schools), they are inherently hierarchical. A hierarchical model could have helped us answer the question of whether a particular construct has the same meaning at the individual and classroom levels. Ignoring this nested structure can give rise to problems of bias within-group homogeneity (Fraser 1998). We do think the clustering effect in this study is negligible due to the number of schools involved in the study. A final limitation of this study was that all data were from self-reports and thus subject to social desirability biases.

The outcome of this chapter is one of the indications of the problems associated with the importation of Western instruments into non-Western countries. We conclude that, cross-cultural educational researchers should be conscious of the problems of construct importation and adaptation, such as, item translation—content overlap, acquiescence, reading difficulty, reverse items, similarly worded items, items that are presented sequentially, construct, method, and item bias that could affect the results of studies. We believe it is important that cross-cultural educational researchers acquire both a theoretical understanding of these issues and a practical ability to address them using *MPLUS* or some other SEM software. Furthermore, cross-cultural educational researchers should pay attention to the construct validity and interpretation of the study outcome. Failing, importation of Western constructs into non-Western countries may lead to inferences that are not valid.

This research has laid a solid foundation for future mathematics belief research in Ghana by making readily available a selection of valid, reliable and applicable questionnaires for researchers, teachers and policy makers.

## References


Ainley, J., Kos, J., & Nicholas, M. (2008). *Participation in science, mathematics and technology in Australian education* (ACER research monographs). Retrieved from http://research.acer.edu.au/acer_monographs/4

Asante, K. O. (2012). Secondary students' attitudes towards mathematics. *IFE PsychologIA, 20*(1), 121–133.

Bodomo, A. (1996). On language and development in Africa: The case of Ghana. *Nordic Journal of African Studies, 5*(2), 31–51.

Bofah, E. M., & Hannula, M. S. (2014). Structural equation modelling: Testing for the factorial validity, replication and measurement invariance of students' views on mathematics. In *SAGE research methods cases.* London: SAGE Publications, Ltd. doi:http://dx.doi.org/10.4135/978144627305014529518

Bofah, E. A., & Hannula, M. S. (submitted). Students' views on mathematics in single-sex and coed classrooms in Ghana. *International Journal of Mathematical Education in Science and Technology.*

Brown, T. A. (2006). *Confirmatory factor analysis for applied research.* New York: Guilford Press.

Brown, T. A., & Moore, M. T. (2012). Confirmatory factor analysis. In R. H. Hoyle (Ed.), *Handbook of structural equation modeling* (pp. 361–379). London: Guilford.

Byrne, B. M. (1991). The Maslach Burnout Inventory: Validating factorial structure and invariance across intermediate, secondary, and university educators. *Multivariate Behavioural Research, 26*(4), 583–605.

Byrne, B. M. (1993). The Maslach Burnout Inventory: Testing for factorial validity and invariance across elementary, intermediate and secondary teachers. *Journal of Occupation and Organizational Psychology, 66*, 197–212.

Byrne, B. M. (2012). *Structural equation modeling with Mplus: Basic concepts, applications, and programming.* New York: Routledge.

Chen, F. F. (2007). Sensitivity of goodness of fit indexes to lack of measurement invariance. *Structural Equation Modeling, 14*(3), 464–504.

Chen, F. F. (2008). What happens if we compare chopsticks with forks? The impact of making inappropriate comparisons in cross-cultural research. *Journal of Personality and Social Psychology, 95*, 1005–1018.

Cheung, F. M. (1996). The assessment of psychopathology in Chinese societies. In M. H. Bond (Ed.), *Handbook of Chinese psychology* (pp. 393–411). Hong Kong: Oxford University Press.

Curran, P. J., West, S. G., & Finch, J. F. (1995). The robustness of test statistics to nonnormality and specification error in confirmatory factor analysis. *Psychological Methods, 1*(1), 16–29.

Edelen, M. O., & Reeve, B. B. (2007). Applying item response theory (IRT) modeling to questionnaire development, evaluation, and refinement. *Quality of Life Research, 16*, 5–18.

Eshun, B. A. (2004). Sex-differences in attitude of students towards mathematics in secondary schools. *Mathematics Connection, 4*, 1–13.

Fennema, E., & Sherman, J. A. (1976). Fennema-Sherman mathematics attitudes scales. *JSAS Catalogue of Selected Documents in Psychology, 6*, 31.

Fraser, B. J. (1998). Science learning environments: Assessment, effects and determinants. In B. J. Fraser & K. G. Tobin (Eds.), *International handbook of science education* (pp. 527–564). Dordrecht: Kluwer.

Geisinger, K. F. (1994). Cross-cultural normative assessment: Translation and adaptation issues influencing the normative interpretation of assessment instruments. *Psychological Assessment, 6*(4), 304–312.

Geldhof, G. J., Preacher, K. J., & Zyphur, M. J. (2014). Reliability estimation in a multilevel confirmatory factor analysis framework. *Psychological Methods, 19*(1), 72–91.

Ghana Education Service. (2013). *Educational statistics: Secondary education statistics*. Accra: Ghana Education Service. Available from: http://www.ges.gov.gh. Accessed on 11 Oct 2013.

Grewal, R., Cote, J. A., & Baumgartner, H. (2004). Multicollinearity and measurement error in structural equation models: Implications for theory testing. *Marketing Science, 23*(4), 519–529.

Hair, J. F., Black, W. C., Babin, B. J., & Anderson, R. E. (2010). *Multivariate data analysis* (7th ed.). Upper Saddle River: Prentice Hall.

Hannula, M. S. (2012, July 8–15). *Emotions in problem solving*. A paper presented in 12th international congress on mathematical education, COEX, Seoul, Korea. Available at: http://www.icme12.org/upload/submission/1983_F.pdf

Hannula, M. S., & Laakso, J. (2011). The structure of mathematics related beliefs, attitudes and motivation among Finnish grade 4 and grade 8 students. In B. Ubuz (Ed.), *Proceedings of the 35th conference of the international group for the psychology of mathematics education* (Vol. 1). Ankara: PME.

Hannula, M. S., Kaasila, R., Laine, A., & Pehkonen, E. (2005). The structure of student teacher's view of mathematics at the beginning of their studies. In M. Bosch (Ed.), *Proceedings of the fourth congress of European research in mathematics education (CERME 4)* (pp. 205–214). Sant Feliu de Guíxols: Fundemi IQS – Universitat Ramon Llull.

Henson, R. K., & Roberts, J. K. (2006). Use of exploratory factor analysis in published research: Common errors and some comment on improved practice. *Educational and Psychological Measurement, 66*, 393–416.

Jöreskog, K. G. (1993). Testing structural equation models. In K. A. Bollen & L. J. Scott (Eds.), *Testing structural equation models* (pp. 294–316). Newbury Park: Sage.

Kaldo, I., & Hannula, M. S. (2012). Structure of university students' view of mathematics in Estonia. *Nordic Studies in Mathematics Education, 17*(2), 5–26.

Kloosterman, P., & Stage, F. K. (1992). Measuring beliefs about mathematical problem solving. *School Science and Mathematics, 92*, 109–115.

Leung, K., & Zhang, J. X. (1995). Systemic considerations: Factors facilitating and impeding the development of psychology in developing countries. *International Journal of Psychology, 30*, 693–706.

Lin, Y. H., Chen, C. Y., & Chiu, P. K. (2005). Cross-cultural research and back-translation. *The Sport Journal, 8*(4), 1–5.

MacCallum, R. C., Rosnowski, M., & Necowitz, L. B. (1992). Model modifications in covariance structure analysis: The problem of capitalization on chance. *Psychological Bulletin, 111*, 490–504.

Malik, K. (2013). *Human development report 2013*. New York: United Nations Development Programme.

Marsh, H. W. (2007). *Self-concept theory, measurement and research into practice: The role of self-concept in educational psychology*. Leicester: British Psychological Society.

Marsh, H. W., Parada, R. H., & Ayotte, V. (2004). A multidimensional perspective of relations between self-concept (self description questionnaire II) and adolescent mental health (youth self-report). *Psychological Assessment, 16*(1), 27–41.

Marsh, H. W., Muthén, B., Asparouhov, A., Lüdtke, O., Robitzsch, A., Morin, A. J., et al. (2009). Exploratory structural equation modeling, integrating CFA and EFA: Application to students' evaluations of university teaching. *Structural Equation Modeling, 16*, 439–476.

Marsh, H. W., Liem, G. A., Martin, A. J., Morin, A. J., & Nagengast, B. (2011). Methodological measurement fruitfulness of exploratory structural equation modeling (ESEM): New approaches to key substantive issues in motivation and engagement. *Journal of Psychoeducational Assessment, 29*, 322.

Marsh, H. W., Abduljabbar, A. S., Abu-Hilal, M. M., Morin, A. J., Abdelfattah, F., Leung, C. K., et al. (2012). Factorial, convergent, and discriminant validity of TIMSS math and science motivation measures: A comparison of Arab and Anglo-Saxon countries. *Journal of Educational Psychology, 105*(1), 108–128.

Mason, C. H., & Perreault, W. D., Jr. (1991). Collinearity, power, and interpretation of multiple regression analysis. *Journal of Marketing Research, 28*(3), 268–280.

McLeod, D. B. (1992). Research on affect in mathematics education: A reconceptualization. In D. A. Grouws (Ed.), *Handbook of research on mathematics teaching and learning* (pp. 575–596). New York: Macmillan.

Metsämuuronen, J. (2012a). Challenges of the Fennema-Sherman test in the international comparisons. *International Journal of Psychological Studies, 4*(3), 1–22.

Metsämuuronen, J. (2012b). Comparison of mental structures of eighth-graders in different countries on the basis of Fennema-Sherman test. *International Journal of Psychological Studies, 4*(4), 1–17.

Mullis, I. V., Martin, M. O., Foy, P., & Arora, A. (2012). *TIMSS 2011 international results in mathematics*. Chestnut Hill: TIMSS & PIRLS International Study Center, Boston College.

Muthén, L. K., & Muthén, B. O. (1998–2012). *Mplus user's guide* (7th ed.). Los Angeles: Muthén & Muthén.

Novick, M. R., & Lewis, C. (1967). Coefficient alpha and the reliability of composite measurements. *Psychometrika, 32*, 1–13. doi:10.1007/BF02289400.

Nurmi, A., Hannula, M. S., Maijala, H., & Pehkonen, E. (2003). On pupils' self-confidence in mathematics: Gender comparisons. In N. A. Pateman, B. J. Dougherty, & J. Zilliox (Eds.), *Proceedings of the 27th conference of the international group for the psychology of mathematics education 3* (pp. 453–460), Hawaii, USA.

Nyala, J. I. (2008). Sex-differences in attitude towards mathematics of junior high school students in Ghana. *Edo Journal of Counselling, 1*(1), 137–161.

O'Connor, B. P. (2000). SPSS and SAS programs for determining the number of components using parallel analysis and Velicer's MAP test. *Behavior Research Methods, Instruments, & Computers, 32*, 396–402.

Official Statistics of Finland. (2013). *Upper secondary general school education*. Helsinki: Statistics Finland. Available from: http://www.stat.fi/til/lop/2012/lop_2012_2013-06-12_tie_001_en.html. Accessed on 11 Oct 2013.

Op't Eynde, P., De Corte, E., & Verschaffel, L. (2002). Framing students' mathematics-related beliefs: A quest for conceptual clarity and a comprehensive categorization. In G. C. Leder, E. Pehkonen, & G. Törner (Eds.), *Beliefs: A hidden variable in mathematics education?* (pp. 13–37). Dordrecht: Kluwer Academic Publishers.

Op't Eynde, P., De Corte, E., & Verschaffel, L. (2006). Epistemic dimensions of students' mathematics-related belief systems. *International Journal of Educational Research, 45*, 57–70.
Pietilä, A. (2002). The ile of mathematics experiences in forming pre-service elementary teachers' views of mathematics. In A. Cockburn & E. Nardi (Eds.), *Proceeding of the 26th conference of the international group for the psychology of mathematics education* (pp. 57–64). Nowich: University of East Anglia.
Raykov, T. (2012). Scale construction and development using structural equation modeling. In R. H. Hoyle (Ed.), *Handbook of structural equation modeling* (pp. 472–492). New York: The Guilford.
Raykov, T., & Marcoulides, G. A. (2010). *Introduction to psychometric theory* (pp. 37–58). New York: Taylor & Francis.
Rhemtulla, M., Brosseau-Liard, P. É., & Savalei, V. (2012). When can categorical variables be treated as continuous? A comparison of robust continuous and categorical SEM estimation methods under suboptimal conditions. *Psychological Methods, 17*(3), 354–373.
Richardson, F., & Suinn, R. (1972). The mathematics anxiety rating scale: Psychometric data. *Journal of Counseling Psychology, 19*, 551–554.
Roesken, B., Hannula, M. S., Pehkonen, E., Kaasila, R., & Laine, A. (2007). Identifying dimensions of students' view of mathematics. In *Proceedings of the fifth congress of the European Society for Research in Mathematics Education* (*CERME-5*) (pp. 349–358), Larnaca, Cyprus.
Roesken, B., Hannula, M. S., & Pehkonen, E. (2011). Dimensions of students' view of themselves as learners of mathematics. *ZDM – International Journal of Mathematics Education, 43*, 497–506.
Rutkowski, L., & Rutkowski, D. (2010). Getting it 'better': The importance of improving background questionnaires in international large-scale assessment. *Journal of Curriculum Studies, 42*(3), 411–430.
Sass, D. A. (2011). Testing measurement invariance and comparing latent factor means within a confirmatory factor analysis framework. *Journal of Psychoeducational Assessment, 29*, 299.
Sass, D. A., & Schmitt, T. A. (2011). Introduction to the special issue: Moving beyond traditional psychometric approaches. *Journal of Psychoeducational Assessment, 29*, 299.
Schmitt, N. (1996). Uses and abuses of coefficient alpha. *Psychological Assessment, 8*, 350–353.
Schmitt, T. A. (2011). Current methodological considerations in exploratory and confirmatory factor analysis. *Journal of Psychoeducational Assessment, 29*, 304–321.
Schmitt, T. A., & Sass, D. A. (2011). Rotation criteria and hypothesis testing for exploratory factor analysis: Implications for factor pattern loadings and interfactor correlations. *Educational and Psychological Measurement, 71*, 95–113.
Tapia, M., & Marsh, G. E. (2000). Attitudes toward mathematics instrument: An investigation with middle school students. *Academic Exchange Quarterly, 8*, 28–44.
UNESCO Institute for Statistics. (2011). *Global education digest 2011: Special focus on secondary education*. Montreal: UNESCO Institute for Statistics.
Van de Vijver, F. J. R., & Leung, K. (2000). Methodological issues in psychological research on culture. *Journal of Cross-Cultural Psychology, 31*(1), 33–51.
Van De Vijver, F. J. R., & Leung, K. (2011). Equivalence and bias: A review of concepts, models, and data analytic procedures. In D. Matsumoto, & F. J. R. Van De Vijver, *Cross-cultural research methods in psychology* (pp. 17–45). Cambridge: Cambridge University Press.

# Reaction to Section 3: Some Methodological Reflections on Studies of Mathematical Affect

**Kenneth Ruthven**

**Abstract** This commentary reviews methodological aspects of nine studies of mathematical affect. It notes the popularity of the Likert-item questionnaire as a means of data collection and examines some of the complexities of data analysis. The studies illustrate why the level of specificity at which the affective constructs measured by such an instrument are constituted, as well as their underlying structure, needs to be subject to investigation. Equally, instruments of this type cannot be assumed to be transposable between different cultural contexts and may themselves be strongly shaped by particular value positions. Adapting an existing instrument of this type or constructing a new one is a challenging task if the resulting questionnaire is to be adequately validated. In pursuit of such validation, greater use might be made of other techniques for purposes both of initial formulation and of subsequent triangulation.

**Keywords** Research methodology • Mathematical affect • Likert-item questionnaires • Instrument validation

## Introduction

The five chapters which make up this section on methodological issues in affect research report a total of nine studies. Table 1 provides an overview of some basic features of each chapter and the studies it reports.

The first three chapters (Andrà, Bofah and Hannula, Chen and Leung) feature relatively broad models of affect, whereas the last two are much more specific in their foci (Kuntze and Dreher, Tsamir et al.). Only two of the studies concern students, at the upper secondary (Bofah and Hannula) and undergraduate university (Andrà Study A) levels. The other seven all focus on pre-service (Andrà Study B, Kuntze and Dreher Study B) or in-service (Chen and Leung, Kuntze and Dreher

K. Ruthven (✉)
Faculty of Education, University of Cambridge, Cambridge, UK
e-mail: kr18@cam.ac.uk

 
B. Pepin, B. Roesken-Winter (eds.), *From beliefs to dynamic affect systems in mathematics education*, Advances in Mathematics Education,
DOI 10.1007/978-3-319-06808-4_19

**Table 1** Overview of chapters and studies to be discussed

| Chapter | Focal aspect of affect | Study | Type of participant | Type of affect instrument | Focal substantive/ methodological issue |
|---|---|---|---|---|---|
| Andrà | Various aspects of student mathematical affect | Study A | Undergraduate university students of STEM subjects | Likert-item questionnaire | Predicting mathematical ability from mathematical affect variables included in questionnaire |
| | | Study B | Pre-service secondary teachers of mathematics | Written narrative<br>Likert-item questionnaire | Relationship between content of narratives and aspects of affect included in questionnaire |
| Bofah and Hannula | Various aspects of student mathematical affect | | Upper-secondary students of mathematics | Likert-item questionnaire | Testing/exploration of generalizability of affect constructs and structure across cultures |
| Chen and Leung | Teacher beliefs about mathematics and about mathematics instruction | | In-service lower-secondary teachers of mathematics | Likert-item questionnaire<br>Semi-structured interview<br>Structured lesson observation | Comparing methods of analysing teacher beliefs and interpreting differences between their results |
| Kuntze and Dreher | Teacher craft knowledge relating to pupil motivation and learning | Study A | In-service upper-secondary teachers of mathematics | Open response questionnaire plus Likert-type rating | Place of affect in teacher representations of successful teaching |
| | | Study B | Pre-service teachers of mathematics | Likert-item questionnaire plus open response section | Teacher evaluation of the contribution of pictorial images to pupil motivation and learning |
| | | Study C | Pre-service and in-service teachers of mathematics | Written narrative | Teacher evaluation of contribution of pictorial images to pupil motivation and learning |
| Tsamir et al. | Teacher self-efficacy in mathematics teaching and in mathematics | Study A | In-service pre-school teachers | Likert-item questionnaire | Degree of specificity or generality of teacher self-efficacy in mathematics teaching |
| | | Study B | In-service pre-school teachers | Likert-item questionnaire | Relationship between teacher self-efficacy in mathematics and mathematical knowledge |

Study A, Tsamir et al. Studies A and B) teachers, or a combination (Kuntze and Dreher Study C).

In terms of method, all but one of these studies employ questionnaires involving Likert-type ratings. In four of these studies (Andrà Study A, Bofah and Hannula, Tsamir et al. Studies A and B) this is the only type of affect instrument used; in two of these studies, questionnaires link open-response and Likert-type items (Kuntze and Dreher Studies A and B); while in the remaining two studies, Likert-type questionnaires are triangulated against other types of affect instrument (Andrà Study B, Chen and Leung). The other types of affect instrument featuring in this collection of studies are written narrative (Andrà Study B, Kuntze and Dreher Study C) and semi-structured interview and structured lesson observation (Chen and Leung).

Having picked out these trends from the table, let me now discuss each of the chapters and studies before making some concluding comments.

## Andrà: Networking Methodologies in the Field of Affect

The chapter by Andrà sketches two studies which examine relations between different forms of evidence about student characteristics. The first of these studies involved building a decision-tree model from a range of dichotomised variables. The target variable was formed from an attainment measure, based on student responses to a multiple-choice test, by splitting it at the threshold value used to decide whether or not students should be directed to a tutorial course in mathematics. The 20 potential predictor variables were each derived by splitting at its median value an attitudinal measure elicited from students through a Likert-type instrument. The attitudinal dichotomies were sifted to build a relatively simple decision-tree model to predict the target allocation dichotomy. This illustrates a technique through which measures of affect and attainment can be integrated in a predictive model.

The second study involved comparing indicative words and phrases from narrative accounts written by prospective mathematics teachers concerning their relationship to mathematics with the substantive themes of the 20 attitudinal variables employed in the first study. The selected examples of such linkages draw interesting connections which have a degree of plausibility, but many depend on relatively high inference and expansive interpretation. Stronger arguments and more powerful findings could be established by developing an analysis which is more exhaustive in coverage of the available narratives and which develops a more explicit protocol to guide and justify claims about relationships between narrative material and attitudinal variables. One way of stimulating development of a more explicit protocol might be to compare the application of different approaches to analysis: for example, the contrast between starting from open coding based on constant comparison across narrative material, as against analytic coding of narrative material based on operational definitions of the 20 attitudinal themes. Equally, such claims would be strengthened if the interpretations made of written narratives could be triangulated against questionnaire responses.

The particular methodological contribution of this chapter, then, is to highlight some strengths and weaknesses of two types of instrumentation for assessing affect: one type based on Likert-style ratings, the other grounded in narrative accounts. The chapter notes that Likert-style questionnaires provide a systematic means of eliciting responses on a range of issues specified by the researcher, whereas the issues that emerge through narrative accounts are controlled more by the respondent and may also be more serendipitous. In addition, while the analysis of correlations between questionnaire items provides a means of identifying patterns of association which can be speculatively linked to causal relations, written narratives may represent respondents' views of such relations in greater depth and cast light on their causal attributions.

## Bofah and Hannula: Factorial Structure of Views on Mathematics

The chapter by Bofah and Hannula reports a study which examined the transposability of an instrument for assessing upper-secondary school students' views of mathematics between the contrasting cultural contexts of Finland and Ghana. Development of the instrument had drawn on established ideas within the field, and been guided by empirical findings about the factorial structure of a pool of relevant Likert-type questionnaire items. The rationale for the resulting multi-factorial instrument conceived five of its seven dimensions as relating to the "cognitive beliefs" of a student in relation to mathematics (designated *ability*, *success*, *difficulty*, *teacher quality* and *family encouragement*), with the final two dimensions conceived as relating to the "motivation" of a student towards mathematics (designated *effort*), and a student's "emotions" relating to mathematics (designated *enjoyment*).

The particular methodological contribution of this chapter is to highlight the potential for linguistic and cultural differences to disrupt the functioning of instruments of this type when they are transposed from one educational system and its local language and culture to another, referring to several earlier studies which have reported such findings. It is not altogether surprising, then, that confirmatory factor analysis found that the pattern of responses in the Ghanaian sample did not conform to the structure established from the Finnish sample. The ensuing exploratory factor analysis of the Ghanaian responses suggested that a reduction of the data to 4- rather than 7-dimensions was most appropriate. Nevertheless, the Finnish and Ghanaian models do share factors of *family encouragement* (based on the same items) and *teacher quality* (based, for the Ghanaian sample, on a subset of the items defining it for the Finnish sample).

It is notable that the four highest-loading items on the *teacher quality* subscale are phrased in ways that are manifestly, and often explicitly, evaluative, whereas the evaluative nuance of the other two items is much more implicit: for example, adding "well" at the end of "The teacher explains the studied topics" would make it more explicitly evaluative. Modifying items in this way would arguably boost both their loading on the factor and the reliability of the subscale (which proved extremely low for the Ghanaian sample). This type of clear evaluative signalling may be particularly important where respondents are answering a questionnaire in a language which is an additional one for them, as English would be for most Ghanaian students.

Related to this linguistic aspect, the study found that two items, which had not loaded on any factor for the Finnish sample, were associated with the leading factor for the Ghanaian sample. As a native speaker of English resident in Britain, I found myself unable to interpret one of these items with any confidence: "My eagerness to study mathematics is seasonal". Likewise, another of the items that loaded on this leading factor required inference on my part, based on my knowledge that many languages do not afford the semantic distinction between "make" and "do": "I have made it well in mathematics". This well illustrates how, even when respondents may appear to be reading the "same" item in the "same" language, it may be expressed in a contextually specific and culturally nuanced way which makes interpretation difficult or different elsewhere.

Finally, even native speakers can have difficulties in interpreting statements involving more complex vocabulary or syntax. For example, while the *family encouragement* construct proved viable in Ghana, it had lower reliability than desirable. One of the items, "My family encourages me to study mathematics" is a model of a simple direct statement. However, "The importance of competence in mathematics has been emphasized at my home" has a complex subject clause followed by a passive verb, and employs both Latinate vocabulary and nominalisation: the result is a sentence which is relatively difficult to interpret compared, say, to "At my home, people emphasise that being able to do mathematics is important".

## Chen and Leung: Affect Research Focusing on Teachers' Beliefs

The chapter by Chen and Leung presents approaches to the collection and analysis of evidence about the beliefs of teachers of mathematics about their subject and its teaching, referring to a study of this type conducted in a context of curricular reform.

The first of these approaches involved use of a Likert-type questionnaire incorporating established scales concerned with "beliefs about mathematics" and with "beliefs about mathematics instruction". Originally developed as a way of measuring the extent to which teachers had made the shift "from a program emphasizing formal mathematical content to a program emphasizing the creative, investigative nature of mathematics" and "from an authoritarian, teacher-dominated classroom, to a child-centred classroom", each scale operationalises an ideologically polarised conception of the focal phenomenon along a "formal-informal dimension" to which the respondent is invited to acquiesce.

At best, the first set of items allows the respondent sufficient free play to treat the view that mathematics "include[es] elements of originality and creativity and [is] characterized by the existence of choices" as not opposed to the view that mathematics "is based on fixed, established forms and requires scrupulous adherence to rule". If sufficient respondents exploit this free play and the resulting data is analysed for its dimensionality, what was envisaged as a unidimensional scale running between the two poles will expand to a two-dimensional space in which each pole creates one facet. Likewise, the second set of items gives the respondent sufficient free play to express a view of the mathematics teacher as someone who may have occasion both to "encourage[] self-discovery and independence from memorized rules" as well as to "define[] and explain[] procedures for students".

The second approach involved analysis of linked observational and interview evidence. One notable requirement was that "lesson(s) to be observed should deal with a completely new mathematical topic, instead of being review or exercise-oriented". This suggests that the focus of this research might more accurately be described as "beliefs about mathematics instruction" in "lesson(s)… deal[ing] with a completely new mathematical topic". Analysis of the observational data was guided by a seven-facet analytic scheme based on a model of the components of a reform-oriented mathematics classroom; with each of these facets defined by a rubric describing desired teacher behaviours. Again, then, this was a strongly normative analysis.

The semi-structured interview protocol set a clear agenda for eliciting teachers' views about mathematics and about mathematics teaching using questions phrased in open and neutral terms. In this respect, then, the approach to interviewing was far more accommodating to respondents' perspectives than the other forms of data collection. However, while relatively little is said about the way in which the resulting interview data was analysed, the impression is given that this was guided by prior analytic schemes that researchers brought to the data: "although mathematics teachers may not describe their personal views of the nature of mathematics in terms of the Platonist, instrumentalist, or social constructivist (problem-solving) positions, such categories may be employed in the analysis of teachers' beliefs"; "although individual teachers may simultaneously hold more than one kind of view… his or her views could be approximately characterized as one particular kind according to their dominant inclination".

One can, of course, argue that the strongly normative approach taken by this study is appropriate where the aim of research is to evaluate change in teacher thinking and practice towards alignment with the precepts of a reform model. As often happens in such cases, teachers had learned to identify with the discourse of reform: "the majority of the subjects held informal beliefs about mathematics and mathematics instruction, which seemed to show a relatively high level of consistency between teachers' mathematics beliefs and the underlying philosophy of the reform-oriented curriculum". However, as also often happens, espousing the reform discourse had not helped teachers to find ways to enact it in their teaching (at least not in the manner envisaged by the reformers and researchers): "as indicated by… classroom observation and interview… most case teachers' mathematics beliefs in both contexts were close to… traditional views". As the chapter itself notes, "almost all [contextual] factors failed to contribute to the realization of the conditions for conceptual change." One could put this more strongly: how adequately had the reformers conceived a viable trajectory from existing practice to robust enaction of the reform ideals, and identified and carried through the crucial contextual changes necessary?

The methodological contribution of this chapter, then, is to highlight choices, explicit or implicit, that are made in developing a conceptual framework for studies of affect and in instrumenting data collection and analysis. In such work, the beliefs of researchers (and those that commission or influence their research) are no less in play than those of the people that they study. And what may appear to be disparities between personal beliefs expressed in different contexts, or between beliefs espoused and beliefs enacted, may prove to be conditioned by the differing goals of actors in these contexts, and the different constraints on them.

## Kuntze and Dreher: Teachers' Views of Affect and Learning

The chapter by Kuntze and Dreher reports three studies which share an interest in how teachers' craft knowledge for teaching mathematics takes account of student affect. Motivating these studies is a concern that too strong attention to this affective dimension on the part of teachers may interfere with other important aspects of their teaching.

The first study used a semi-structured questionnaire in which in-service upper-secondary teachers were invited to nominate up to six criteria for evaluating the quality of mathematical instruction and to rate their level of importance. Semantic analysis of these criteria indicated that "aspects of motivation/affect" were proposed by over 80 % of the teachers, making this the most widely nominated category, with over 75 % of respondents rating such aspects as very important. It is interesting also to note that the category with the next highest proportion of teachers (over 40 %) rating it very important was concerned with "activation of students". So, contrary to popular stereotype, a not inconsiderable proportion of upper-secondary teachers appear to be rather student focused and attentive to affective issues. Other aspects which were nominated and rated important or very important by a good proportion (over 40 %) of teachers included "structuring/adapting contents", the category that appears most closely related to subject matter; "use of preferred teaching methods", the category that appears most closely related to subject pedagogy; with "presentation" more ambiguous between subject matter and pedagogy, and finally "discipline" which might be taken as a general aspect of pedagogy. The interpretability of these categories for the reader would have been higher had some representative exemplars for each been provided, perhaps as a supporting table.

The second study presented pre-service teachers with four fraction tasks intended for pupils, each featuring some form of visual image. For each task, respondents were asked to rate six statements expressing opinions about various aspects of the pedagogical value of the image. For two of the tasks the images had been selected on the basis of their potential to be cognitively productive ("providing… insight") rather than affectively positive ("motivating"); in the other two, the reverse. Analysis of the data confirmed, first, that responses to each item were structured in terms of distinct cognitive and affective dimensions; and, second, that respondents did indeed rate the tasks with images chosen to be affectively positive (Tasks 1 and 3) more highly on that dimension than those not chosen on this basis (Tasks 2 and 4). However, only one of the images chosen by the researchers as cognitively productive received an affirmative mean rating on this score from the respondents, which was also the highest mean rating (Task 4), whereas the other received the lowest mean rating (Task 2), suggesting that what the researchers saw as its cognitive qualities were apparent to few pre-service teachers. To better understand the situation, response patterns concerning the cognitive rating of each of the tasks were identified through clustering of respondents, and interpreted in the light of later comments in open responses. One of these response patterns (Cluster 1) matched the viewpoint of the researchers, grounded in the "providing insight" perspective. The next response pattern (Cluster 2) reversed the viewpoint of the researchers, apparently due to domination by the "motivating" perspective, since this ostensibly "providing insight" pattern for this cluster was almost identical to the one that they produced for "motivating". A further response pattern (Cluster 3) is dominated by a more specific "revealing misconceptions" perspective which fits one "providing insight" task (Task 4) but not the other (Task 2): the presence of this cluster helps to explain the low rating of Task 2. A valuable contribution of this study, then, is to illustrate these techniques of data collection and analysis, which proved relatively illuminating of broad patterns in the questionnaire responses.

The final study analyses three examples of teachers' written narratives evaluating the record of a teacher's use, for the purpose of addressing a pupil misconception, of a diagram which shows a kangaroo leaping on a number line. Essentially, these cases were chosen to illustrate three archetypical perspectives: one dominated by the "motivating" aspect of the image; one dominated by the "providing insight" aspect; and one incorporating attention to both aspects.

The main contribution of this chapter, then, is in demonstrating the salience of affective issues for teachers, notably through the broader first study. Equally, the second and third studies confirm the concern expressed at the start of the chapter that some teachers, both pre-service and in-service, may focus more strongly on affective issues at the expense of cognitive aspects, at least as regards the role of images in task design and use.

## Tsamir, Tirosh, Levenson, Tabach, and Barkai: Teachers' Knowledge and Self-Efficacy

The chapter by Tsamir, Tirosh, Levenson, Tabach and Barkai reports two cognate studies which examined, respectively, the specificity of teachers' mathematical-pedagogical self-efficacy, and the association between teachers' mathematical self-efficacy and their mathematical knowledge.

The motivation for the first study is a lack of agreement in the research literature on the specificity of self-efficacy. At one extreme, it is suggested, some researchers treat self-efficacy as action-specific or problem-specific; at the other extreme, certain researchers treat it in highly generic terms. This first study, then, examines preschool teachers' ratings of four Likert-type items involving verbal descriptions of competence in the form "I am capable of designing tasks to *promote/evaluate* children's knowledge of *cones/cylinders*". The sound functioning of these items depends on respondents sharing, both amongst themselves and with the researchers, substantially common or similar external referents that calibrate their interpretations of what it means to design tasks of the types referred to in the items. As the chapter notes, other studies have often provided more explicit referents to anchor the self-assessment of respondents. The researchers compared pairs of items controlled for task function or type of shape, showing that the mean ratings for the *promote* items are slightly higher than those for the *evaluate* items, and those for the *cylinder* items slightly higher than those for the *cone* items.

The researchers also formed aggregate measures by combining ratings from pairs of items involving the same task-type or shape-type. Indeed, regardless of whether the mean scores for each item were similar, the aggregate of each respondent's ratings across the four items would provide a more generic index of self-efficacy in designing tasks to promote and evaluate children's knowledge of cones and cylinders, adequate to compare respondents at this domain level. However, to better understand the underlying structure of teacher self-efficacy and its fundamental level of specificity, it would be valuable to examine how strongly the basic item variables correlate, indicating whether the aggregate measure is more an index than a scale. It seems, then, that rather than there being an absolute level of specificity to which a construct of self-efficacy must be tied, such constructs can reasonably be defined at various levels, with structure at the basic level clarifying how more generic constructs at a higher level should be treated. This study, then, serves well to illuminate some of the complexities which arise in conceptualising and operationalising measures of teacher self-efficacy, in particular as these relate to the specificity or generality of such a construct.

The motivation for the second study is the question of how teacher self-assessment of their own knowledge of a mathematical topic (i.e. a component of their mathematical self-efficacy) compares with a researcher assessment of that knowledge. In this study, preschool teachers rated pairs of verbal statements concerning their capacity to identify examples and non-examples of particular types of geometric shape. They were asked to make this self-assessment without any indication of the terms in which the researchers might conceive examples and non-examples of these shapes. As the chapter discusses, the researchers faced a real dilemma here: while they could have reversed the order in which the self-efficacy and knowledge instruments were presented so that the latter then served to illuminate the meaning of the examples and non-examples subsequently referred to in the former, doing so would risk teachers' self-efficacy assessments being overly and immediately influenced by their experience of taking the test.

The chapter explains how the study used the notion of prototype to guide its choice of examples and non-examples in the knowledge test. Here it is particularly instructive to note the everyday epistemology which underpins the researchers' labelling of shapes. For instance, in Fig. 5, the examples of a cone are, first, the archetypical "cone" which provides the benchmark against which the reoriented "upside-down cone" and "cone lying down" are defined. Likewise, many non-examples are framed as truncated versions of archetypical shapes: "cone with its top cut off", "cylinder cut on a slant". At the preschool level, one imagines that sensitivity in handling the way in which informal everyday and formal mathematical epistemologies interact and are sometimes in tension (and, more broadly, in recognising and managing the contrast between prototypical and definitional reasoning) represents an important component of pedagogical content knowledge.

This study also illustrates some difficulties that may arise in analysing the degree of association between self-efficacy and knowledge. First, because most teachers performed well on all the knowledge scales, the mean scores were (as the chapter acknowledges) high: indeed, they were sufficiently close to the ceiling value to suggest that the measures were likely to have poor discrimination. Likewise, the mean self-efficacy scores for each shape were also high, as well as being formed by combining only two 4-point Likert ratings. So neither type of measure could be expected to be a strong discriminator across this sample. A second problem in the cases of 2-D shapes was the small number of teachers for which data was available (N between 17 and 19). These factors undoubtedly limited the power of the statistical tests: it is not surprising, then, that no significant correlation was found between self-efficacy and knowledge measures for 2-D figures. However, the number of subjects for the 3-D figures was greater (N between 62 and 63) and so the power of these tests correspondingly higher. Nevertheless, the mean knowledge score for cones (93 %) was closer to the ceiling value than that for cylinders (87 %), suggesting that discrimination may be poorer in the former case: this should be borne in mind in interpreting the statistically significant association found with respect to cylinders (and 3-D figures overall) but not cones.

These same factors undoubtedly remained in play when dichotomised versions of the self-efficacy and knowledge variables for the 3-D shapes were crosstabulated in the final part of the analysis. To further complicate interpretation of results, the splitting of variables at the mean (rather than, for example, around the median) produced some partitions which were very uneven within variables and unbalanced between them, constraining the patterns of association achievable within a cross-tabulation. Bearing these considerations in mind, for example, the large number of cases falling into the high knowledge and low self-efficacy cell in the cones cross-tabulation is not surprising. Equally, here the study posits a "phenomenon of [teachers] being able to correctly identify figures but yet not being aware of this knowledge": however, even were such an observation to be more than an artefact of method or a result of chance variation, it would not be surprising if modest preschool teachers were reluctant to rate their mathematical expertise highly, particularly just before taking an unseen test devised by researchers in mathematics education! This is simply another side of the dilemma, referred to earlier, over how to sequence the self-efficacy and knowledge instruments.

In view of these complications, then, this second study cannot be considered a dependable test of the relationship between teacher self-efficacy and knowledge, as the discussion section of the chapter acknowledges. Indeed, this study provides a good illustration of just how challenging it can be to investigate such variables and the relationship between them.

## Conclusion

As displayed in Table 1 and noted earlier, a single method pervades the studies reported in these chapters: the Likert-item questionnaire. While the possibility of combining a Likert response format with forms of open response is illustrated by two of the Kuntze and Dreher studies, in the other studies this technique is employed in a standard way. Clearly, then, this is an unusually convenient technique which serves researchers well in generating results. Equally, as the chapters by Bofah and Hannula and Kuntze and Dreher illustrate, a wide range of more advanced statistical techniques are available to help identify pattern and structure in such results. However, as the study by Tsamir et al. highlights, the level of specificity at which the affective constructs measured by such an instrument are constituted, as well as their underlying structure, needs to be subject to investigation. Equally, as the studies by Bofah and Hannula and Chen and Leung show, instruments of this type cannot be assumed to be transposable between different cultural contexts and may themselves be strongly shaped by particular value positions. Adapting an existing instrument of this type or constructing a new one is a challenging task if the resulting questionnaire is to be adequately validated. In pursuit of such validation, greater use might be made of other techniques for purposes both of initial formulation and of subsequent triangulation, such as methods of narrative as employed by Andrà and Kuntze and Dreher, or of interview and observation used only by Chen and Leung.

# What Counts, When? – Reflections on Beliefs, Affect, Attitude, Orientations, Habits of Mind, Grain Size, Time Scale, Context, Theory, and Method

**Alan H. Schoenfeld**

**Abstract** Research in the "affective domain" is densely populated with overlapping constructs, partially commensurate methods, and somewhat contradictory findings. One productive consequence of such confusion, as seen in this volume, has been recent work on networking theories – attempts to document and build on commonalities in what appear on the surface to be very different perspectives and methods. Such work is "close to the ground," seeking to bind together and build upon the various strands of extant work. Here I take much more of a bird's-eye view. My first goal is to characterize what might be explainable, and what theory and methods might be productive in producing rigorous explanations of people's beliefs/affect/values/preferences/habits of mind shape their in-the-moment decision making. My second goal is to address questions of what it takes to have a positive impact on people's beliefs/affect/values/preferences/habits of mind.

## What Counts? (Or, "Watch What I Do, Not What I Say.")

In this section I focus on people's beliefs and actions over relatively short time frames, such as teaching a lesson or trying to solve a mathematics problem. I begin with some examples to set the stage for my comments.

*Example 1* When I was in high school a newspaper conducted a telephone poll, asking people to indicate whether or not they read a series of magazines. The results were fascinating. A very large percentage of those polled indicated that they read a highbrow magazine that had a relatively low circulation – about as many as indicated that they read a common newsweekly.

A.H. Schoenfeld
University of California, Berkeley, CA, USA

 
B. Pepin, B. Roesken-Winter (eds.), *From beliefs to dynamic affect systems in mathematics education*, Advances in Mathematics Education,
DOI 10.1007/978-3-319-06808-4

*Example 2* In my early problem solving work (see, e.g., Schoenfeld 1985, 1989), I gave students simple construction problems to solve, thinking that their geometric knowledge would enable them to solve the problems in short order. I was shocked by the conjectures they made, which seemed grounded in guesswork and were almost "knowledge free." I pursued the issue, and in later years had student work some proof problems that provided the information that the students needed for the construction problem before they worked them. The students solved the proof problems without difficulty. Then, in working the construction problems, they made conjectures that flatly contradicted what they had just proven.

*Example 3* Example 2 is neither unique nor atypical; (apparent) contradictions abound in what students say and do in mathematical contexts. Carpenter et al. (1983) commented as follows when discussing the secondary school mathematics results from the Third National Assessment of Educational Progress:

> [Students] felt very strongly that mathematics always gives a rule to follow to solve problems. Yet, they felt just as strongly that knowing how to solve a problem is as important as getting a solution and that knowing why an answer is correct is as important as getting the correct answer. (p. 656)

*Example 4* Mrs. "Oublier" (Cohen 1990) said she was teaching in a manner consistent with the new "reform" standards in mathematics.

> She eagerly embraced change, rather than resisting it. She found new ideas and materials that worked in her classroom, rather than resisting innovation. Mrs. O sees her class as a success for the new mathematics framework. Though her revolution began while the framework was still being written, it was inspired by many of the same ideas. She reports that her math teaching has wound up where the framework intends it to be. (p. 312)

And yet…

> She used the new materials, but used them as though mathematics contained only right and wrong answers. She has revised the curriculum to help students understand math, but she conducts the class in ways that discourage exploration of students' understanding. (p. 312)

*Example 5* Mrs. Oublier was hardly an outlier. Schraw and Olafson (2002) describe three epistemological world views held by teachers: realist, relativist, and contextualist. Using questionnaire data, they show how there are consistent patterns of people's beliefs about knowledge, curriculum, pedagogy, assessment, and the roles of teacher and student, given their alignment with one or another epistemological world view. So far, so good. But … there was at least a 70 % overlap between people holding the contextualist and relativist positions, and a 25 % overlap between realist and contextualist. In short, the epistemological positions are made to seem distinct, but people say they hold more than one. More important, just as in the case of Mrs. Oublier, the reported beliefs didn't translate into classroom actions: a very large proportion of those who talk a contextualist/relativist (read "constructivist" or "reform") game wind up teaching in a manner more consistent with the traditional "there's a right and wrong way to do things, here's how to do it right" model.

*Example 6* Speer (2005) interviewed two collegiate calculus instructors. Both described their teaching by saying that they used questioning strategies because they believed such strategies would help their students develop rich understandings of the mathematical content. Yet, their teaching looked radically different.

*Example 7* Are you a racist? I thought not. Hardly anyone is, according to what they say.

## Discussion

What do I take from these examples?

*Conclusion 1* It's what people do that matters, not what they say. Moreover, understanding why people (specifically, students and teachers) do what they do is what really counts. If we want students to become effective problem solvers, we need to know why they make the productive or unproductive choices they make while engaged in problem solving. If we want teachers to become more effective at producing students who are good mathematical thinkers, we need to know why they make the productive or unproductive choices they make while engaged in teaching.

Needless to say, the study of beliefs and affect is central to this issue. In what follows, my comments are shaped by the wish to understand the role of belief and affect in shaping people's decisions and actions. I will note here, and address in the meta-level discussion below, that there are methodological entailments to this perspective.

*Conclusion 2* People lie. They lie to others, they lie to themselves. You can *never* take what people say (in interviews, in response to questionnaires, etc.) at face value. To put this another way …

*Conclusion 3* Statements of beliefs do not predict actions. If you're interested in what people do, what they say they believe is of limited value. In particular, while artfully constructed beliefs questionnaires may be of some value in some contexts (e.g., in pointing to the coherence of what people say about what they believe, or in documenting that certain percentages of a given population profess to hold certain beliefs), they are of little use in helping us understand why people do what they do. See the meta-level discussion below.

*Conclusion 4* Grain size matters when discussing beliefs. In example 6, both instructors wanted their students to "understand the subject matter." But for one instructor, such understanding meant being able to answer straightforward procedural test questions; his questioning strategies were oriented toward making sure his students knew what to do when working such questions. For the other instructor, such understanding meant seeing how everything fits together – and his questions were aimed at getting students to develop rich understandings of the underlying concepts. Thus, beliefs about what constitutes understanding are the right level of grain size for understanding these instructors' choices, not "I want my students to understand the mathematics."

*Conclusion 5* Clusters of beliefs, not simply individual beliefs, shape behavior (See also Aguirre and Speer 2000). In example 6, "I believe in asking questions to elicit student understanding" and "this is what I think is important about mathematical understanding" and "this is what I think is most important about the content in the example they are working on" all shape the individual teachers' in-the-moment decision making. For Mrs. Oublier, "math problems have one right answer" and "students need specific methods for obtaining the right answer" and "listening to incorrect procedures may confuse students" may all have contributed to her decision making.

*Conclusion 6* Different beliefs will get activated (or have different levels of activation) in different contexts. The teacher who believes that student scores on standardized tests are critically important may or may not act on that belief early in the school year because there is not yet pressure to perform – but, classroom practices may be seriously modified during the month prior to a major standardized assessment. Activations can vary from moment to moment. Thus, for example, a teacher may firmly believe that students should understand why a certain procedure works the way it does; but if five minutes remain in the class period, the teacher may decide to defer explanation because other things need to be done. In fact, the contradictory statements in example 3 are most likely the result of students invoking different contexts for their answers: the belief that "mathematics always gives a rule to follow to solve problems" may come from their experience, while "knowing how to solve a problem is as important as getting a solution" and "knowing why an answer is correct is as important as getting the correct answer" may reflect the internalization of the rhetoric they have heard over the years.

*Conclusion 7* Beliefs can be subtle, and it takes work to unearth them. Example 2 is one indication. What happened was that students behaved strangely: they failed to use certain knowledge that I knew they had. But unearthing the underlying belief structures – that (a) the students believed that proofs only confirmed what was already known, and were not therefore productive tools for discovery, and (b) that constructions were empirical activities to be judged by empirical standards, and thus that formal proof knowledge was irrelevant to the construction process – took some years.[1] The same is the case for teaching beliefs, such as the idea that the main task of school math is "answer getting" – so that work stops when students have arrived at a correct answer. Similarly, a teacher's belief that "ordinary students should be given rules and procedures to solve problems, because they will get confused otherwise" may not be apparent at first – but it has very serious consequences. Different beliefs are activated in different contexts ("ordinary" and "honors" classes), resulting in different instructional activities.

## Meta-level Discussion

I was going to begin this section with the statement "My orientation is practical, but it has theoretical and methodological entailments." But, as I began to type it, I realized that the statement "My orientation is theoretical, but it also has practical and methodological entailments" is equally true.

[1] I note that once the beliefs are understood, their impact is easy to observe. However, identifying them in the first place can be a challenge.

As the examples given above make clear, what "counts" are people's decisions and actions. Beliefs (and values, and affect, and preferences, and habits of mind – what I call "orientations" as an umbrella category) are critically important, but their main importance is in how they shape people's decisions and actions.

Here is where issues of theory and method become central. The question is, what does it mean to "explain" behavior? One's response to this question will vary according to one's disciplinary training – habits of mind matter! My particular orientation is toward the perspective of the sciences, where one strives for theoretical explanations at a level of mechanism ("this is *how* things work") and one tests one's theoretical explanations by building models. If the models work (in that the predictions they make are consistent with the behavior one is trying to explain), then at least one's explanations are consistent with observed reality; if they fail to work, then one has clear evidence that one's ideas need refinement.[2]

This approach was fundamental to my work on decision making, as elaborated in my book *How We Think* (Schoenfeld 2010). The goal was to provide a theory that explained how and why people made the decisions they did, in the midst of complex activities such as problem solving, teaching, or medical practice. In that theory, the key theoretical constructs were *orientations* (an abstraction of beliefs including aspects of affect such as preferences, values, and habits of mind) *goals* (just what is the individual trying to achieve?), *resources* (what does the individual know, what tools are available?), and a mechanism for choosing among options given any particular constellation of orientations, goals, and resources. That's the theory. The mechanism for theory testing was modeling. If the theory was right, then it should be possible to take a number of examples of teaching or other behavior; to identify the salient orientations, goals, and resources; and build and "run" a model.

In short, it worked. *How We Think* offers detailed models of three radically different teaching episodes, and an array of substantiating evidence that strongly suggest that the theory is robust and applies widely. But that is only part of the story. In the next section I expand the scope of the discussion.

## The Big Picture

Up to this point, what I have discussed has been somewhat narrowly framed – the question being, "what are the roles of beliefs/affect/habits of mind/orientations in shaping people's in-the-moment decision making, and how do they operate?" Now I wish to open up the scope, to discuss theoretical and pragmatic issues related to making change. The question for this part of my chapter is, "what are productive ways to think about the growth and change of beliefs/values/affect/habits of mind/orientations? Here the time scale, rather then being micro (an hour or two subject to careful modeling), is macro – months or years if we are truly looking to see changes

[2] Note that one can take this stance without having to take a position on the nature of falsifiability (the philosophical question of whether theoretical claims must be falsifiable). Framing one's ideas so that they can be falsified is a good way to make scientific progress.

in people's orientations. And here, much of what I will say echoes two main themes that permeate this volume:

1. The web of influences on "individual" beliefs/values/affect/habits of mind/orientations is deeply cultural.
2. The challenges of facilitating meaningful change

## The Web of Influences on "Individual" Beliefs/Values/Affect/Habits of Mind/Orientations Is Deeply Cultural.

An emphasis on the fundamental role of culture in shaping "inner" experiences is most overt in Radford's chapter, but I would argue that it appears strongly in, for example, the chapters by Rolka and Roesken-Winter and by Skott as well. Let me try to link them here, and discuss the implications.

I take as a given Radford's claim that much of which we take to be highly personal is, in fact, culturally mediated. If you doubt this, I suggest you look at the web site http://www.culinaryschools.org/cuisine/10-disgusting-delicacies/, "The 10 Most Disgusting Delicacies to Try Before You Die." As the web site notes, food that is considered utterly disgusting by some people is taken as a challenge ("culinary thrill-seeking") by some people, and – most important – "time-honored traditions for others." A particular dish of food may be likely to make you salivate if you have a particular sociocultural background, and may make you likely to gag if you have another.

What this means, of course, is that all of the constructs that are at the heart of this book – beliefs, affect, values, habits of mind, "personal" preferences and orientations – are culturally influenced in fundamental ways. To take just one example, consider the fact that in the U.S., most people believe in something like a "math gene" – that either you're good at mathematics or you're not, and there's not much you can do about it. In contrast, in some Asian nations, the assumption is that how good you are at mathematics is a function of how hard you work at it. With that in mind, think about issues of self-efficacy. It stands to reason that those who believe that mathematical talent is innate rather than malleable will score lower on measures of mathematical self-efficacy than those who believe that "talent" is a matter of effort. There is, thus, a cultural component to this ostensibly individual trait.

That said, I think it is necessary to unpack "culture." Or, perhaps better, to use a term that is less laden with problematic connotations, such as the assumption that cultures are homogeneous. A concept that has proven useful in the identity literature is "community of practice" (Lave and Wenger 1991). In what follows I will take an expansive view of the notion, where there are no geographical boundaries to such communities. And, I will argue, that there are nested communities of practice as well as overlapping ones.

Consider teaching. One way to view Shulman's (2005) notion of "signature pedagogies" is that different communities of teaching practice (e.g., those teaching in K-12 or college, in medical school, in law school, in colleges of engineering) have developed particular, internally consistent forms of pedagogy. That is, members of these communities of teaching practice teach in certain ways. But, the whole point of *How We Think* is that teaching decisions and actions are not "just" actions; they are a function of knowledge, goals, and orientations. Practitioners of any "signature" pedagogy, being members of that extended community of pedagogical practice, have been enculturated into a set of pedagogical assumptions (that is, beliefs and orientations) that reflect the underpinnings of that pedagogical approach. If someone is employing a signature pedagogy, then there is a good chance that they share many of the beliefs and orientations tied to that pedagogy. This is important if you want to think about teacher change, in that it helps to identify the orientations teachers are likely to have (and that may need to shift).

But there is more. While Shulman refers to the signature pedagogy of K-12 instruction,[3] Stigler and Hiebert's (1999) book *The teaching gap* shows that if you look more closely, there is significant cultural variation within such teaching. Specifically, Stigler and Hiebert reveal that there is much greater between-country variance than within-country variance in pedagogical practices. That is, Japan, Germany and the U.S. can be considered extended communities of teaching practice, each with its own signature pedagogy and associated orientations. (In the Japanese TIMSS lessons, evoking student reasoning and building on that reasoning is a core part of the pedagogy. The U.S. lessons are much more of the "demonstrate and practice" type. This reflects underlying beliefs and orientations regarding the nature of learning. See also Clarke et al. 2006). Japanese Lesson Study employs a series of technical terms to describe aspects of pedagogical practice for which there are no comparable nouns in much of the Western world. Consider for example, "kikanshido":

> Japanese word, literally means "in-between desk instruction." During kikanshido, the classroom teacher observes how each student is solving the problem, consider in what sequence the various solution strategies may be shared and discussed, and provide appropriate support to individual students. During kikanshido, however, the teacher does not spend too much time with a single student as a major goal of this phase is for the teacher to know how all students are approaching the problem. (Downloaded March 7, 2014, from http://hrd.apec.org/index.php/Glossary_of_Lesson_Study_Terms.)

As above, this aspect of Japanese signature pedagogy is distinctive (it is as different from the pedagogy of "demonstrate and practice" as case methods are different from lectures). And, the belief structures underlying that practice reflect a perspective on learning that is deeply engrained and fundamentally different from the belief structures underlying the pedagogy of "demonstrate and practice." This is one reason that it has not been easy to adopt lesson study in the U.S.

Differentiation doesn't stop at the national level. It can be argued that within the U.S., "standards-based" and "traditional" teaching reflect different underlying orientations. Indeed, different school districts and schools may have distinctive communities of practice. Horn (2007) documents a case where the teachers' practices – coherent at the school level but very different between schools – reflected very different orientations regarding students' innate capacity and their potential for growth.

---

[3] To be sure, at the grain size of analysis that Shulman employs, K-12 instruction (in contrast to the pedagogies in law and medical schools) represents a distinctive form of pedagogy.

All of this matters because there is a strong relationship between orientations and communities of practice; identifying those communities may help us to understand which orientations people hold. And understanding those orientations is a key to change, the topic of my final comments.

## The Challenges of Facilitating Meaningful Change

Let me end where I began. From my perspective, what counts is behavior – and in teaching, it means teaching in ways that result in students' developing deep understandings of the subject matter. I will speak for myself, but I think my sentiments are shared.

First: we have, individually and collectively, some ideas about productive teaching practices.[4]

Second: A great deal of current classroom practice does not match our ideas about productive teaching practices. We would like to change this.

Third: As has been noted, decision-making is a complex function of resources, goals and orientations. Thus, addressing orientations – that is, beliefs, affect, values, habits of mind, and "personal" preferences is essential. This volume represents progress in our understandings of this complex construct.

Here in conclusion I will say some things about change. I will use the term orientations as an umbrella term, referring to the spectrum of beliefs, affect, values, habits of mind, and preferences discussed above.

*Orientations develop slowly – typically over periods of years.* Think about the national pedagogical styles, as discussed above. There is what has been called the "apprenticeship of observation": in a culture where a pedagogical approach is prevalent, people come to accept that approach as being "natural." Or, consider beliefs such as "learning mathematics consists of memorizing the methods that are used to solve particular classes of problems." Such beliefs (as evidenced by the 1983 NAEP exam: Carpenter et al. 1983) are abstracted from one's experience with mathematics through the years. They are not the only possible beliefs of course: with different experiences, students could abstract the belief that "mathematics is a discipline of sense-making." But either way, the formation of such orientations takes time – and the resulting orientations can be very robust.

*People are often unaware of orientations that drive their behavior.* Students may not be aware of holding the beliefs that "all problems can be solved in five minutes or less," or that "proof knowledge is irrelevant when one is working on a construction problem." Teachers may not be consciously aware of the beliefs that

[4] In my case the ideas are quite explicit: Schoenfeld (2013) describes five dimensions of mathematically productive classrooms. By the time this volume appears in print, readers will be able to find research reviews, analytic tools, and professional development documents in support of these ideas on the web sites of the Algebra Teaching study (http://ats.berkeley.edu/) and the Mathematics Assessment Project (http://map.mathshell.org/).

underpin their use of teaching strategies such as "demonstrate and practice," or the reasons they shy away from using open-ended problems. Yet, those beliefs may well shape their behavior – and because the underlying beliefs are not recognized, change will be that much more difficult.

There are two corollaries to these two points.

*Changes in beliefs and practices will take place slowly.* If it takes months or years to develop a belief – on the basis of experience and/or cultural immersion – it stands to reason that it will take comparably long for those beliefs and practices to change. Being told that "mathematics is a sense-making discipline" or "proof is a useful tool for mathematical discovery" is not enough; students need to *experience* mathematics as sense-making and *use* proof as a tool for discover in order for their beliefs to change in a robust way. Similarly, teachers can't be told that "students will develop understandings through problem solving": they need to experience the fact that it can happen in their classrooms, and be supported in it. The case of Mrs. Oublier (Cohen 1990) demonstrates what happens when changes are rhetorical. While Mrs. Oublier believed she had changed her teaching (and in some ways she had), a successful change in her teaching required both a set of new teaching techniques and a constellation of orientations (about mathematics, about problem solving, about what students could do, about classroom practices) in order to take hold. True change, resulting in different orientations and practices, will come slowly.

*Change will be hastened if orientations become the conscious focus of attention.* It is difficult to change something if you don't know you should be attending to it. If there is an underlying reason for the practices we engage in, then reflecting on those reasons may give us reason to problematize the practices. (See, Arcavi and Schoenfeld 2008, for an example.)

The study of change will also call for tools and methods that differ from those discussed in the first part of the chapter. It will be useful to develop theory and methods related to teachers' developmental trajectories – a theory of change, in response to changing context. Charting the evolution of beliefs and affect will certainly be a part of this effort.

In sum, the path toward the improvement of mathematics teaching will be long and slow. But, it is worth it – and the contributions in this volume show that we are making progress toward that goal.

## References

Aguirre, J., & Speer, N. (2000). Examining the relationship between beliefs and goals in teacher practice. *Journal of Mathematical Behavior, 18*(3), 327–356.

Arcavi, A., & Schoenfeld, A. H. (2008). Using the unfamiliar to problematize the familiar. *Canadian Journal of Science, Mathematics, and Technology Education, 8*(3), 280–295.

Carpenter, T. P., Lindquist, M. M., Matthews, W., & Silver, E. A. (1983). Results of the third NAEP mathematics assessment: Secondary School. *Mathematics Teacher, 76*(9), 652–659.

Clarke, D., Keitel, C., & Shimizu, Y. (2006). *Mathematics classrooms in twelve countries: The insider's perspective*. Rotterdam: Sense Publishers.

Cohen, D. K. (1990). A revolution in one classroom: the case of Mrs. Oublier. *Educational Evaluation and Policy Analysis, 12*(3), 311–329.
Horn, I. S. (2007). Fast kids, slow kids, lazy kids: Framing the mismatch problem in math teachers' conversations. *Journal of the Learning Sciences, 16*(1), 37–79.
Lave, J., & Wenger, E. (1991). *Situated learning: Legitimate peripheral participation*. Cambridge: Cambridge University Press.
Schoenfeld, A. H. (1985). *Mathematical problem solving*. New York: Academic.
Schoenfeld, A. H. (1989). Explorations of students' mathematical beliefs and behavior. *Journal for Research in Mathematics Education, 20*(4), 338–355.
Schoenfeld, A. H. (2010). *How we think: A theory of goal-oriented decision making and its educational applications*. New York: Routledge.
Schoenfeld, A. H. (2013). Classroom observations in theory and practice. *ZDM – The International Journal of Mathematics Education, 45*, 607–621.
Schraw, G., & Olafson, L. J. (2002). Teachers' epistemological worldviews and educational practices. *Issues in Education, 8*(2), 99–148.
Shulman, L. S. (2005). Signature pedagogies in the professions. *Daedalus, 134*(Summer), 52–59.
Speer, N. M. (2005). Issues of methods and theory in the study of mathematics teachers' professed and attributed beliefs. *Educational Studies in Mathematics, 58*, 361–391.
Stigler, J., & Hiebert, J. (1999). *The teaching gap*. New York: Free Press.

# Index




B. Pepin, B. Roesken-Winter (eds.), *From beliefs to dynamic affect systems in mathematics education*, Advances in Mathematics Education,
DOI 10.1007/978-3-319-06808-4

F



G



I



L



M



N



O



P

MIX
Papier aus verantwortungsvollen Quellen
Paper from responsible sources
FSC® C105338

If you have any concerns about our products,
you can contact us on
**ProductSafety@springernature.com**

In case Publisher is established outside the EU,
the EU authorized representative is:
**Springer Nature Customer Service Center GmbH**
**Europaplatz 3, 69115 Heidelberg, Germany**

Printed by Libri Plureos GmbH
in Hamburg, Germany